APPLIED SOIL MECHANICS

APPLIED SOIL MECHANICS

with ABAQUS Applications

SAM HELWANY

BICENTENNIAL
1807
WILEY
2007
BICENTENNIAL

JOHN WILEY & SONS, INC.

Library of Congress Cataloging-in-Publication Data:

Helwany, Sam, 1958-
   Applied soil mechanics with ABAQUS applications / Sam Helwany.
       p. cm.
   Includes index.
   ISBN 978-0-471-79107-2 (cloth)
   1. Soil mechanics. 2. Finite element method. 3. ABAQUS. I. Title.
   TA710.H367 2007
   624.1'5136–dc22
                                          2006022830

To the memory
of my parents

CONTENTS

PREFACE

The purpose of this book is to provide civil engineering students and practitioners with simple basic knowledge on how to apply the finite element method to soil mechanics problems. This is essentially a soil mechanics book that includes traditional soil mechanics topics and applications. The book differs from traditional soil mechanics books in that it provides a simple and more flexible alternative using the finite element method to solve traditional soil mechanics problems that have closed-form solutions. The book also shows how to apply the finite element method to solve more complex geotechnical engineering problems of practical nature that do not have closed-form solutions.

In short, the book is written mainly for undergraduate students, to encourage them to solve geotechnical engineering problems using both traditional engineering solutions and the more versatile finite element solutions. This approach not only teaches the concepts but also provides means to gain more insight into geotechnical engineering applications that reinforce the concepts in a very profound manner. The concepts are presented in a basic form so that the book can serve as a valuable learning aid for students with no background in soil mechanics. The main prerequisite would be strength of materials (or equivalent), which is a prerequisite for soil mechanics in most universities.

General soil mechanics principles are presented for each topic, followed by traditional applications of these principles with longhand solutions, which are followed in turn by finite element solutions for the same applications, and then both solutions are compared. Further, more complex applications are presented and solved using the finite element method.

The book consist of nine chapters, eight of which deal with traditional soil mechanics topics, including stresses in semi-infinite soil mass, consolidation, shear strength, shallow foundations, lateral earth pressure, deep foundations (piles), and seepage. The book includes one chapter (Chapter 2) that describes several elastic and elastoplastic material models, some of which are used within the framework of the finite element method to simulate soil behavior, and that includes a generalized three-dimensional linear elastic model, the Cam clay model, the cap model and Lade's model. For undergraduate teaching, one can include a brief description of the essential characteristics and parameters of the Cam clay model and the cap model without much emphasis on their mathematical derivations.

Over 60 solved examples appear throughout the book. Most are solved longhand to illustrate the concepts and then solved using the finite element method embodied in a computer program: ABAQUS. All finite element examples are solved using ABAQUS. This computer program is used worldwide by educators and engineers to solve various types of civil engineering and engineering mechanics problems. One of the major advantages of using this program is that it is capable of solving most geotechnical engineering problems. The program can be used to tackle geotechnical engineering problems involving two- and three-dimensional configurations that may include soil and structural elements, total and effective stress analysis, consolidation analysis, seepage analysis, static and dynamic (implicit and explicit) analysis, failure and post-failure analysis, and a lot more. Nevertheless, other popular finite element or finite difference computer programs specialized in soil mechanics can be used in conjunction with this book in lieu of ABAQUS—obviously, this depends on the instructor's preference.

The PC Education Version of ABAQUS can be obtained via the internet so that the student and practitioner can use it to rework the examples of the book and to solve the homework assignments, which can be chosen from those end-of-chapter problems provided. Furthermore, the input data for all examples can be downloaded from the book's website (www.wiley.com/college/helwany). This can be very useful for the student and practitioner, since they can see how the input should be for a certain problem, then can modify the input data to solve more complex problems of the same class.

I express my deepest appreciation to the staff at John Wiley & Sons Publishing Company, especially Mr. J. Harper, Miss K. Nasdeo, and Miss M. Torres for their assistance in producing the book. I am also sincerely grateful to Melody Clair for her editing parts of the manuscript.

Finally, a very special thank you to my family, Alba, Eyad, and Omar, and my brothers and sisters for their many sacrifices during the development of the book.

CHAPTER 1

PROPERTIES OF SOIL

1.1 SOIL FORMATION

Soil is a three-phase material consisting of solid particles, water, and air. Its mechanical behavior is largely dependent on the size of its solid particles and voids. The solid particles are formed from physical and chemical weathering of rocks. Therefore, it is important to have some understanding of the nature of rocks and their formation.

A *rock* is made up of one or more minerals. The characteristics of a particular rock depend on the minerals it contains. This raises the question: What is a mineral? By definition, a *mineral* is a naturally occurring inorganic element or compound in a solid state. More than 4000 different minerals have been discovered but only 10 elements make up 99% of Earth's crust (the outer layer of Earth): oxygen (O), silicon (Si), aluminum (Al), iron (Fe), calcium (Ca), sodium (Na), potassium (K), magnesium (Mg), titanium (Ti), and hydrogen (H). Most of the minerals (74%) in Earth's crust contain oxygen and silicon. The silicate minerals, containing oxygen and silicon, comprise 90% of all rock-forming minerals. One of the interesting minerals in soil mechanics is the clay mineral *montmorillonite* (an expansive clay), which can expand up to 15 times its original volume if water is present. When expanding, it can produce pressures high enough to damage building foundations and other structures.

Since its formation, Earth has been subjected to continuous changes caused by seismic, volcanic, and climatic activities. Moving from the surface to the center of Earth, a distance of approximately 6370 km, we encounter three different layers. The top (outer) layer, the *crust*, has an average thickness of 15 km and an average density of 3000 kg/m^3. By comparison, the density of water is 1000 kg/m^3 and that of iron is 7900 kg/m^3. The second layer, the *mantle*, has an average thickness

1

of 3000 km and an average density of 5000 kg/m^3. The third, the *core*, contains primarily nickel and iron and has an average density of 11,000 kg/m^3.

Within the crust, there are three major groups of rocks:

1. *Igneous rocks*, which are formed by the cooling of magma. Fast cooling occurs above the surface, producing igneous rocks such as basalt, whereas slow cooling occurs below the surface, producing other types of igneous rocks, such as granite and dolerite. These rocks are the ancestors of sedimentary and metamorphic rocks.

2. *Sedimentary rocks*, which are made up of particles and fragments derived from disintegrated rocks that are subjected to pressure and cementation caused by calcite and silica. Limestone (chalk) is a familiar example of a sedimentary rock.

3. *Metamorphic rocks*, which are the product of existing rocks subjected to changes in pressure and temperature, causing changes in mineral composition of the original rocks. Marble, slate, and schist are examples of metamorphic rocks.

Note that about 95% of the outer 10 km of Earth's crust is made up of igneous and metamorphic rocks, and only 5% is sedimentary. But the exposed surface of the crust contains at least 75% sedimentary rocks.

Soils Soils are the product of physical and chemical weathering of rocks. Physical weathering includes climatic effects such as freeze–thaw cycles and erosion by wind, water, and ice. Chemical weathering includes chemical reaction with rainwater. The particle size and the distribution of various particle sizes of a soil depend on the weathering agent and the transportation agent.

Soils are categorized as gravel, sand, silt, or clay, depending on the predominant particle size involved. Gravels are small pieces of rocks. Sands are small particles of quartz and feldspar. Silts are microscopic soil fractions consisting of very fine quartz grains. Clays are flake-shaped microscopic particles of mica, clay minerals, and other minerals. The average size (diameter) of solid particles ranges from 4.75 to 76.2 mm for gravels and from 0.075 to 4.75 mm for sands. Soils with an average particle size of less than 0.075 mm are either silt or clay or a combination of the two.

Soils can also be described based on the way they were deposited. If a soil is deposited in the vicinity of the original rocks due to gravity alone, it is called a *residual soil*. If a soil is deposited elsewhere away from the original rocks due to a transportation agent (such as wind, ice, or water), it is called a *transported soil*.

Soils can be divided into two major categories: cohesionless and cohesive. *Cohesionless soils*, such as gravelly, sandy, and silty soils, have particles that do not adhere (stick) together even with the presence of water. On the other hand, *cohesive soils* (clays) are characterized by their very small flakelike particles, which can attract water and form plastic matter by adhering (sticking) to each other. Note that whereas you can make shapes out of wet clay (but not too wet) because of its cohesive characteristics, it is not possible to do so with a cohesionless soil such as sand.

1.2 PHYSICAL PARAMETERS OF SOILS

Soils contain three components: solid, liquid, and gas. The *solid components* of soils are the product of weathered rocks. The *liquid component* is usually water, and the *gas component* is usually air. The gaps between the solid particles are called *voids*. As shown in Figure 1.1*a*, the voids may contain air, water, or both. Let us discuss the soil specimen shown in Figure 1.1*a*. The total volume (V) and the total weight (W) of the specimen can be measured in the laboratory. Next, let us separate the three components of the soil as shown in Figure 1.1*b*. The solid particles are gathered in one region such that there are no voids in between, as shown in the figure (this can only be done theoretically). The volume of this component is V_s and its weight is W_s. The second component is water, whose volume is V_w and whose weight is W_w. The third component is the air, which has a volume V_a and a very small weight that can be assumed to be zero. Note that the volume of voids (V_v) is the sum of V_a and V_w. Therefore, the total volume is $V = V_v + V_s = V_a + V_w + V_s$. Also, the total weight $W = W_w + W_s$.

In the following we present definitions of several basic soil parameters that hold important physical meanings. These basic parameters will be used to obtain relationships that are useful in soil mechanics.

The void ratio e is the proportion of the volume of voids with respect to the volume of solids:

$$e = \frac{V_v}{V_s} \tag{1.1}$$

The porosity n is given as

$$n = \frac{V_v}{V} \tag{1.2}$$

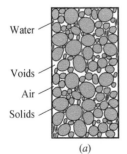

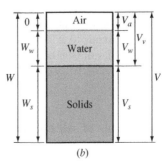

FIGURE 1.1 (*a*) Soil composition; (*b*) phase diagram.

Note that

$$e = \frac{V_v}{V_s} = \frac{V_v}{V - V_v} = \frac{V_v/V}{V/V - V_v/V} = \frac{n}{1 - n} \tag{1.3}$$

or

$$n = \frac{e}{1 + e} \tag{1.4}$$

The degree of saturation is defined as

$$S = \frac{V_w}{V_v} \tag{1.5}$$

Note that when the soil is fully saturated, all the voids are filled with water (no air). In that case we have $V_v = V_w$. Substituting this into (1.5) yields $S = 1$ (or 100% saturation). On the other hand, if the soil is totally dry, we have $V_w = 0$; therefore, $S = 0$ (or 0% saturation).

The moisture content (or water content) is the proportion of the weight of water with respect to the weight of solids:

$$\omega = \frac{W_w}{W_s} \tag{1.6}$$

The water content of a soil specimen is easily measured in the laboratory by weighing the soil specimen first to get its total weight, W. Then the specimen is dried in an oven and weighed to get W_s. The weight of water is then calculated as $W_w = W - W_s$. Simply divide W_w by W_s to get the moisture content, (1.6).

Another useful parameter is the specific gravity G_s, defined as

$$G_s = \frac{\gamma_s}{\gamma_w} = \frac{W_s/V_s}{\gamma_w} \tag{1.7}$$

where γ_s is the unit weight of the soil solids (not the soil itself) and γ_w is the unit weight of water ($\gamma_w = 9.81$ kN/m^3). Note that the specific gravity represents the relative unit weight of solid particles with respect to water. Typical values of G_s range from 2.65 for sands to 2.75 for clays.

The unit weight of soil (the bulk unit weight) is defined as

$$\gamma = \frac{W}{V} \tag{1.8}$$

and the dry unit weight of soil is given as

$$\gamma_d = \frac{W_s}{V} \tag{1.9}$$

Substituting (1.6) and (1.9) into (1.8), we get

$$\gamma = \frac{W}{V} = \frac{W_s + W_w}{V} = \frac{W_s + \omega W_s}{V} = \frac{W_s(1 + \omega)}{V} = \gamma_d(1 + \omega)$$

or

$$\gamma_d = \frac{\gamma}{1 + \omega} \tag{1.10}$$

Let us assume that the volume of solids V_s in Figure 1.1b is equal to 1 unit (e.g., 1 m^3). Substitute $V_s = 1$ into (1.1) to get

$$e = \frac{V_v}{V_s} = \frac{V_v}{1} \rightarrow V_v = e \tag{1.11}$$

Thus,

$$V = V_s + V_v = 1 + e \tag{1.12}$$

Substituting $V_s = 1$ into (1.7) we get

$$G_s = \frac{\gamma_s}{\gamma_w} = \frac{W_s/V_s}{\gamma_w} = \frac{W_s/1}{\gamma_w} \rightarrow W_s = \gamma_w G_s \tag{1.13}$$

Substitute (1.13) into (1.6) to get

$$W_w = \omega W_s = \omega \gamma_w G_s \tag{1.14}$$

Finally, substitute (1.12), (1.13), and (1.14) into (1.8) and (1.9) to get

$$\gamma = \frac{W}{V} = \frac{W_s + W_w}{V} = \frac{\gamma_w G_s + \omega \gamma_w G_s}{1 + e} = \frac{\gamma_w G_s(1 + \omega)}{1 + e} \tag{1.15}$$

and

$$\gamma_d = \frac{W_s}{V} = \frac{\gamma_w G_s}{1 + e} \tag{1.16}$$

Another interesting relationship can be obtained from (1.5):

$$S = \frac{V_w}{V_v} = \frac{W_w/\gamma_w}{V_v} = \frac{\omega \gamma_w G_s/\gamma_w}{e} = \frac{\omega G_s}{e} \rightarrow eS = \omega G_s \tag{1.17}$$

Equation (1.17) is useful for estimating the void ratio of saturated soils based on their moisture content. For a saturated soil $S = 1$ and the value of G_s can be assumed (2.65 for sands and 2.75 for clays). The moisture content can be obtained from a simple laboratory test (described earlier) performed on a soil specimen taken from the field. An approximate in situ void ratio is calculated as $e = \omega G_s \approx (2.65 - 2.75)\omega$.

For a fully saturated soil, we have $e = \omega G_s \rightarrow G_s = e/\omega$. Substituting this into (1.15), we can obtain the following expression for the saturated unit weight:

$$\gamma_{sat} = \frac{\gamma_w G_s (1 + \omega)}{1 + e} = \frac{\gamma_w [G_s + \omega e/\omega]}{1 + e} = \frac{\gamma_w (G_s + e)}{1 + e} \tag{1.18}$$

Example 1.1 A 0.9-m^3 soil specimen weighs 17 kN and has a moisture content of 9%. The specific gravity of the soil solids is 2.7. Using the fundamental equations (1.1) to (1.10), calculate (a) γ, (b) γ_d, (c) e, (d) n, (e) V_w, and (f) S.

SOLUTION: Given: $V = 0.9$ m^3, $W = 17$ kN, $\omega = 9\%$, and $G_s = 2.7$.

(a) From the definition of unit weight, (1.8):

$$\gamma = \frac{W}{V} = \frac{17\,\text{kN}}{0.9\,\text{m}^3} = 18.9\ \text{kN/m}^3$$

(b) From (1.10):

$$\gamma_d = \frac{\gamma}{1 + \omega} = \frac{18.9\ \text{kN/m}^3}{1 + 0.09} = 17.33\ \text{kN/m}^3$$

(c) From (1.9):

$$\gamma_d = \frac{W_s}{V} \rightarrow W_s = \gamma_d V = 17.33\ \text{kN/m}^3 \times 0.9\ \text{m}^3 = 15.6\ \text{kN}$$

From the phase diagram (Figure 1.1b), we have

$$W_w = W - W_s = 17\ \text{kN} - 15.6\ \text{kN} = 1.4\ \text{kN}$$

From (1.7):

$$G_s = \frac{\gamma_s}{\gamma_w} = \frac{W_s/V_s}{\gamma_w} \rightarrow V_s = \frac{W_s}{\gamma_w G_s} = \frac{15.6\ \text{kN}}{9.81\ \text{kN/m}^3 \times 2.7} = 0.5886\ \text{m}^3$$

Also, from the phase diagram (Figure 1.1b), we have

$$V_v = V - V_s = 0.9\ \text{m}^3 - 0.5886\ \text{m}^3 = 0.311\ \text{m}^3$$

From (1.1) we get

$$e = \frac{V_v}{V_s} = \frac{0.311\ \text{m}^3}{0.5886\ \text{m}^3} = 0.528$$

(d) Equation (1.2) yields

$$n = \frac{V_v}{V} = \frac{0.311\,\text{m}^3}{0.9\,\text{m}^3} = 0.346$$

(e) From the definition of the unit weight of water,

$$V_w = \frac{W_w}{\gamma_w} = \frac{1.4\,\text{kN}}{9.81\,\text{kN/m}^3} = 0.143\,\text{m}^3$$

(f) Finally, from (1.5):

$$S = \frac{V_w}{V_v} = \frac{0.143\,\text{m}^3}{0.311\,\text{m}^3} = 0.459 = 45.9\%$$

1.2.1 Relative Density

The compressibility and strength of a granular soil are related to its relative density D_r, which is a measure of the compactness of the soil grains (their closeness to each other). Consider a uniform sand layer that has an in situ void ratio e. It is possible to tell how dense this sand is if we compare its in situ void ratio with the maximum and minimum possible void ratios of the same sand. To do so, we can obtain a sand sample from the sand layer and perform two laboratory tests (ASTM 2004: Test Designation D-4253). The first laboratory test is carried out to estimate the maximum possible dry unit weight $\gamma_{d-\max}$ (which corresponds to the minimum possible void ratio $e_{\min}$) by placing a dry sand specimen in a container with a known volume and subjecting the specimen to a surcharge pressure accompanied with vibration. The second laboratory test is performed to estimate the minimum possible dry unit weight $\gamma_{d-\min}$ (which corresponds to the maximum possible void ratio $e_{\max}$) by pouring a dry sand specimen very loosely in a container with a known volume. Now, let us define the relative density as

$$D_r = \frac{e_{\max} - e}{e_{\max} - e_{\min}} \tag{1.19}$$

This equation allows us to compare the in situ void ratio directly with the maximum and minimum void ratios of the same granular soil. When the in situ void ratio e of this granular soil is equal to $e_{\min}$, the soil is at its densest possible condition and D_r is equal to 1 (or $D_r = 100\%$). When e is equal to $e_{\max}$, the soil is at its loosest possible condition, and its D_r is equal to 0 (or $D_r = 0\%$). Note that the dry unit weight is related to the void ratio through the equation

$$\gamma_d = \frac{G_s \gamma_w}{1 + e} \tag{1.20}$$

It follows that

$$\gamma_{d-\max} = \frac{G_s \gamma_w}{1 + e_{\min}} \quad \text{and} \quad \gamma_{d-\min} = \frac{G_s \gamma_w}{1 + e_{\max}}$$

1.3   MECHANICAL PROPERTIES OF SOIL

Soil engineers usually classify soils to determine whether they are suitable for particular applications. Let us consider three borrow sites from which we need to select a soil that has the best compaction characteristics for a nearby highway embankment construction project. For that we would need to get details about the *grain-size distribution* and the *consistency* of each soil. Then we can use available charts and tables that will give us the exact type of each soil. From experience and/or from available charts and tables we can determine which of these soils has the best compaction characteristics based on its classification.

Most soil classification systems are based on the grain-size distribution curve and the Atterberg limits for a given soil. The grain-size analysis is done using sieve analysis on the coarse portion of the soil (> 0.075 mm in diameter), and using hydrometer analysis on the fine portion of the soil (< 0.075 mm in diameter). The consistency of soil is characterized by its Atterberg limits as described below.

1.3.1   Sieve Analysis

A set of standardized sieves is used for the analysis. Each sieve is 200 mm in diameter and 50 mm in height. The opening size of the sieves ranges from 0.075 mm for sieve No. 200 to 4.75 mm for sieve No. 4. Table 1.1 lists the designation of each sieve and the corresponding opening size. As shown in Figure 1.2, a set of sieves stacked in descending order (the sieve with the largest opening size is on top) is secured on top of a standardized shake table. A dry soil specimen is then

TABLE 1.1   Standard Sieve Sizes

Sieve No.	Opening Size (mm)
4	4.75
10	2.00
20	0.85
40	0.425
60	0.250
80	0.180
100	0.15
120	0.125
140	0.106
170	0.090
200	0.075

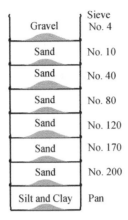

FIGURE 1.2 Typical set of U.S. standard sieves.

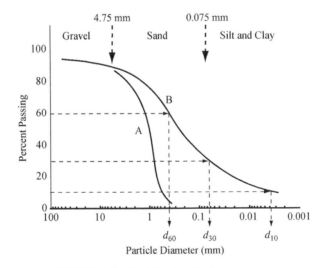

FIGURE 1.3 Particle-size distribution curve.

shaken through the sieves for 10 minutes. As shown in Figure 1.3, the percent by weight of soil passing each sieve is plotted as a function of the grain diameter (corresponding to a sieve number as shown in Table 1.1). It is customary to use a logarithmic horizontal scale on this plot.

Figure 1.3 shows two grain-size distribution curves, A and B. Curve A represents a *uniform soil* (also known as *poorly graded soil*) that includes a narrow range of particle sizes. This means that the soil is not well proportioned, hence the expression "poorly graded soil." In this example, soil A is uniform coarse sand. On the other hand, curve B represents a *nonuniform soil* (also known as *well-graded soil*) that includes a wide spectrum of particle sizes. In this case the soil is well proportioned—it includes gravel, sand (coarse, medium, and fine), and silt/clay.

There are two useful indicators, C_u and C_c, that can be obtained from the grain-size distribution curve. C_u is the uniformity coefficient, defined as $C_u = d_{60}/d_{10}$, and C_c is the coefficient of gradation, defined as $C_c = d_{30}^2/(d_{10}d_{60})$. Here d_{10}, d_{30}, and d_{60} are the grain diameters corresponding respectively to 10%, 30%, and 60% passing, as shown in Figure 1.3. For a well-graded sand the value of the coefficient of gradation should be in the range $1 \leq C_c \leq 3$. Also, higher values of the uniformity coefficient indicate that the soil contains a wider range of particle sizes.

1.3.2 Hydrometer Analysis

Sieve analysis cannot be used for clay and silt particles because they are too small (<0.075 mm in diameter) and they will be suspended in air for a long time during shaking. The grain-size distribution of the fine-grained portion that passes sieve No. 200 can be obtained using hydrometer analysis. The basis of hydrometer analysis is that when soil particles are dispersed in water, they will settle at different velocities because of their different sizes. Assuming that soil particles are perfect spheres dispersed in water with a viscosity η, Stokes' law can be used to relate the terminal velocity v of a particle to its diameter D:

$$v = \frac{\rho_s - \rho_w}{18\eta} D^2 \tag{1.21}$$

in which ρ_s is the density of soil particles and ρ_w is the density of water. Equation (1.21) indicates that a larger particle will have a greater terminal velocity when dropping through a fluid.

In the hydrometer laboratory test (ASTM 2004) a dry soil specimen weighing 50 g is mixed thoroughly with water and placed in a graduated 1000-mL glass flask. A floating instrument called a *hydrometer* (Figure 1.4) is placed in the flask to measure the specific gravity of the mixture in the vicinity of the hydrometer center. In a 24-hour period the time t and the corresponding depth L are recorded. The measured depth (see Figure 1.4) is correlated with the amount of soil that is still in suspension at time t. From Stokes' law, (1.21), it can be shown that the diameter of the largest soil particles still in suspension is given by

$$D = \sqrt{\frac{18\eta}{[(\rho_s/\rho_w) - 1]\gamma_w} \frac{L}{t}} \tag{1.22}$$

in which γ_w is the unit weight of water. From the hydrometer readings (L versus t) and with the help of (1.22), one can calculate the percent of finer particles and plot a gradation curve. The part of curve B (Figure 1.3) with particle diameter smaller than 0.075 mm is obtained from a hydrometer test.

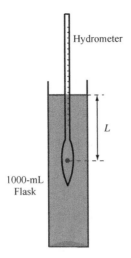

Hydrometer

L

1000-mL
Flask

FIGURE 1.4   Hydrometer test.

1.4   SOIL CONSISTENCY

Clays are flake-shaped microscopic particles of mica, clay minerals, and other minerals. Clay possesses a large specific surface, defined as the total surface of clay particles per unit mass. For example, the specific surfaces of the three main clay minerals; kaolinite, illite, and montmorillonite, are 15, 80, and 800 m^2/g, respectively. It is mind-boggling that just 1 g of montmorillonite has a surface of 800 m^2! This explains why clays are fond of water. It is a fact that the surface of a clay mineral has a net negative charge. Water, on the other hand, has a net positive charge. Therefore, the clay surface will bond to water if the latter is present. A larger specific surface means more absorbed water. As mentioned earlier, montmorillonite can increase 15-fold in volume if water is present, due to its enormous specific surface. Montmorillonite is an expansive clay that causes damage to adjacent structures if water is added (rainfall). It also shrinks when it dries, causing another type of damage to structures. Illite is not as expansive, due to its moderate specific surface. Kaolinite is the least expansive.

It is clear that the moisture (water) content has a great effect on a clayey soil, especially in terms of its response to applied loads. Consider a very wet clay specimen that looks like slurry (fluid). In this *liquid state* the clay specimen has no strength (i.e., it cannot withstand any type of loading). Consider a potter's clay specimen that has a moderate amount of moisture. This clay is in its *plastic state* because in this state we can actually make shapes out of the clay knowing that it will not spring back as elastic materials do. If we let this plastic clay dry out for a short time (i.e., so that it is not totally dry), it will lose its plasticity because if we try to shape it now, many cracks will appear, indicating that the clay is in its *semisolid state*. If the specimen dries out further, it reaches its *solid state*, where it becomes exceedingly brittle.

Atterberg limits divide the four states of consistency described above. These three limits are obtained in the laboratory on reconstituted soil specimens using the techniques developed early in the twentieth century by a Swedish scientist. As shown in Figure 1.5, the *liquid limit* (LL) is the dividing line between the liquid and plastic states. LL corresponds to the moisture content of a soil as it changes from the plastic state to the liquid state. The *plastic limit* (PL) is the moisture content of a soil when it changes from the plastic to the semisolid state. The *shrinkage limit* (SL) is the moisture content of a soil when it changes from the semisolid state to the solid state. Note that the moisture content in Figure 1.5 increases from left to right.

1.4.1 Liquid Limit

The liquid limit is obtained in the laboratory using a simple device that includes a shallow brass cup and a hard base against which the cup is bumped repeatedly using a crank-operated mechanism. The cup is filled with a clay specimen (paste), and a groove is cut in the paste using a standard tool. The liquid limit is the moisture content at which the shear strength of the clay specimen is so small that the soil "flows" to close the aforementioned groove at a standard number of blows (ASTM 2004: Designation D-4318).

1.4.2 Plastic Limit

The plastic limit is defined as the moisture content at which a soil crumbles when rolled down into threads 3 mm in diameter (ASTM 2004: Designation D-4318). To do that, use your hand to roll a round piece of clay against a glass plate. Being able to roll a moist piece of clay is an indication that it is now in its plastic state (see Figure 1.5). By rolling the clay against the glass, it will lose some of its moisture moving toward its semisolid state, as indicated in the figure. Crumbling of the thread indicates that it has reached its semisolid state. The moisture content of the thread at that stage can be measured to give us the plastic limit, which is the verge between the plastic and semisolid states.

1.4.3 Shrinkage Limit

In its semisolid state, soil has some moisture. As a soil loses more moisture, it shrinks. When shrinking ceases, the soil has reached its solid state. Thus, the

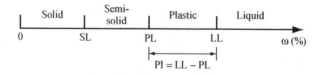

FIGURE 1.5 Atterberg limits.

moisture content at which a soil ceases to shrink is the shrinkage limit, which is the verge between the semisolid and solid states.

1.5 PLASTICITY CHART

A useful indicator for the classification of fine-grained soils is the *plasticity index* (PI), which is the difference between the liquid limit and the plastic limit (PI = LL − PL). Thus, PI is the range within which a soil will behave as a plastic material. The plasticity index and the liquid limit can be used to classify fine-grained soils via the Casagrande (1932) empirical plasticity chart shown in Figure 1.6. The line shown in Figure 1.6 separates silts from clays. In the plasticity chart, the liquid limit of a given soil determines its plasticity: Soils with LL ≤ 30 are classified as low-plasticity clays (or low-compressibility silts); soils with 30 < LL ≤ 50 are medium-plasticity clays (or medium-compressibility silts); and soils with LL > 50 are high-plasticity clays (or high-compressibility silts). For example, a soil with LL = 40 and PI = 10 (point A in Figure 1.6) is classified as silt with medium compressibility, whereas a soil with LL = 40 and PI = 20 (point B in Figure 1.6) can be classified as clay of medium plasticity.

To determine the state of a natural soil with an in situ moisture content ω, we can use the *liquidity index* (LI), defined as

$$LI = \frac{\omega - PL}{LL - PL} \tag{1.23}$$

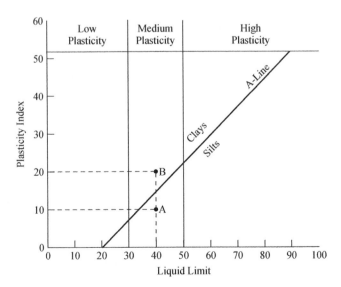

FIGURE 1.6 Plasticity chart.

For heavily overconsolidated clays, $\omega < $ PL and therefore LI < 0, and the soil is classified as nonplastic (i.e., brittle); if $0 \leq$ LI ≤ 1 (i.e., PL $< \omega <$ LL), the soil is in its plastic state; and if LI > 1 (i.e., $\omega >$ LL), the soil is in its liquid state.

Another useful indicator is the *activity* A of a soil (Skempton, 1953):

$$A = \frac{PI}{\% \text{ clay fraction } (< 2\,\mu\text{m})} \tag{1.24}$$

In this equation the clay fraction (clay content) is defined as the weight of the clay particles ($< 2\,\mu$m) in a soil. The activity is a measure of the degree of plasticity of the clay content of the soil. Typical activity values for the main clay minerals are: kaolinite, $A = 0.3$ to 0.5; illite, $A = 0.5$ to 1.2; and montmorillonite, $A = 1.5$ to 7.0.

1.6 CLASSIFICATION SYSTEMS

The two most widely used classification systems are the American Association of State Highway and Transportation Officials (AASHTO) and the Unified Soil Classification System (USCS). Our discussion here will involve only the USCS system.

The Unified Soil Classification System (ASTM 2004: Designation D-2487) classifies soils based on their grain-size distribution curves and their Atterberg limits. As shown in Table 1.2, a soil is called *coarse-grained* if it has less than 50% passing sieve No. 200. Soils in this group can be sandy soils (S) or gravelly soils (G). It follows that a soil is called *fine-grained* if it has more than 50% passing sieve No. 200. Soils in this group include inorganic silts (M), inorganic clays (C), or organic silts and clays (O). The system uses the symbol W for well-graded soils, P for poorly graded soils, L for low-plasticity soils, and H for high-plasticity soils. The combined symbol GW, for example, means well-graded gravel, SP means poorly graded sand, and so on. Again, to determine the exact designation of a soil using the Unified Soil Classification System, you will need to have the grain-size distribution curve and the Atterberg limits of that soil. Then you can use Table 1.2 to get the soil symbol.

Example 1.2   Using the Unified Soil Classification System, classify a soil that has 95% passing a No. 10 sieve; 65% passing No. 40; and 30% passing No. 200. The soil has a liquid limit of 25 and a plastic limit of 15.

SOLUTION: The soil has 30% passing a No. 200 sieve, therefore, it is a coarse-grained soil according to the first column in Table 1.2. The soil has 95% passing a No. 10 sieve, so it must have at least 95% passing a No. 4 sieve (No. 4 has a larger opening size than No. 10). This means that the soil has less than 5% gravel (see Figure 1.7). According to the second column in Table 1.2, the soil is classified as sand. But since it has 30% fines, it is a sandy soil with fines according to the third column in Table 1.2.

TABLE 1.2 Unified Soil Classification System (adapted from Das 2004)

		Criteria		Symbol
Coarse-grained soils: less than 50% passing No. 200 sieve	Gravel: more than 50% of coarse fraction retained on No. 4 sieve	Clean gravels: less than 5% fines	$C_u \geq 4$ and $1 \leq C_c \leq 3$	GW
			$C_u < 4$ and/or $1 > C_c > 3$	GP
		Gravels with fines: more than 12% fines	PI < 4 or plots below A line (Fig. 1.6)	GM
			PI > 7 and plots on or above A line (Fig. 1.6)	GC
	Sands: 50% or more of coarse fraction passes No. 4 sieve	Clean sands: less than 5% fines	$C_u \geq 6$ and $1 \leq C_c \leq 3$	SW
			$C_u < 6$ and/or $1 > C_c > 3$	SP
		Sands with fines: more than 12% fines	PI < 4 or plots below A line (Fig. 1.6)	SM
			PI > 7 and plots on or above A line (Fig. 1.6)	SC
Fine-grained soils: 50% or more passing No. 200 sieve	Silts and clays: LL < 50	Inorganic	PI > 7 and plots on or above A line (Fig. 1.6)	CL
			PI < 4 or plots below A line (Fig. 1.6)	ML
		Organic	$\dfrac{\text{LL(oven dried)}}{\text{LL(not dried)}} < 0.75$	OL
	Silts and clays: LL $\geq$ 50	Inorganic	PI plots on or above A line (Fig. 1.6)	CH
			PI plots below A line (Fig. 1.6)	MH
		Organic	$\dfrac{\text{LL(oven dried)}}{\text{LL(not dried)}} < 0.75$	OH
Highly organic soils	Primarily organic matter, dark in color, and organic odor			Pt

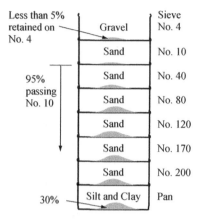

FIGURE 1.7   Particle-size distribution for Example 1.2.

The plasticity index of the soil is PI = LL − PL = 25 − 15 = 10 > 7. Also, the point with LL = 25 and PI = 10 plots above the A line in Figure 1.6. Therefore, the soil is classified as SC = clayey sand according to the fourth and fifth columns in Table 1.2.

1.7 COMPACTION

Compaction involves applying mechanical energy to partially saturated soils for densification purposes. The densification process brings soil particles closer to each other, thus decreasing the size of the voids by replacing air pockets with soil solids. Theoretically, we can achieve 100% saturation by replacing all air pockets by soil solids if we apply enough mechanical energy (compaction), but that is practically impossible. With proper compaction, the soil becomes stronger (higher shear strength), less compressible when subjected to external loads (i.e., less future settlement), and less permeable, making the soil a good construction material for highway embankments, ramps, earth dams, dikes, backfill for retaining walls and bridge abutments, and many other applications.

Soils are compacted in layers (called *lifts*) with each layer being compacted to develop a final elevation and/or shape. Compaction machines such as smooth rollers, pneumatic rollers, and sheepsfoot rollers are generally used for this purpose. The compaction energy generated by a compactor is proportional to the pressure applied by the compactor, its speed of rolling, and the number of times it is rolled (number of passes). Usually, a few passes are needed to achieve the proper dry unit weight, provided that the proper moisture content is used for a particular soil. The required field dry density is 90 to 95% of the maximum dry density that can be achieved in a laboratory compaction test (standard proctor test or modified proctor test: ASTM 2004: Designation D-698 and D-1557) carried out on the same soil.

The *standard proctor test* is a laboratory test used to determine the maximum dry unit weight and the corresponding optimum moisture content for a given compaction energy and a given soil. The soil specimen is obtained from the borrow site, which is usually an earthcut that is close to the construction site. The soil is first dried and crushed and then mixed with a small amount of water in a uniform manner. The resulting moisture content should be well below the natural moisture content of the soil. The Proctor test involves placing the moist soil in three equal layers inside an extended mold (removable extension). The inside volume of the mold (without the extension) is exactly 1000 cm^3. Each soil layer is compacted using 25 blows from a 2.5-kg hammer. Each blow is applied by raising the hammer 305 mm and releasing it (free fall). The 25 blows are distributed uniformly to cover the entire surface of each layer. After compacting the third layer, the mold extension is removed and the soil is carefully leveled and weighed. Knowing the weight W of the moist soil and its volume V, we can calculate the unit weight as $\gamma = W/V$. A small sample is taken from the compacted soil and dried to measure the moisture content ω. Now we can calculate the dry unit weight of the compacted soil as $\gamma_d = \gamma/(1 + \omega)$. Once this is done, the soil sample is crushed and added to the remainder of the soil in the mixing pan. The moisture content of the soil is increased (1 to 2%) by adding more water. The test is repeated in the same manner as described above. We need to repeat the test at least four to five times to establish a compaction curve such as the one shown in Figure 1.8.

The compaction curve shown in Figure 1.8 provides the relationship between the dry unit weight and the moisture content for a given soil subjected to a specific compaction effort. It is noted from the figure that the dry unit weight increases

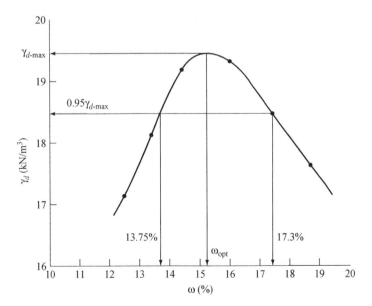

FIGURE 1.8 Compaction curve.

as the moisture content increases until the maximum dry unit weight is reached. The moisture content associated with the maximum dry unit weight is called the *optimum water content*. As shown in the figure, when the moisture content is increased beyond the optimum water content, the dry unit weight decreases. This is caused by the water that is now occupying many of the voids and making it more difficult for the soil to compact further. Note that the degree of saturation corresponding to the optimum moisture content is 75 to 80% for most soils (i.e., 75 to 80% of the voids are filled with water).

The compaction curve provides valuable information: the maximum dry unit weight and the optimum water content that can be conveyed to the compaction contractor by specifying the required relative compaction, RC, defined as

$$RC = \frac{\gamma_{d-\text{field}}}{\gamma_{d-\text{max}}} \times 100\% \qquad (1.25)$$

in which $\gamma_{d-\text{field}}$ is the required dry unit weight of the compacted soil and $\gamma_{d-\text{max}}$ is the maximum dry unit weight obtained from the laboratory compaction test. Usually, the required relative compaction is 90 to 95%. This is because it is very difficult (and costly) to achieve a field dry unit weight that is equal to the maximum dry unit weight obtained from the laboratory compaction test.

It is not enough to specify the relative compaction RC alone. We need to specify the corresponding moisture content that must be used in the field to achieve a specific RC. This is due to the nature of the bell-shaped compaction curve that can have two different moisture contents for the same dry unit weight. Figure 1.8 shows that we can use either $\omega = 13.75\%$ or $\omega = 17.3\%$ to achieve a dry unit weight $\gamma_{d-\text{field}} = 18.5$ kN/m^3, which corresponds to RC = 95%.

In general, granular soils can be compacted in thicker layers than silt and clay. Granular soils are usually compacted using kneading, tamping, or vibratory compaction techniques. Cohesive soils usually need kneading, tamping, or impact. It is to be noted that soils vary in their compaction characteristics. Soils such as GW, GP, GM, GC, SW, SP, and SM (the Unified Soil Classification System, Table 1.2) have good compaction characteristics. Other soils, such as SC, CL, and ML, are characterized as good to poor. Cohesive soils with high plasticity or organic contents are characterized as fair to poor. At any rate, the quality of field compaction needs to be assured by measuring the in situ dry unit weight of the compacted soil at random locations. Several test methods can be used for this purpose:

1. *The sand cone method* (ASTM 2004: Designation D-1556) requires that a small hole be excavated in a newly compacted soil layer. The soil removed is weighed (W) and its moisture content (ω) is determined. The volume (V) of the hole is determined by filling it with Ottawa sand that has a known unit weight. The field dry unit weight can be calculated as $\gamma_{d-\text{field}} = \gamma/(1 + \omega)$, in which γ is calculated as $\gamma = W/V$.

2. There is a method similar to the sand cone method that determines the volume of the hole by filling it with oil (instead of sand) after sealing the surface

of the hole with a thin rubber membrane. This method is called the *rubber balloon method* (ASTM 2004: Designation D-2167).

3. *The nuclear density method* uses a low-level radioactive source that is inserted, via a probe, into the center of a newly compacted soil layer. The source emits rays through the compacted soil that are captured by a sensor at the bottom surface of the nuclear density device. The intensity of the captured radioactivity is inversely proportional to soil density. The apparatus is calibrated using the sand cone method for various soils, and it usually provides reliable estimates of moisture content and dry unit weight. The method provides fast results, allowing the user to perform a large number of tests in a short time.

PROBLEMS

1.1 Refer to Figure 1.3. For soil B: **(a)** determine the percent finer than sieves No. 4, 10, 100, and 200; **(b)** determine d_{10}, d_{30}, and d_{60}; **(c)** calculate the uniformity coefficient; and **(d)** calculate the coefficient of gradation.

1.2 Refer to the phase diagram shown in Figure 1.9. In this phase diagram it is assumed that the total volume of the soil specimen is 1 unit. Show that **(a)** $\gamma_d = G_s\gamma_w(1 - n)$, and **(b)** $\gamma = G_s\gamma_w(1 - n)(1 + \omega)$.

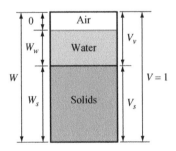

FIGURE 1.9

1.3 For a moist soil specimen, the following are given: $V = 0.5 \ m^3$, $W = 9.5$ kN, $\omega = 7.3\%$, and $G_s = 2.7$. Determine the bulk unit weight γ, the dry unit weight γ_d, the void ratio e, the porosity n, and the degree of saturation S.

1.4 The field unit weight of a compacted soil is $\gamma = 17.5$ kN/m^3, and its moisture content is $\omega = 7\%$. Calculate the relative density of the compacted soil knowing that its $e_{max} = 0.9$, $e_{min} = 0.5$, and $G_s = 2.7$.

1.5 A moist soil has $G_s = 2.65$, $\gamma = 20$ kN/m^3, and $\omega = 15.2\%$. Calculate its dry unit weight γ_d, void ratio e, porosity n, and degree of saturation S.

1.6 A 1.3-m^3 soil specimen weighs 25.7 kN and has a moisture content of 11%. The specific gravity of the soil solids is 2.7. Using the fundamental equations, (1.1) to (1.10), calculate its bulk unit weight γ, dry unit weight γ_d, void ratio e, porosity n, volume of water V_w, and degree of saturation S.

1.7 Four standard Proctor tests were performed on a clayey soil with the following results:

Bulk unit weight (kN/m^3)	Moisture content (%)
20.6	13.3
21.96	14.4
22.5	16
21.7	18.5

Plot the compaction curve and obtain the maximum dry unit weight and the corresponding optimum moisture content.

1.8 The same soil as that in Figure 1.8 is used to construct an embankment. It was compacted using a moisture content below the optimum moisture content and a relative compaction of 95%. What is the compacted unit weight of the soil in the field? If the specific gravity of the soil solids is 2.68, what are the soil's in situ porosity and degree of saturation?

CHAPTER 2

ELASTICITY AND PLASTICITY

2.1 INTRODUCTION

Predicting soil behavior by constitutive equations that are based on experimental findings and embodied in numerical methods such as the finite element method is a significant aspect of soil mechanics. This allows engineers to solve various types of geotechnical engineering problems, especially problems that are inherently complex and cannot be solved using traditional analysis without making simplifying assumptions that may jeopardize the value of the analytical solution.

Soils are constituted of discrete particles, and most soil models assume that the forces and displacements within these particles are represented by continuous stresses and strains. It is not the intention of most soil models to predict the behavior of the soil mass based on the behavior of soil particles and the interaction among particles due to a given loading regime. Rather, these stress–strain constitutive laws are generally fitted to experimental measurements performed on specimens that include a large number of particles.

In this chapter we present three elastoplastic soil models. These models must be calibrated with the results of laboratory tests performed on representative soil samples. Usually, a minimum of three conventional triaxial compression tests and one isotropic consolidation (compression) test are needed for any of the three models. Undisturbed soil specimens are obtained from the field and tested with the assumption that they represent the average soil behavior at the location from which they were obtained. The triaxial tests should be performed under conditions that are similar to the in situ conditions. This includes soil density, the range of stresses, and the drainage conditions in the field (drained loading versus undrained loading conditions). Also, the laboratory tests should include unloading–reloading cycles to characterize the elastic parameters of the soil.

Before presenting the plasticity models we discuss briefly some aspects of the elasticity theory. The theory of elasticity is used to calculate the elastic strains that occur prior to yielding in an elastoplastic material. First we present the stress matrix, then present the generalized Hooke's law for a three-dimensional stress condition, a uniaxial stress condition, a plane strain condition, and a plane stress condition.

2.2 STRESS MATRIX

The stress state at a point A within a soil mass can be represented by an *infinitesimal* (very small) cube with three stress components on each of its six sides (one normal and two shear components), as shown in Figure 2.1. Since point A is under static equilibrium (assuming the absence of body forces such as the self-weight), only nine stress components from three planes are needed to describe the stress state at point A. These nine stress components can be organized into the *stress matrix*:

$$\begin{pmatrix} \sigma_{11} & \tau_{12} & \tau_{13} \\ \tau_{21} & \sigma_{22} & \tau_{23} \\ \tau_{31} & \tau_{32} & \sigma_{33} \end{pmatrix} \tag{2.1}$$

where σ_{11}, σ_{22}, and σ_{33} are the normal stresses (located on the diagonal of the stress matrix) and τ_{12}, τ_{21}, τ_{13}, τ_{31}, τ_{23}, and τ_{32} are the shear stresses. The shear stresses across the diagonal are identical (i.e., $\tau_{12} = \tau_{21}$, $\tau_{13} = \tau_{31}$ and $\tau_{23} = \tau_{32}$) as a result of static equilibrium (to satisfy moment equilibrium). This arrangement

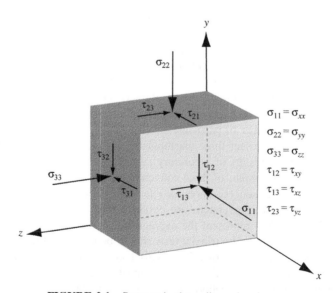

FIGURE 2.1   Stresses in three-dimensional space.

of the nine stress components is also known as the *stress tensor*. The subscripts 1, 2, and 3 are used here instead of the subscripts x, y, and z, respectively (see Figure 2.1).

The subscripts used for the nine stress components $\sigma_{\alpha\beta}$ and $\tau_{\alpha\beta}$ have the following meaning: α is the direction of the surface normal upon which the stress acts, and β is the direction of the stress component. For example, the shear stress component τ_{13} in Figure 2.1 is acting on a plane whose normal is parallel to the x-axis ($1 \equiv x$), and the shear stress component itself is parallel to the z-axis ($3 \equiv z$).

2.3 ELASTICITY

In the following we present the three-dimensional generalized Hooke's law suited for isotropic linear elastic materials in three-dimensional stress conditions. The generalized Hooke's law will be applied to the uniaxial stress condition (one-dimensional), the plane strain condition (two-dimensional), and the plane stress condition (also two-dimensional). Hooke's law is not appropriate for soils because soils are neither linear elastic nor isotropic. Nevertheless, sometimes we idealize soils as being linear elastic and isotropic materials—only then can we use Hooke's law to estimate the elastic strains associated with applied stresses within a soil mass.

2.3.1 Three-Dimensional Stress Condition

The simplest form of linear elasticity is the isotropic case. Being isotropic means that the elastic moduli, such as E and v, are orientation independent. This means, for example, that E_{11}, E_{22}, and E_{33} are identical and they are all equal to E (Young's modulus). The stress–strain relationship of the linear elastic isotropic case is given by

$$
\begin{Bmatrix} \sigma_{11} \\ \sigma_{22} \\ \sigma_{33} \\ \tau_{12} \\ \tau_{13} \\ \tau_{23} \end{Bmatrix} = \frac{E}{(1+v)(1-2v)} \begin{bmatrix} 1-v & v & v & 0 & 0 & 0 \\ v & 1-v & v & 0 & 0 & 0 \\ v & v & 1-v & 0 & 0 & 0 \\ 0 & 0 & 0 & 1-2v & 0 & 0 \\ 0 & 0 & 0 & 0 & 1-2v & 0 \\ 0 & 0 & 0 & 0 & 0 & 1-2v \end{bmatrix}
$$

$$
\times \begin{Bmatrix} \varepsilon_{11} \\ \varepsilon_{22} \\ \varepsilon_{33} \\ \varepsilon_{12} \\ \varepsilon_{13} \\ \varepsilon_{23} \end{Bmatrix} \tag{2.2}
$$

The elastic properties are defined completely by Young's modulus, E, and Poisson's ratio, v. Equation (2.2) is also known as the *generalized Hooke's law*. Recall that Hooke's law for the one-dimensional (uniaxial) stress condition is $\sigma = E\varepsilon$. This equation has the same general form as (2.2). It will be shown below that (2.2) reduces to $\sigma = E\varepsilon$ for the uniaxial stress condition.

Equation (2.2) can be inverted to yield

$$
\begin{Bmatrix} \varepsilon_{11} \\ \varepsilon_{22} \\ \varepsilon_{33} \\ \varepsilon_{12} \\ \varepsilon_{13} \\ \varepsilon_{23} \end{Bmatrix} = \begin{bmatrix} 1/E & -v/E & -v/E & 0 & 0 & 0 \\ -v/E & 1/E & -v/E & 0 & 0 & 0 \\ -v/E & -v/E & 1/E & 0 & 0 & 0 \\ 0 & 0 & 0 & 1/2G & 0 & 0 \\ 0 & 0 & 0 & 0 & 1/2G & 0 \\ 0 & 0 & 0 & 0 & 0 & 1/2G \end{bmatrix} \begin{Bmatrix} \sigma_{11} \\ \sigma_{22} \\ \sigma_{33} \\ \tau_{12} \\ \tau_{13} \\ \tau_{23} \end{Bmatrix}
$$

$$(2.3)$$

In this equation, the shear modulus, G, can be expressed in terms of E and v as $G = E/2(1 + v)$.

2.3.2 Uniaxial Stress Condition

The stress condition resulting from an axial stress σ_{11} (tension) applied to a steel rebar can be thought of as a uniaxial stress condition (Figure 2.2). In a uniaxial stress condition we have $\sigma_{22} = \sigma_{33} = \tau_{12} = \tau_{13} = \tau_{23} = 0$, and $\sigma_{11} \neq 0$. Substituting into (2.3), we get

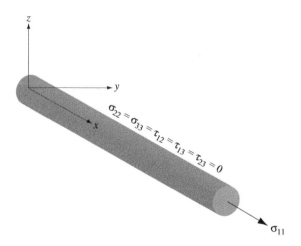

FIGURE 2.2   Uniaxial stress condition.

$$
\begin{Bmatrix} \varepsilon_{11} \\ \varepsilon_{22} \\ \varepsilon_{33} \\ \varepsilon_{12} \\ \varepsilon_{13} \\ \varepsilon_{23} \end{Bmatrix} =
\begin{bmatrix}
1/E & -v/E & -v/E & 0 & 0 & 0 \\
-v/E & 1/E & -v/E & 0 & 0 & 0 \\
-v/E & -v/E & 1/E & 0 & 0 & 0 \\
0 & 0 & 0 & 1/2G & 0 & 0 \\
0 & 0 & 0 & 0 & 1/2G & 0 \\
0 & 0 & 0 & 0 & 0 & 1/2G
\end{bmatrix}
\begin{Bmatrix} \sigma_{11} \\ 0 \\ 0 \\ 0 \\ 0 \\ 0 \end{Bmatrix}
$$

This reduces to two equations:

$$
\varepsilon_{11} = \frac{1}{E} \sigma_{11} \tag{2.4}
$$

and

$$
\varepsilon_{22} = \varepsilon_{33} = \frac{-v}{E} \sigma_{11} \tag{2.5}
$$

substituting (2.4) into (2.5) yields

$$
\varepsilon_{22} = \varepsilon_{33} = \frac{-v}{E} \sigma_{11} = -v \frac{\varepsilon_{11}}{\sigma_{11}} \sigma_{11} = -v \varepsilon_{11}
$$

or

$$
v = \frac{-\varepsilon_{33}}{\varepsilon_{11}} \rightarrow \varepsilon_{33} = -v \varepsilon_{11}
$$

This equation indicates that as the axial stress causes the steel rebar to extend in the axial direction, the rebar becomes slimmer (negative ε_{33}), due to Poisson's effect.

2.3.3 Plane Strain Condition

The plane strain assumption is frequently used in geotechnical analysis of soil structures that are very long in one dimension while having a uniform cross section with finite dimensions. Figure 2.3 illustrates a soil embankment that is long in the z-direction while having a uniform cross section with finite dimensions in the x–y plane. In this case we can assume a plane strain condition in which the strains along the z-axis are assumed to be nil (i.e., $\varepsilon_{33} = \varepsilon_{13} = \varepsilon_{23} = 0$). The seemingly three-dimensional embankment problem reduces to a two-dimensional plane problem in which the cross section of the embankment, in the x–y plane, is assumed to represent the entire embankment. Now, let us substitute $\varepsilon_{33} = \varepsilon_{13} = \varepsilon_{23} = 0$ into (2.2):

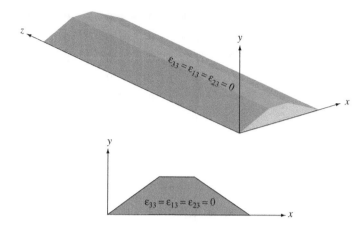

FIGURE 2.3 Plane strain condition.

$$
\begin{Bmatrix} \sigma_{11} \\ \sigma_{22} \\ \sigma_{33} \\ \tau_{12} \\ \tau_{13} \\ \tau_{23} \end{Bmatrix} = \frac{E}{(1+v)(1-2v)} \begin{bmatrix} 1-v & v & v & 0 & 0 & 0 \\ v & 1-v & v & 0 & 0 & 0 \\ v & v & 1-v & 0 & 0 & 0 \\ 0 & 0 & 0 & 1-2v & 0 & 0 \\ 0 & 0 & 0 & 0 & 1-2v & 0 \\ 0 & 0 & 0 & 0 & 0 & 1-2v \end{bmatrix}
$$

$$
\times \begin{Bmatrix} \varepsilon_{11} \\ \varepsilon_{22} \\ 0 \\ \varepsilon_{12} \\ 0 \\ 0 \end{Bmatrix}
$$

or

$$
\begin{Bmatrix} \sigma_{11} \\ \sigma_{22} \\ \tau_{12} \end{Bmatrix} = \frac{E}{(1+v)(1-2v)} \begin{bmatrix} 1-v & v & 0 \\ v & 1-v & 0 \\ 0 & 0 & 1-2v \end{bmatrix} \begin{Bmatrix} \varepsilon_{11} \\ \varepsilon_{22} \\ \varepsilon_{12} \end{Bmatrix} \tag{2.6}
$$

Inverting (2.6), we get

$$
\begin{Bmatrix} \varepsilon_{11} \\ \varepsilon_{22} \\ \varepsilon_{12} \end{Bmatrix} = \frac{1+v}{E} \begin{bmatrix} 1-v & v & 0 \\ v & 1-v & 0 \\ 0 & 0 & 1 \end{bmatrix} \begin{Bmatrix} \sigma_{11} \\ \sigma_{22} \\ \tau_{12} \end{Bmatrix} \tag{2.7}
$$

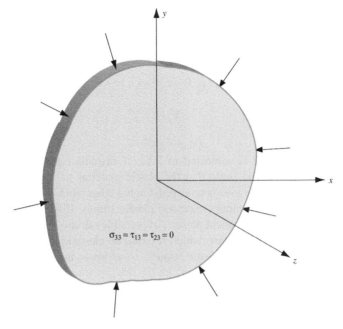

$$\sigma_{33} = \tau_{13} = \tau_{23} = 0$$

FIGURE 2.4   Plane stress condition.

2.3.4 Plane Stress Condition

In the plane stress condition the stresses in the z-direction are assumed negligible (i.e., $\sigma_{33} = \tau_{13} = \tau_{23} = 0$; see Figure 2.4). Substituting these stresses into (2.3), we have

$$
\begin{Bmatrix} \varepsilon_{11} \\ \varepsilon_{22} \\ \varepsilon_{33} \\ \varepsilon_{12} \\ \varepsilon_{13} \\ \varepsilon_{23} \end{Bmatrix} =
\begin{bmatrix}
1/E & -\nu/E & -\nu/E & 0 & 0 & 0 \\
-\nu/E & 1/E & -\nu/E & 0 & 0 & 0 \\
-\nu/E & -\nu/E & 1/E & 0 & 0 & 0 \\
0 & 0 & 0 & 1/2G & 0 & 0 \\
0 & 0 & 0 & 0 & 1/2G & 0 \\
0 & 0 & 0 & 0 & 0 & 1/2G
\end{bmatrix}
\begin{Bmatrix} \sigma_{11} \\ \sigma_{22} \\ 0 \\ \tau_{12} \\ 0 \\ 0 \end{Bmatrix}
\tag{2.8}
$$

Therefore,

$$
\begin{Bmatrix} \varepsilon_{11} \\ \varepsilon_{22} \\ \varepsilon_{12} \end{Bmatrix} = \frac{1}{E}
\begin{bmatrix}
1 & -\nu & 0 \\
-\nu & 1 & 0 \\
0 & 0 & 1+\nu
\end{bmatrix}
\begin{Bmatrix} \sigma_{11} \\ \sigma_{22} \\ \tau_{12} \end{Bmatrix}
\tag{2.9}
$$

Inverting (2.9), we get

$$
\left\{
\begin{array}{c}
\sigma_{11} \\
\sigma_{22} \\
\tau_{12}
\end{array}
\right\}
=
\frac{E}{1 - \nu^2}
\left[
\begin{array}{ccc}
1 & \nu & 0 \\
\nu & 1 & 0 \\
0 & 0 & 1 - \nu
\end{array}
\right]
\left\{
\begin{array}{c}
\varepsilon_{11} \\
\varepsilon_{22} \\
\varepsilon_{12}
\end{array}
\right\}
\qquad (2.10)
$$

2.4   PLASTICITY

When an elastic material is subjected to load, it sustains elastic strains. Elastic strains are reversible in the sense that the elastic material will spring back to its undeformed condition if the load is removed. On the other hand, if a plastic material is subjected to a load, it sustains elastic and plastic strains. If the load is removed, the material will sustain permanent plastic (irreversible) strains, whereas the elastic strains are recovered. Hooke's law, which is based on elasticity theory, is sufficient (in most cases) to estimate the elastic strains. To estimate the plastic strains, one needs to use plasticity theory.

Plasticity theory was originally developed to predict the behavior of metals subjected to loads exceeding their elastic limits. Similar models were developed later to calculate the irreversible strains in concrete, soils, and polymers. In this chapter we present three plasticity models for soils that are frequently used in geotechnical engineering applications. It is customary in plasticity theory to decompose strains into elastic and plastic parts. A plasticity model includes (1) a yield criterion that predicts whether the material should respond elastically or plastically due to a loading increment, (2) a strain hardening rule that controls the shape of the stress–strain response during plastic straining, and (3) a plastic flow rule that determines the direction of the plastic strain increment caused by a stress increment.

2.5   MODIFIED CAM CLAY MODEL

Researchers at Cambridge University formulated the first critical-state models for describing the behavior of soft soils: the Cam clay and modified Cam clay models (Roscoe and Burland, 1968; Schofield and Wroth, 1968). Both models are capable of describing the stress–strain behavior of soils; in particular, the models can predict the pressure-dependent soil strength and the compression and dilatancy (volume change) caused by shearing. Because the models are based on critical-state theory, they both predict unlimited soil deformations without changes in stress or volume when the critical state is reached. The following description is limited to the modified Cam clay model.

Soil is composed of solids, liquids, and gases. The Cam clay model assumes that the voids between the solid particles are filled only with water (i.e., the soil is fully saturated). When the soil is loaded, significant irreversible (plastic) volume changes occur, due to the water that is expelled from the voids. Realistic prediction of these deformations is crucial for many geotechnical engineering problems. Formulations of the modified Cam clay model are based on plasticity theory, through which it is possible to predict realistically volume changes due to various types of loading.

In critical-state theory, the state of a soil specimen is characterized by three parameters: mean effective stress p', deviator stress (shear stress) q, and void ratio, e. The mean effective stress can be calculated in terms of the principal effective stresses σ_1', σ_2', and σ_3' as

$$p' = \frac{\sigma_1' + \sigma_2' + \sigma_3'}{3} \tag{2.11}$$

and the shear stress is defined as

$$q = \frac{1}{\sqrt{2}}\sqrt{(\sigma_1' - \sigma_2')^2 + (\sigma_2' - \sigma_3')^2 + (\sigma_1' - \sigma_3')^2} \tag{2.12}$$

For the consolidation stage of a consolidated–drained triaxial compression test, we have $\sigma_1' = \sigma_2' = \sigma_3'$, where σ_3' is the confining pressure; therefore,

$$p' = \frac{\sigma_1' + \sigma_2' + \sigma_3'}{3} = \frac{3\sigma_3'}{3} = \sigma_3' \tag{2.13}$$

and

$$q = \frac{1}{\sqrt{2}}\sqrt{(\sigma_1' - \sigma_2')^2 + (\sigma_2' - \sigma_3')^2 + (\sigma_1' - \sigma_3')^2} = 0 \tag{2.14}$$

For the shearing stage of a triaxial compression test we have $\sigma_1' \neq \sigma_2' = \sigma_3'$; therefore,

$$p' = \frac{\sigma_1' + \sigma_2' + \sigma_3'}{3} = \frac{\sigma_1' + 2\sigma_3'}{3} \tag{2.15}$$

and

$$q = \frac{1}{\sqrt{2}}\sqrt{(\sigma_1' - \sigma_2')^2 + (\sigma_2' - \sigma_3')^2 + (\sigma_1' - \sigma_3')^2} = \frac{1}{\sqrt{2}}\sqrt{2(\sigma_1' - \sigma_3')^2} = \sigma_1' - \sigma_3' \tag{2.16}$$

Note that in a triaxial stress condition the shear stress $q = \sigma_1' - \sigma_3'$ is termed the *deviator stress*, $\Delta\sigma_d (= \sigma_1' - \sigma_3')$.

The effective stress path of a triaxial test represents the locus of the effective stress state in the p'–q plane. The effective stress path can be calculated easily from the results of a triaxial test using (2.15) and (2.16). For a consolidated–drained triaxial test, the effective stress path is a straight line whose slope is defined as

$$\text{slope} = \frac{\Delta q}{\Delta p'}$$

Noting that σ_3' is constant and using (2.16) we have

$$\Delta q = \Delta\sigma_1' - \Delta\sigma_3' = \Delta\sigma_1' - 0 = \Delta\sigma_1'$$

From (2.15) we have

$$\Delta p' = \frac{\Delta\sigma'_1 + 2\Delta\sigma'_3}{3} = \frac{\Delta\sigma'_1 + 0}{3} = \frac{\Delta\sigma'_1}{3}$$

Therefore,

$$\text{slope} = \frac{\Delta\sigma'_1}{\Delta\sigma'_1/3} = 3$$

2.5.1 Normal Consolidation Line and Unloading–Reloading Lines

The consolidation characteristics of a soil can be measured in the laboratory using a one-dimensional consolidation test or isotropic consolidation test. A *one-dimensional consolidation test* involves a cylindrical soil specimen confined in a rigid ring and subjected to normal pressure (Chapter 4). The normal pressure is increased in stages, each ending when the excess pore water pressure generated by the pressure increment has ceased. The results of a one-dimensional consolidation test are usually presented in the e–log σ'_v plane as shown in Figure 2.5a, where e is the void ratio and σ'_v is the vertical effective stress.

In reference to Figure 2.5a, let us define a *preconsolidation pressure*, σ'_c, as the maximum past pressure exerted on the clay specimen. A *normally consolidated* (NC) *clay* is defined as a clay that has a present (in situ) vertical effective stress σ'_0 equal to its preconsolidation pressure σ'_c. An *overconsolidated* (OC) *clay* is defined as a clay that has a present vertical effective stress less than its preconsolidation pressure. Finally, define an *overconsolidation ratio* (OCR) as

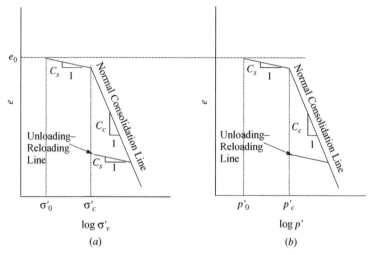

FIGURE 2.5 Idealized consolidation curves: (*a*) one-dimensional consolidation; (*b*) isotropic consolidation (compression).

the ratio of the preconsolidation pressure to the present vertical effective stress (OCR $= \sigma'_c/\sigma'_0$). Imagine a clay layer that was subjected to a constant pressure of 100 kPa caused by a thick layer of sand. Then the sand layer is removed and the pressure exerted is totally gone. In this case the preconsolidation pressure is $\sigma'_c = 100$ kPa—it is the maximum past pressure exerted. Let us assume that the present vertical effective stress in the middle of the clay layer is $\sigma'_0 = 50$ kPa. The present vertical effective stress, 50 kPa, is less than the preconsolidation pressure, 100 kPa. Therefore, the clay is overconsolidated. The overconsolidation ratio of this clay is OCR $= \sigma'_c/\sigma'_0 = 100$ kPa/50 kPa $= 2$.

The preconsolidation pressure is a soil parameter that can be obtained from the e–log σ'_v curve deduced from the results of a one-dimensional consolidation test (Figure 2.5a). The preconsolidation pressure is located near the point where the e–log σ'_v curve changes in slope. Other consolidation parameters, such as the compression index (C_c) and the swelling index (C_s), are also obtained from an e–log σ'_v curve. The *compression index* is the slope of the loading portion in the e–log σ'_v plane, and the *swelling index* is the slope of the unloading portion, as indicated in the figure.

An *isotropic consolidation* (compression) *test* can also be performed to obtain the consolidation characteristics of soils. The test consists of a cylindrical soil specimen subjected to an all-around confining pressure during which the specimen is allowed to consolidate. The confining pressure is increased in increments, each of which ends when the excess pore water pressure generated by the stress increment has ceased. The void ratio versus mean effective stress relationship in a semilogarithmic plane (e–log p') is obtained from the changes in volume at the end of each loading stage of the isotropic consolidation test. The mean effective stress in an isotropic consolidation test is $p' = (\sigma'_1 + \sigma'_2 + \sigma'_3)/3 = (\sigma'_3 + \sigma'_3 + \sigma'_3)/3 = \sigma'_3$, where σ'_3 is the confining pressure. An example of an e–log p' curve is shown in Figure 2.5b.

In the derivation of the modified Cam clay model it is assumed that when a soil sample is consolidated under isotropic stress conditions ($p' = \sigma'_1 = \sigma'_2 = \sigma'_3$), the relationship between its void ratio (e) and ln p' (natural logarithm of p') is a straight line. This line is the normal consolidation line shown in Figure 2.6. Also, there exists a set of straight unloading–reloading (swelling) lines that describe the unloading–reloading behavior of the soft soil in the e–ln p' plane, as shown in the figure. Note that λ is the slope of the normal consolidation line in the e–ln p' plane and κ is the slope of the unloading–reloading line in the same plane.

Consider a soil specimen that is subjected initially to a mean effective stress $p'_A = 1$ kPa and has a void ratio $e_A = e_N = 0.92$ in an isotropic consolidation test (Figure 2.6). This condition is represented by point A in the figure. In the modified Cam clay model, when the mean effective stress is increased to $p' = 30$ kPa, for example, the stress condition in the e–ln p' plane will move down the normal consolidation line from point A to point B. If the sample is unloaded back to $p' = 1$ kPa, point B will not move back to point A; instead, it will move up the unloading–reloading line to point C, at which the soil has a smaller void ratio $e_C = 0.86$.

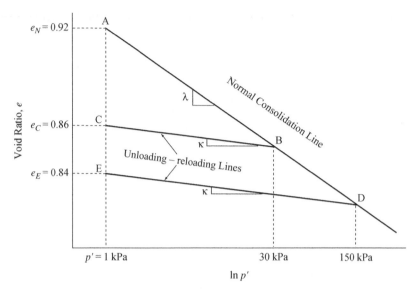

FIGURE 2.6 Consolidation curve in the void ratio versus mean effective stress (natural logarithm of p') plane.

If the sample is then reloaded to a stress $p' = 150$ kPa, point C will first move down the unloading–reloading line to point B, at which $p' = 30$ kPa. When $p' = 30$ kPa is reached, the stress condition will change course and move down the normal consolidation line to point D. If the sample is again unloaded to $p' = 1$ kPa, point D will move up the unloading–reloading line to point E, at which the soil will have a void ratio $e_E = 0.84$.

In the e–$\ln p'$ plane (Figure 2.6), the normal consolidation line is defined by the equation

$$e = e_N - \lambda \ln p' \tag{2.17}$$

The normal consolidation line exists in the e–p' plane as shown in Figure 2.7; therefore, its equation in the p'–q plane is $q = 0$.

In the e–$\ln p'$ plane the equation for an unloading–reloading line has the form

$$e = e_C - \kappa \ln p' \tag{2.18}$$

and it has the form $q = 0$ in the p'–q plane.

The material parameters λ, κ, and e_N are unique for a particular soil. λ is the slope of the normal consolidation line and the critical-state line (which is described below) in the e–$\ln p'$ plane, κ is the slope of the unloading–reloading line in the e–$\ln p'$ plane, and e_N is the void ratio on the normal consolidation line at unit mean effective stress (point A in Figure 2.6). Thus, e_N is dependent on the pressure

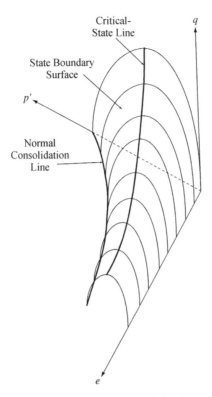

Critical-
State Line

State Boundary
Surface

p'

Normal
Consolidation
Line

q

e

FIGURE 2.7   State boundary surface of the Cam clay model.

units used in the e–ln p' plot. Note that the unloading–reloading line BC is fully defined by its slope κ and by the coordinates of either point B or point C.

Any point along the normal consolidation line represents the stress state of a normally consolidated (NC) soil. Also, any point along an unloading–reloading line represents an overconsolidated stress state. Take, for example, point C (Figure 2.6), at which $p'_C = 1$ kPa and $e_C = 0.86$. This point has an overconsolidated stress state since the soil was previously subjected to a preconsolidation pressure of $p'_B = 30$ kPa. In fact, the overconsolidation ratio at point C is OCR $= p'_B/p'_C = 30$ kPa/1 kPa $= 30$, which means that the soil at point C is heavily overconsolidated.

2.5.2   Critical-State Line

Applying an increasing shear stress on a soil sample in a triaxial test, for example, will eventually lead to a state in which further shearing can occur without changes in volume, as shown in Figure 2.8, known as the *critical-state condition*. The *critical-state line* (CSL) (Figures 2.9 and 2.10) is a presentation of the critical state condition. The critical-state line in e–p'–q space is shown in Figure 2.7. To obtain the critical-state line we need to perform consolidated–drained (CD)

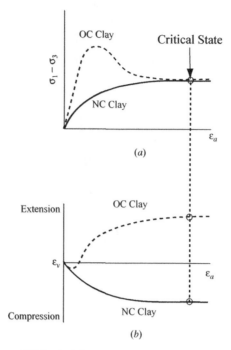

(a)

(b)

FIGURE 2.8   Critical-state definition.

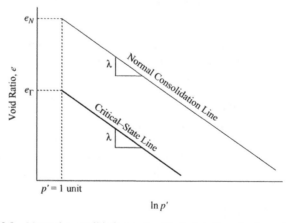

FIGURE 2.9   Normal consolidation and critical-state lines in the e–$\ln p'$ plane.

or consolidated–undrained (CU) triaxial compression tests on representative soil specimens. From the test results we can obtain the critical-state friction angle of the soil by drawing the effective-stress Mohr's circles that represent a critical-state stress condition, such as the one shown in Figure 2.8. Next, we draw the effective-stress Mohr–Coulomb failure criterion, which is a straight line tangent to

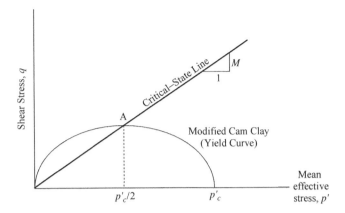

FIGURE 2.10   Yield surface of a Cam clay model in the q–p' plane.

the effective-stress Mohr's circles. The slope of this line is the critical-state friction angle ϕ'. The slope of the critical-state line in the p'–q plane, M (Figure 2.10), can be calculated as

$$M = \frac{6 \sin \phi'}{3 - \sin \phi'} \tag{2.19}$$

In reference to Figure 2.10, the critical-state line has the following equation in the p'–q plane:

$$q_f = M p'_f \tag{2.20}$$

where p'_f is the mean effective stress at failure and q_f is the shear stress at failure (i.e., the shear strength). Equation (2.20) is the failure criterion used in the modified Cam clay model. This failure criterion bears the same meaning as the Mohr–Coulomb failure criterion $\tau_f = c' + \sigma' \tan \phi'$, where τ_f is the shear stress at failure and σ' is the effective normal stress. Equation (2.19) can be obtained by comparing the effective-stress Mohr–Coulomb failure criterion with (2.20), where c' is assumed to be zero in the Mohr–Coulomb failure criterion (true for sands and soft clays).

The critical-state line is parallel to the normal consolidation line in the e–ln p' plane, as shown in Figure 2.9. The equation of the critical-state line in this plane is given as

$$e_f = e_\Gamma - \lambda \ln p' \tag{2.21}$$

where e_f is the void ratio at failure and e_Γ is the void ratio of the critical-state line at $p' = 1$ kPa (or any other unit). Note that the parameters e_N and e_Γ (Figure 2.9) are related by the equation

$$e_\Gamma = e_N - (\lambda - \kappa) \ln 2 \tag{2.22}$$

Because of (2.22), either e_N or e_Γ needs to be provided. The other parameter can be calculated using (2.22).

2.5.3 Yield Function

In the p'–q plane, the modified Cam clay yield surface is an ellipse given by

$$\frac{q^2}{p'^2} + M^2\left(1 - \frac{p'_c}{p'}\right) = 0 \tag{2.23}$$

Figure 2.10 shows an elliptical yield surface corresponding to a preconsolidation pressure p'_c. The parameter p'_c controls the size of the yield surface and is different for each unloading–reloading line. The parameter p'_c is used to define the hardening behavior of soil. The soil behavior is elastic until the stress state of the soil specimen (p', q) hits the yield surface. Thereafter, the soil behaves in a plastic manner. Note that the critical-state line intersects the yield surface at point A, located at the crown of the ellipse (and thus has the maximum q). Also note that the p'-coordinate of the intersection point (point A) is $p'_c/2$. Figure 2.7 presents the yield surface of the modified Cam clay model in e–p'–q three-dimensional space, termed the *state boundary surface*. Note that the size of the yield surface decreases as the void ratio increases.

2.5.4 Hardening and Softening Behavior

Consider a soil specimen that is isotropically consolidated to a mean effective stress p'_c and then is unloaded slightly to p'_0, as shown in Figure 2.11a. Here p'_c is the preconsolidation pressure and p'_0 is the present pressure. The size of the initial yield surface is determined by p'_c, as shown in the figure. Note that the soil is lightly overconsolidated such that OCR $= p'_c/p'_o < 2$.

Next, let us start shearing the soil specimen under drained conditions. The effective stress path of this consolidated–drained (CD) triaxial test is shown in the figure as a straight line with a 3 : 1 slope (3 vertical to 1 horizontal). If the stress path touches the initial yield surface to the right of the point at which the CSL intersects the yield surface, hardening behavior, accompanied by compression, will occur. This side of the yield surface is the *wet side* as indicated in Figure 2.11a.

During shearing, the soil specimen sustains only elastic strains within the initial yield surface. When the stress state of the soil touches the yield surface, the specimen will sustain plastic strains as well as elastic strains. The yield surface will expand (hardening), causing further plastic strains, until the stress state of the specimen touches the critical state line at point F, where failure occurs; the soil will continue to distort without changes in shear stress or volume. Figure 2.11b shows the stress–strain hardening behavior for normally consolidated and lightly overconsolidated clays.

Consider another soil specimen that is isotropically consolidated to a mean effective stress p'_c and then unloaded to p'_0 such that the specimen is heavily

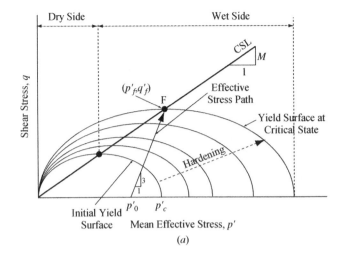

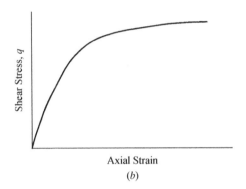

FIGURE 2.11 Cam clay hardening behavior: (*a*) evolution of a yield surface during hardening; (*b*) stress–strain curve with strain hardening.

overconsolidated (OCR > 2). Note that p'_c is the preconsolidation pressure and p'_o is the present pressure. As shown in Figure 2.12*a*, the size of the initial yield surface is governed by p'_c.

When shearing the soil specimen under drained conditions, the effective stress path is a straight line making a 3 : 1 slope, as shown in Figure 2.12*a*. In the case of heavily overconsolidated clay, the stress path traverses the initial yield surface, to the left of the point at which the CSL intersects the yield surface, inducing softening behavior accompanied by dilatancy (expansion). This side of the yield surface is the *dry side*, as indicated in Figure 2.12*a*.

The heavily overconsolidated soil specimen sustains only elastic strains within the initial yield surface. Note that the effective stress path traverses the critical-state line before touching the initial yield surface and without causing failure in the soil specimen elastic behavior. When the effective stress path touches the yield surface,

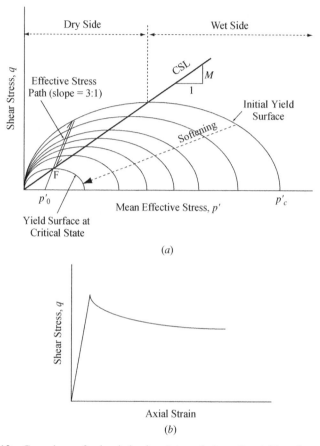

FIGURE 2.12 Cam clay softening behavior: (*a*) evolution of a yield surface during softening; (*b*) stress–strain curve with strain softening.

the yield surface will contract (softening), causing further plastic strains, until the stress state of the specimen touches the critical-state line again at point F, where failure occurs. Figure 2.12*b* shows the stress–strain softening behavior for heavily overconsolidated clays.

2.5.5 Elastic Moduli for Soil

Assuming that the elastic response of soil is isotropic and elastic, we need to have two elastic moduli to define its elastic stiffness completely. The elastic material constants commonly used to relate stresses to strains are Young's modulus E, shear modulus G, Poisson's ratio v, and bulk modulus K. These moduli are related—if you know two, you can calculate the other two. For example, if you know K and

v, you can calculate E and G using

$$E = 3K(1 - 2v) \tag{2.24}$$

and

$$G = \frac{3K(1 - 2v)}{2(1 + v)} \tag{2.25}$$

respectively. The elastic behavior of soil is nonlinear and stress dependent. Therefore, the elastic moduli need to be presented in incremental form.

For soils modeled using the modified Cam clay model, the bulk modulus K is stress dependent (i.e., K is not a constant). The bulk modulus depends on the mean effective stress p', void ratio e_0, and unloading–reloading line slope κ. The following equation can be obtained easily from the equation of the unloading–reloading line, (2.18), which describes the elastic behavior of soil:

$$K = \frac{(1 + e_0)p'}{\kappa} \tag{2.26}$$

Substituting (2.26) into (2.24) and (2.25), respectively, we can obtain

$$E = \frac{3(1 - 2v)(1 + e_0)p'}{\kappa} \tag{2.27}$$

and

$$G = \frac{3(1 - 2v)(1 + e_0)p'}{2(1 + v)\kappa} \tag{2.28}$$

Note that E and G are also not constants. They are a function of the mean effective stress p', void ratio e_0, unloading–reloading line slope κ, and Poisson's ratio v. In (2.27) and (2.28) we can, for simplicity, assume a constant Poisson's ratio.

2.5.6   Summary of Modified Cam Clay Model Parameters

Overconsolidation Ratio   The stress state of a soil can be described by its current mean effective stress p'_0, void ratio e, and yield stress p'_c (preconsolidation pressure). The ratio of preconsolidation pressure to current mean effective stress is the overconsolidation ratio (OCR):

$$\text{OCR} = \frac{p'_c}{p'_0} \tag{2.29}$$

$\text{OCR} = 1$ indicates a normal consolidation state; a state in which the maximum mean effective stress experienced previously by the soil is equal to the current mean effective stress. $\text{OCR} > 1$ indicates an overconsolidated stress state where the preconsolidation pressure is greater than the present mean effective pressure.

Slope M of the Critical-State Line The slope M of the CSL in the $p'–q$ plane can be calculated from the internal friction angle ϕ' obtained from triaxial tests results at failure:

$$M = \frac{6 \sin \phi'}{3 - \sin \phi'}$$

Alternatively, the at-failure stresses from triaxial tests results can be plotted in the $p'–q$ plane. The data points can be best fitted with a straight line whose slope is M.

λ *and* κ Slopes λ and κ of the normal consolidation and unloading–reloading lines in the e–$\ln p'$ plane are related to the compression index C_c and swelling index C_s obtained from an isotropic consolidation test:

$$\lambda = \frac{C_c}{\ln 10} = \frac{C_c}{2.3}$$

and

$$\kappa = \frac{C_s}{\ln 10} = \frac{C_s}{2.3}$$

2.5.7 Incremental Plastic Strains

In their derivation of the modified Cam clay formulations, Roscoe and Burland (1968) assumed that the work done on a soil specimen by a load q, p', is given by

$$dW = p' \, d\varepsilon_v^p + q \, d\varepsilon_s^p \tag{2.30}$$

where $d\varepsilon_v^p$ is the plastic (irreversible) volumetric strain increment and $d\varepsilon_s^p$ is the plastic shear strain increment (also irreversible). In a triaxial stress state, $d\varepsilon_v^p$ is equal to $d\varepsilon_1^p + 2d\varepsilon_3^p$ and $d\varepsilon_s^p$ is equal to $\frac{2}{3}(d\varepsilon_1^p - d\varepsilon_3^p)$. Also, p' and q are given by (2.15) and (2.16), respectively.

Roscoe and Burland (1968) derived an associated *plastic flow rule* given by

$$\frac{d\varepsilon_v^p}{d\varepsilon_s^p} = \frac{M^2 - \eta^2}{2\eta} \tag{2.31}$$

where $\eta = q/p'$ is the stress ratio. Note that η is equal to M when $q = q_f$ and $p' = p'_f$ (at failure). In the modified Cam clay model the plastic strain increment resulting from a load increment $d\eta = dq/dp'$, shown in Figure 2.13a, is normal to the yield surface as shown in Figure 2.13b. This is referred to as the *normality rule*.

Next, we present equations for strain increments caused by a stress increment, $d\eta = dq/dp'$. These equations allow us to calculate the plastic volumetric strain increment $d\varepsilon_v^p$, the elastic volumetric strain increment $d\varepsilon_v^e$, the plastic shear strain increment $d\varepsilon_s^p$, and the elastic shear strain increment $d\varepsilon_s^e$. Note that the total volumetric strain increment is given as

$$d\varepsilon_v = d\varepsilon_v^e + d\varepsilon_v^p \tag{2.32}$$

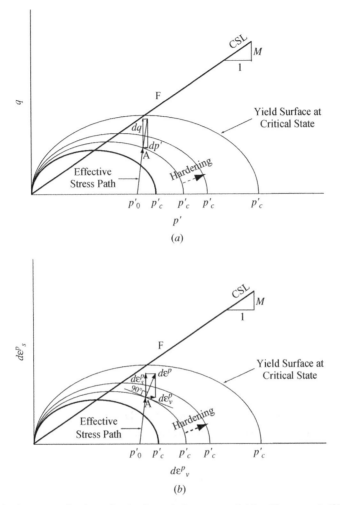

FIGURE 2.13 Determination of a plastic strain increment: (*a*) load increment; (*b*) direction of the plastic strain increment (normality rule).

and the total shear strain increment is given as

$$d\varepsilon_s = d\varepsilon_s^e + d\varepsilon_s^p \qquad (2.33)$$

For simplicity, the critical-state theory assumes that no recoverable energy is associated with shear distortion (i.e., $d\varepsilon_s^e = 0$). Thus, (2.33) is reduced to $d\varepsilon_s = d\varepsilon_s^p$. It is important to note that the following equations are given in incremental forms and thus need to be used in an incremental manner: The load must be applied in small increments and the corresponding strain increments are calculated. The strain increments are accumulated to give us the total strain.

Desai and Siriwardane (1984) presented the following equations for volumetric and shear strain increments:

Volumetric strains The plastic volumetric strain increment

$$d\varepsilon_v^p = \frac{\lambda - \kappa}{1 + e}\left(\frac{dp'}{p'} + \frac{2\eta\, d\eta}{M^2 + \eta^2}\right) \tag{2.34}$$

The elastic volumetric strain increment

$$d\varepsilon_v^e = \frac{\kappa}{1 + e}\frac{dp'}{p'} \tag{2.35}$$

Thus, the total volumetric strain increment:

$$d\varepsilon_v = \frac{\lambda}{1 + e}\left[\frac{dp'}{p'} + \left(1 - \frac{\kappa}{\lambda}\right)\frac{2\eta\, d\eta}{M^2 + \eta^2}\right] \tag{2.36}$$

Shear strains

$$d\varepsilon_s = d\varepsilon_s^p = \frac{\lambda - \kappa}{1 + e}\left(\frac{dp'}{p'} + \frac{2\eta\, d\eta}{M^2 + \eta^2}\right)\frac{2\eta}{M^2 - \eta^2} \tag{2.37}$$

or

$$d\varepsilon_s = d\varepsilon_s^p = d\varepsilon_v^p\frac{2\eta}{M^2 - \eta^2} \tag{2.38}$$

2.5.8 Calculations of the Consolidated–Drained Stress–Strain Behavior of a Normally Consolidated Clay Using the Modified Cam Clay Model

Using (2.34) to (2.38), we can predict the stress–strain behavior of a normally consolidated (NC) clay. The stress is applied using a given stress path, and the corresponding strains are calculated. The effective stress path in a consolidated–drained triaxial test is a straight line making a slope of 3 in the p'–q plane, as described earlier. Equations (2.34) to (2.38) are given in incremental form. Thus, incremental strains caused by stress increments should be calculated. The increments are then added together to calculate the total strains.

Figure 2.14a describes the behavior of a lightly overconsolidated soil specimen in a CD test. Initially, the stress state of the specimen is located on the NCL, as indicated by point 1 in Figure 2.14b. The specimen is subjected to an all-around confining pressure (isotropic consolidation), bringing the state of stress to point 2, which corresponds to the preconsolidation pressure p_c', as shown in the figure. The specimen is then unloaded to point 3, which corresponds to the present pressure p_0'. Now we can start shearing the specimen in a drained condition by allowing water to leave during shearing. This means that there will be volume changes in the soil specimen. The effective stress path during drained shearing is a straight

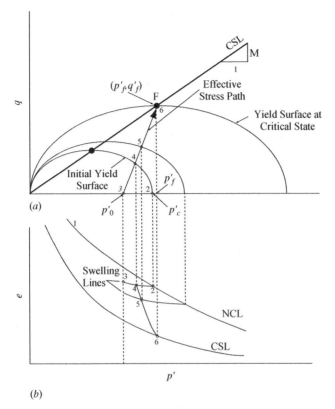

(a)

(b)

FIGURE 2.14 Consolidated drained triaxial test behavior of a lightly overconsolidated clay.

line, with a slope $= 3$, emanating from point 3. Thus, the equation of the effective stress path is given as

$$q_f = 3(p'_f - p'_0) \tag{2.39}$$

The soil specimen will encounter only elastic strains as long as the effective stress state is within the elastic region (i.e., within the initial yield surface). When the effective stress path touches point 4, the soil starts to yield. The effective stress path will maintain its course toward the critical-state line, during which the yield surface grows (strain hardening) until the effective stress path touches the critical-state line at point 6, where failure occurs. Note now how point 1 travels along the NCL to point 2 during isotropic consolidation. Upon unloading, point 2 moves along the swelling line to point 3. When shearing starts, point 3 moves along the same swelling line to point 4. From point 3 to point 4, the soil specimen encounters only elastic strains. As shearing continues beyond the elastic zone, plastic strains ensue and the yield surface grows. During that time, point 4 travels to point 5, located on another swelling line, which corresponds to the new yield surface that passes through point 5. Further loading will cause the yield surface to expand, and point 5 moves to point 6, located on the critical-state line.

The following step-by-step procedure, adopted from Desai and Siriwardane (1984), can be programmed in a spreadsheet to calculate the stress–strain behavior. A similar procedure, used by Budhu (2007), can be used for lightly overconsolidated soils.

2.5.9   Step-by-Step Calculation Procedure for a CD Triaxial Test on NC Clays

Step 1: Choose a stress path for a specific loading condition. For a normally consolidated clay in a consolidated–drained triaxial test, the effective stress path (ESP) has a slope of 3 in the p'–q plane. The stress path starts from $p' = p_0' = \sigma_3'$. Let us apply a small mean effective stress increment $\Delta p'$. Next, let us vary p' from p_0' to p_f' as follows: $p' = p_0'$, $p_0' + \Delta p'$, $p_0' + 2\Delta p'$, $p_0' + 3\Delta p', \ldots, p_f'$, where p_f' is the mean effective stress at failure. The spreadsheet shown in Table 2.1 is used to predict the consolidated–drained triaxial behavior of a normally consolidated clay specimen subjected to a confining pressure $\sigma_3' = 206.7$ kPa. Column (1) in the spreadsheet is reserved for the variation of the effective mean stress. $p_0' = 206.7$ kPa and $\Delta p' = 7.2$ kPa are used. Since $\Delta q = 3\Delta p'$ in the effective stress path of a CD triaxial test, q should vary as follows: $q = 0, 3\Delta p', 6\Delta p', 9\Delta p', \ldots, q_f$, where q_f is the shear stress at failure ($= Mp_f'$). This is shown in column (2).

The mean effective stress at failure p_f' and the shear stress at failure q_f can be calculated easily if we realize that they are at the point where the effective stress path intersects the critical-state line (Figure 2.14a). The equation of the critical-state line is $q_f = Mp_f'$, and the equation of the effective stress path is $q_f = 3(p_f' - p_0')$. Solving these two equations simultaneously, we get

$$p_f' = \frac{3p_0'}{3 - M} \quad \text{and} \quad q_f = \frac{3Mp_0'}{3 - M}$$

In the present example, use $M = 1$.

Step 2: Calculate the stress ratio $\eta = q/p'$ for each row [column (3)].

Step 3: Calculate the increment of stress ratio $d\eta = \eta_i - \eta_{i-1}$, where i is the current row and $i - 1$ is the preceding row [column (4)].

Step 4: Calculate the total volumetric strain increment using (2.36):

$$d\varepsilon_v = \frac{\lambda}{1 + e}\left[\frac{dp'}{p'} + \left(1 - \frac{\kappa}{\lambda}\right)\frac{2\eta\, d\eta}{M^2 + \eta^2}\right] \quad \text{column (5)}$$

In the present example, use $\kappa = 0.026$, $\lambda = 0.174$, and $e_0 = 0.889$.

Step 5: Calculate the total volumetric strain, $(\varepsilon_v)_k = \sum_{i=1}^{i=k}(d\varepsilon_v)_i$, where k is the current row. This is done in column (6).

TABLE 2.1

(1) p' (kPa)	(2) q (kPa)	(3) η	(4) $d\eta$	(5) $d\varepsilon_v$	(6) ε_v	(7) de	(8) e	(9) $d\varepsilon_s$	(10) ε_s	(11) ε_1
0	0	0.00000	0.00000	0.00000	0.00000	0.00000	0.88900	0.00000	0.00000	0.00000
206.7	0.0	0.00000	0.00000	0.00322	0.00322	0.00609	0.88291	0.00000	0.00000	0.00107
213.9	21.7	0.10139	0.10139	0.00472	0.00794	0.00888	0.87403	0.00088	0.00088	0.00353
221.2	43.4	0.19615	0.09476	0.00583	0.01377	0.01093	0.86310	0.00224	0.00313	0.00772
228.4	65.1	0.28491	0.08876	0.00660	0.02038	0.01230	0.85079	0.00393	0.00706	0.01385
235.6	86.8	0.36822	0.08331	0.00708	0.02746	0.01311	0.83768	0.00588	0.01294	0.02209
242.9	108.5	0.44657	0.07835	0.00734	0.03480	0.01349	0.82420	0.00807	0.02101	0.03261
250.1	130.1	0.52039	0.07382	0.00742	0.04222	0.01354	0.81065	0.01056	0.03157	0.04564
257.3	151.8	0.59007	0.06967	0.00739	0.04961	0.01338	0.79727	0.01346	0.04503	0.06157
264.5	173.5	0.65593	0.06586	0.00727	0.05689	0.01307	0.78420	0.01701	0.06204	0.08100
271.8	195.2	0.71829	0.06236	0.00710	0.06399	0.01267	0.77153	0.02161	0.08365	0.10498
279.0	216.9	0.77742	0.05913	0.00690	0.07088	0.01221	0.75932	0.02806	0.11170	0.13533
286.2	238.6	0.83356	0.05614	0.00667	0.07755	0.01174	0.74758	0.03816	0.14986	0.17571
293.5	260.3	0.88694	0.05337	0.00644	0.08399	0.01125	0.73633	0.05691	0.20678	0.23477
300.7	282.0	0.93774	0.05081	0.00620	0.09019	0.01077	0.72557	0.10568	0.31246	0.34252
307.9	303.7	0.98617	0.04842	0.00597	0.09616	0.01030	0.71527	0.58718	0.89964	0.93169

Step 6: Calculate the change of void ratio: $de = (1 + e)d\varepsilon_v$ [column (7)], where e is the void ratio at the beginning of the current increment taken from column (8). For the first row, use $e = e_0$. In the example spreadsheet we used $e_0 = 0.889$. In column (8), calculate the updated void ratio $e_i = e_{i-1} - de$, where i is the current row and $i - 1$ is the preceding row.

Step 7: Calculate the total shear strain increment using (2.37):

$$d\varepsilon_s = d\varepsilon_s^p = \frac{\lambda - \kappa}{1 + e}\left(\frac{dp'}{p'} + \frac{2\eta\, d\eta}{M^2 + \eta^2}\right)\left(\frac{2\eta}{M^2 - \eta^2}\right) \qquad \text{column (9)}$$

Step 8: Calculate the total shear strain: $(\varepsilon_s)_k = \sum_{i=1}^{i=k}(d\varepsilon_s)_i$, where k is the current row. This is done in column (10).

Step 9: Calculate the total axial strain: $\varepsilon_1 = \varepsilon_v/3 + \varepsilon_s$ [column (11)].

The deviator stress versus axial strain behavior predicted for the soil can be plotted using data from columns (2) and (11). Also, the volumetric strain versus axial strain can be plotted using data from columns (6) and (11). Figure 2.15 shows the behavior predicted for this normally consolidated clay when tested in consolidated–drained triaxial compression with a confining pressure of 206.7 kPa. The figure also shows the behavior predicted for the same soil when subjected to confining pressures of 68.9 and 137.8 kPa.

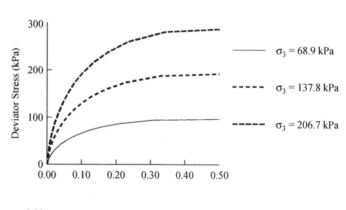

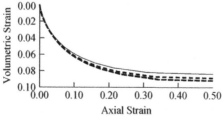

FIGURE 2.15 Predicting the CD triaxial behavior of NC clay using a Cam clay model.

2.5.10   Calculations of the Consolidated–Undrained Stress–Strain Behavior of a Normally Consolidated Clay Using the Modified Cam Clay Model

Figure 2.16a illustrates the behavior of a lightly overconsolidated soil specimen in a CU test. Initially, the stress state of the specimen is located on the NCL as indicated by point 1 in Figure 2.16b. The specimen is subjected to an all-around confining pressure (isotropic consolidation) that brings the state of stress to point 2, which corresponds to the preconsolidation pressure p'_c as shown in Figure 2.16. The specimen is then unloaded to point 3, which corresponds the present pressure p'_0. Now we can start shearing the specimen in an undrained condition by preventing water from leaving the specimen. This means that there will be no volume change. The total stress path during undrained shearing is a straight line, with a slope $= 3$, emanating from point 3. Thus, the equation of the total stress path is give as

$$q = 3(p - p'_0) \tag{2.40}$$

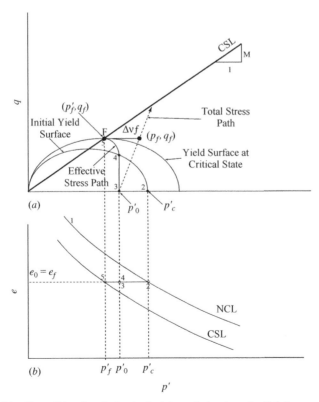

FIGURE 2.16   Consolidated undrained triaxial test behavior of a lightly overconsolidated clay.

The effective stress path during undrained shearing is a vertical line (line 3–4 in Figure 2.16a) in the elastic region within the initial yield surface. If the line were not vertical, there would be volumetric strains resulting from changes in the mean effective stress, and that is not admissible in an undrained condition. At point 4 the soil starts yielding. The effective stress path will turn to the left toward the critical-state line. During that time the yield surface grows (strain hardening) until the effective stress path touches the critical-state line at point 5, where failure occurs. Note how points 2, 3, 4, and 5 travel along a horizontal line, from right to left, in the e–p' plane as shown in Figure 2.16b. If line 2–5 were not horizontal in the e–p' plane, there would be a change in the void ratio, which is not admissible in this undrained condition.

Point 5 in Figure 2.16 describes the stress state at failure. With reference to Figure 2.16b, point 5 has a void ratio e_f (on the critical-state line) which is identical to the initial void ratio e_0 at point 3. The equation of the critical-state line, (2.21), in the e–$\ln p'$ plane is given as

$$e_f = e_\Gamma - \lambda \ln p'$$

At point 5 this equation becomes

$$e_f = e_0 = e_\Gamma - \lambda \ln p'_f$$

Rearranging this equation yields

$$p'_f = \exp\left(\frac{e_\Gamma - e_0}{\lambda}\right) \tag{2.41}$$

where p'_f is the mean effective stress at failure. But

$$q_f = Mp'_f$$

Therefore,

$$q_f = M \exp\left(\frac{e_\Gamma - e_0}{\lambda}\right) \tag{2.42}$$

where q_f is the shear stress at failure.

In Figure 2.16a, the horizontal distance between the total stress path and the effective stress path is the excess pore water pressure. At failure (point 5), we can write

$$\Delta u_f = p_f - p'_f$$

Applying (2.40), the total stress path equation, to the at-failure stress conditions (i.e., at $p = p_f$ and $q = q_f$), we get

$$q_f = 3(p_f - p'_0)$$

or

$$p_f = p'_0 + \frac{q_f}{3} = p'_0 + \frac{Mp'_f}{3}$$

but

$$\Delta u_f = p_f - p'_f = p_0 + \frac{M}{3} p'_f - p'_f = p_0 + \left(\frac{M}{3} - 1 \right) p'_f$$

Substituting (2.41) into the equation above yields

$$\Delta u_f = p_0 + \left(\frac{M}{3} - 1 \right) \exp \left(\frac{e_\Gamma - e_0}{\lambda} \right) \qquad (2.43)$$

Using (2.34) to (2.38), we predict the stress–strain behavior of a normally consolidated (NC) clay in a CU triaxial test. These equations are given in an incremental form. Thus, increments of stress will be applied in accordance with the CU effective stress path, and the resulting strain increments will be calculated. The strain increments are then added together to calculate the total strain.

The following step-by-step procedure, adopted from Budhu (2007), can be programmed in a spreadsheet to calculate the stress–strain behavior. It is to be noted that Budhu's procedure can be used for the more general case of a lightly overconsolidated soil.

2.5.11 Step-by-Step Calculation Procedure for a CU Triaxial Test on NC Clays

Step 1: Calculate the mean effective stress at failure using (2.41):

$$p'_f = \exp \left(\frac{e_\Gamma - e_0}{\lambda} \right)$$

Then choose a small mean effective stress increment, $\Delta p'$, such that $\Delta p' = (p'_0 - p'_f)/N$, where N is an integer. In the spreadsheet example in Table 2.2 we used $N = 14$. You can use a larger value of N for better accuracy. In the first column of the spreadsheet, use $p' = p'_0, p'_0 - \Delta p', p'_0 - 2\Delta p', p'_0 - 3\Delta p', \ldots, p'_f$. The spreadsheet is used to predict the consolidated–undrained triaxial behavior of a normally consolidated clay specimen subjected to a confining pressure $\sigma'_3 = 206.7$ kPa. Column (1) shows the variation in the mean effective stress. The initial mean effective stress $p'_0 = 206.7$ kPa is used along with $\Delta p' = 7.0$ kPa.

Step 2: Update the yield surface for each load increment. Figure 2.17 shows two consecutive yield surfaces corresponding to two consecutive points (points 1 and 2) along the effective stress path of a CU triaxial test. Point 1 corresponds to a mean effective stress p'_1 located on a yield surface with a major axis $(p'_c)_1$. Point 2 corresponds to a mean effective stress p'_2 located on a yield surface

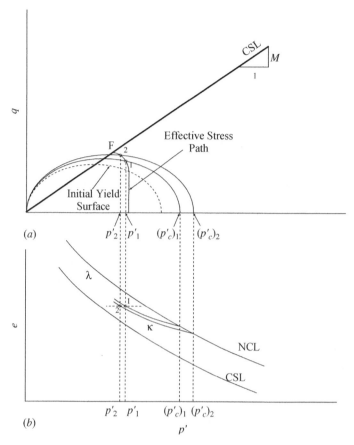

FIGURE 2.17   Evolution of a Cam clay yield surface under CU triaxial stress conditions.

with a major axis $(p'_c)_2$. With the help of Figure 2.17, it can be shown that

$$(p'_c)_2 = (p'_c)_1 \left(\frac{p'_1}{p'_2} \right)^{\kappa/(\lambda-\kappa)} \tag{2.44}$$

In column (2) of Table 2.2, calculate the updated yield surface for each load increment using the general equation

$$(p'_c)_i = (p'_c)_{i-1} \left(\frac{p'_{i-1}}{p'_i} \right)^{\kappa/(\lambda-\kappa)} \tag{2.45}$$

where $(p'_c)_i$ is the preconsolidation pressure in the current increment, $(p'_c)_{i-1}$ the preconsolidation pressure in the preceding increment, p'_i the mean effective stress in the current increment, and p'_{i-1} the mean effective stress in the preceding increment. Because the soil is normally consolidated, use $p'_c = p'_0$ in the first row.

Step 3: In column (3), calculate q for each increment using (2.46):

$$q = Mp' \sqrt{\frac{p'_c}{p'} - 1} \tag{2.46}$$

This equation is obtained from the equation of the yield surface, (2.23).

Step 4: Calculate the elastic volumetric strain increment using (2.35):

$$d\varepsilon^e_v = \frac{\kappa}{1+e} \frac{dp'}{p'} \qquad \text{column (4)}$$

Note that e in this equation is constant ($= e_0$) because of the undrained condition. Note also in Table 2.2 that the elastic volumetric strain increments calculated are all negative because the change in the mean effective stress is negative under undrained triaxial conditions.

Step 5: Calculate the plastic volumetric strain increment:

$$d\varepsilon^p_v = -d\varepsilon^e_v = \frac{-\kappa}{1+e} \frac{dp'}{p'} \qquad \text{column (5)}$$

Again, this is because of the undrained condition (no volume change), in which

$$d\varepsilon_v = d\varepsilon^e_v + d\varepsilon^p_v = 0 \rightarrow d\varepsilon^e_v = -d\varepsilon^p_v$$

Step 6: Calculate the plastic shear strain increment using (2.38):

$$d\varepsilon^p_s = d\varepsilon^p_v \frac{2\eta}{M^2 - \eta^2} \qquad \text{column (6)}$$

Step 7: Calculate the elastic shear strain increment using

$$d\varepsilon^e_s = \frac{\Delta q}{3G} \qquad \text{column (7)}$$

where G is the shear modulus given by (2.28).

In the present example, use $e_0 = 0.889$, $\kappa = 0.026$, and $v = 0.3$ [see (2.28)].

Step 8: Calculate the shear strain increment using (2.33):

$$d\varepsilon_s = d\varepsilon^e_s + d\varepsilon^p_s \qquad \text{column (8)}$$

Step 9: Calculate the total shear strain $(\varepsilon_s)_k = \sum_{i=1}^{i=k} (d\varepsilon_s)_i$, where k is the current row. This is done in column (9).

TABLE 2.2

(1) p' (kPa)	(2) p'_c (kPa)	(3) q (kPa)	(4) $d\varepsilon_v^e$	(5) $d\varepsilon_v^p$	(6) $d\varepsilon_s^p$	(7) $d\varepsilon_s^e$	(8) $d\varepsilon_s$	(9) $\varepsilon_s = \Sigma d\varepsilon_s$	(10) ε_1	(11) p (kPa)	(12) Δu (kPa)
206.70	206.70	0.00	0.00000	0.00000	0.00000	0.00000	0.00000	0.00000	0.00000	206.70	0.00
199.70	207.95	40.47	−0.00047	0.00047	0.00020	0.00195	0.00215	0.00215	0.00215	220.19	20.49
192.70	209.26	56.31	−0.00049	0.00049	0.00032	0.00076	0.00108	0.00323	0.00323	225.47	32.77
185.70	210.63	67.81	−0.00051	0.00051	0.00043	0.00055	0.00099	0.00421	0.00421	229.30	43.60
178.70	212.05	76.95	−0.00053	0.00053	0.00056	0.00044	0.00100	0.00521	0.00521	232.35	53.65
171.70	213.55	84.49	−0.00055	0.00055	0.00072	0.00036	0.00108	0.00630	0.00630	234.86	63.16
164.70	215.11	90.82	−0.00057	0.00057	0.00092	0.00030	0.00122	0.00752	0.00752	236.97	72.27
157.70	216.76	96.19	−0.00060	0.00060	0.00117	0.00026	0.00143	0.00895	0.00895	238.76	81.06
150.70	218.50	100.75	−0.00062	0.00062	0.00153	0.00022	0.00175	0.01070	0.01070	240.28	89.58
143.70	220.33	104.59	−0.00065	0.00065	0.00206	0.00018	0.00224	0.01294	0.01294	241.56	97.86
136.70	222.27	107.80	−0.00069	0.00069	0.00292	0.00015	0.00307	0.01601	0.01601	242.63	105.93
129.70	224.34	110.42	−0.00072	0.00072	0.00459	0.00013	0.00471	0.02072	0.02072	243.51	113.81
122.70	226.53	112.50	−0.00076	0.00076	0.00917	0.00010	0.00927	0.02999	0.02999	244.20	121.50
115.70	228.88	114.06	−0.00081	0.00081	0.07374	0.00007	0.07381	0.10380	0.10380	244.72	129.02

Step 10: Calculate the axial strain ε_1 [column (10)]. Under undrained conditions we have

$$\varepsilon_v = \varepsilon_1 + 2\varepsilon_3 = 0 \rightarrow \varepsilon_1 = -2\varepsilon_3$$

But

$$\varepsilon_s = \tfrac{2}{3}(\varepsilon_1 - \varepsilon_3)$$

Therefore, $\varepsilon_1 = \varepsilon_s$.

Step 11: Calculate the current total mean stress using the equation of the total stress path, (2.40):

$$p = p'_0 + \frac{q}{3} \qquad \text{column (11)}$$

Step 12: Calculate the pore water pressure increment: $\Delta u = p - p'$ [column (12)].

The deviator stress versus axial strain behavior predicted for the soil can be plotted using data from columns (3) and (10), respectively. Also, the excess pore water pressure versus axial strain can be plotted using data from columns (12) and (10), respectively. Figure 2.18 shows the behavior predicted for this normally consolidated clay when tested in consolidated undrained triaxial compression with a confining pressure of 206.7 kPa. The figure also shows the behavior predicted for the same soil when subjected to confining pressures of 68.9 and 137.8 kPa.

2.5.12 Comments on the Modified Cam Clay Model

The modified Cam clay model formulations are based on the triaxial stress condition in which the intermediate and the minor principal stresses are equal ($\sigma_2 = \sigma_3$). It is desirable, however, to express the modified Cam clay model in terms of stress invariants, described below, so that one can use the model in a more generalized manner. This is particularly useful for problems involving three-dimensional stress conditions and plane strain conditions that are common in geotechnical engineering. Here we present a summary of stress and strain invariants that are commonly used in engineering mechanics and soil models. Then we present the extended (generalized) Cam clay model.

2.6 STRESS INVARIANTS

In the following, all stresses are assumed to be *effective stresses*. This means that the value of the excess pore water pressure is known and is subtracted from total stresses to obtain the effective stresses. The stress characteristics equation is

$$\sigma^3 - I_{1\text{-stress}}\sigma^2 + I_{2\text{-stress}}\sigma - I_{3\text{-stress}} = 0 \qquad (2.47)$$

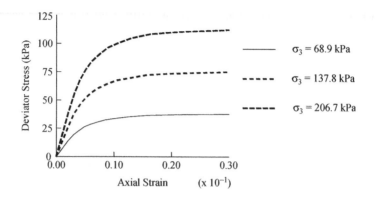

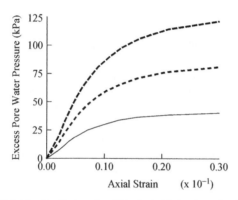

FIGURE 2.18   Predicting the CU triaxial behavior of NC clay using the Cam clay model.

where $I_{\text{1-stress}}$, $I_{\text{2-stress}}$, and $I_{\text{3-stress}}$, the stress invariants, are defined as follows:

$$I_{\text{1-stress}} = \sigma_{11} + \sigma_{22} + \sigma_{33} \tag{2.48}$$

$$I_{\text{2-stress}} = \begin{vmatrix} \sigma_{11} & \tau_{12} \\ \tau_{21} & \sigma_{22} \end{vmatrix} + \begin{vmatrix} \sigma_{22} & \tau_{23} \\ \tau_{32} & \sigma_{33} \end{vmatrix} + \begin{vmatrix} \sigma_{11} & \tau_{13} \\ \tau_{31} & \sigma_{33} \end{vmatrix} \tag{2.49}$$

$$I_{\text{3-stress}} = \begin{vmatrix} \sigma_{11} & \tau_{12} & \tau_{13} \\ \tau_{21} & \sigma_{22} & \tau_{23} \\ \tau_{31} & \tau_{32} & \sigma_{33} \end{vmatrix} \tag{2.50}$$

Three other invariants, J_1, J_2, and J_3, are related to the stress invariants above. These are called *invariants of the stress tensor* and defined as

$$J_1 = I_{\text{1-stress}} = \sigma_{11} + \sigma_{22} + \sigma_{33} \tag{2.51}$$

$$J_2 = \tfrac{1}{2}(I_{\text{1-stress}}^2 - 2I_{\text{2-stress}}) \tag{2.52}$$

$$J_3 = \tfrac{1}{3}(I_{\text{1-stress}}^3 - 3I_{\text{1-stress}}I_{\text{2-stress}} + 3I_{\text{3-stress}}) \tag{2.53}$$

2.6.1 Decomposition of Stresses

The stress matrix can be decomposed into a deviator stress matrix and a hydrostatic stress matrix:

$$
\begin{pmatrix}
\sigma_{11} & \tau_{12} & \tau_{13} \\
\tau_{21} & \sigma_{22} & \tau_{23} \\
\tau_{31} & \tau_{32} & \sigma_{33}
\end{pmatrix}
=
\begin{pmatrix}
S_{11} & \tau_{12} & \tau_{13} \\
\tau_{21} & S_{22} & \tau_{23} \\
\tau_{31} & \tau_{32} & S_{33}
\end{pmatrix}
+
\begin{pmatrix}
p & 0 & 0 \\
0 & p & 0 \\
0 & 0 & p
\end{pmatrix}
\tag{2.54}
$$

where

$$
p = \frac{J_1}{3} = \frac{\sigma_{11} + \sigma_{22} + \sigma_{33}}{3}
\tag{2.55}
$$

On the right-hand side of (2.54), the first matrix represents the deviator stresses in a soil element, and the second matrix represents the hydrostatic stresses. This decomposition is convenient for soil modeling because we can assume that the distortion of soil is caused by deviator stresses and that the soil volume change is caused by hydrostatic stresses.

From (2.54), the deviator stress matrix is calculated as

$$
\begin{pmatrix}
S_{11} & \tau_{12} & \tau_{13} \\
\tau_{21} & S_{22} & \tau_{23} \\
\tau_{31} & \tau_{32} & S_{33}
\end{pmatrix}
=
\begin{pmatrix}
\sigma_{11} & \tau_{12} & \tau_{13} \\
\tau_{21} & \sigma_{22} & \tau_{23} \\
\tau_{31} & \tau_{32} & \sigma_{33}
\end{pmatrix}
-
\begin{pmatrix}
p & 0 & 0 \\
0 & p & 0 \\
0 & 0 & p
\end{pmatrix}
\tag{2.56}
$$

Using the deviator stresses calculated by (2.56), we can define a new set of invariants called *invariants of deviator stresses*. The first invariant of deviator stress is

$$
J_{1D} = S_{11} + S_{22} + S_{33} = 0
\tag{2.57}
$$

The second invariant of deviator stress is

$$
J_{2D} = J_2 - \frac{J_1^2}{6}
\tag{2.58}
$$

Using principal stresses, J_{2D} can be expressed as

$$
J_{2D} = \tfrac{1}{6}[(\sigma_1 - \sigma_2)^2 + (\sigma_2 - \sigma_3)^2 + (\sigma_1 - \sigma_3)^2]
\tag{2.59}
$$

The third invariant of deviator stress is

$$
J_{3D} = J_3 - \tfrac{2}{3}J_1 J_2 + \tfrac{2}{27}J_1^3
\tag{2.60}
$$

Example 2.1   In a triaxial stress state we have

$$
\begin{pmatrix}
\sigma_{11} & \tau_{12} & \tau_{13} \\
\tau_{21} & \sigma_{22} & \tau_{23} \\
\tau_{31} & \tau_{32} & \sigma_{33}
\end{pmatrix}
=
\begin{pmatrix}
\sigma_1 & 0 & 0 \\
0 & \sigma_3 & 0 \\
0 & 0 & \sigma_3
\end{pmatrix}
$$

Calculate (a) the deviator stress matrix using (2.56), and (b) the second invariant of deviator stress using (2.58).

SOLUTION: (a) We start by calculating the hydrostatic pressure p using (2.55):

$$p = \frac{J_1}{3} = \frac{\sigma_1 + 2\sigma_3}{3}$$

From (2.56) we can write

$$\begin{pmatrix} S_{11} & \tau_{12} & \tau_{13} \\ \tau_{21} & S_{22} & \tau_{23} \\ \tau_{31} & \tau_{32} & S_{33} \end{pmatrix} = \begin{pmatrix} \sigma_1 & 0 & 0 \\ 0 & \sigma_3 & 0 \\ 0 & 0 & \sigma_3 \end{pmatrix} - \begin{pmatrix} \dfrac{\sigma_1 + 2\sigma_3}{3} & 0 & 0 \\ 0 & \dfrac{\sigma_1 + 2\sigma_3}{3} & 0 \\ 0 & 0 & \dfrac{\sigma_1 + 2\sigma_3}{3} \end{pmatrix}$$

or

$$\begin{pmatrix} S_{11} & \tau_{12} & \tau_{13} \\ \tau_{21} & S_{22} & \tau_{23} \\ \tau_{31} & \tau_{32} & S_{33} \end{pmatrix} = \begin{pmatrix} \dfrac{2(\sigma_1 - \sigma_3)}{3} & 0 & 0 \\ 0 & \dfrac{\sigma_3 - \sigma_1}{3} & 0 \\ 0 & 0 & \dfrac{\sigma_3 - \sigma_1}{3} \end{pmatrix}$$

Therefore,

$$S_{11} = \frac{2(\sigma_1 - \sigma_3)}{3}$$

and

$$S_{22} = S_{33} = \frac{-(\sigma_1 - \sigma_3)}{3}$$

Note that

$$S_{11} + S_{22} + S_{33} = \frac{2(\sigma_1 - \sigma_3)}{3} + \frac{-(\sigma_1 - \sigma_3)}{3} + \frac{-(\sigma_1 - \sigma_3)}{3} = 0$$

(b) To calculate the second invariant of deviator stress using (2.58), we need to calculate J_1 and J_2:

$$I_{1\text{-stress}} = \sigma_1 + 2\sigma_3$$

or

$$J_1 = \sigma_1 + 2\sigma_3$$

$$I_{2\text{-stress}} = \begin{vmatrix} \sigma_1 & 0 \\ 0 & \sigma_3 \end{vmatrix} + \begin{vmatrix} \sigma_3 & 0 \\ 0 & \sigma_3 \end{vmatrix} + \begin{vmatrix} \sigma_1 & 0 \\ 0 & \sigma_3 \end{vmatrix} = 2\sigma_1\sigma_3 + \sigma_3^2$$

$$J_2 = \frac{1}{2}(I_{1\text{-stress}}^2 - 2I_{2\text{-stress}}) = \frac{1}{2}[(\sigma_1 + 2\sigma_3)^2 - 2(2\sigma_1\sigma_3 + \sigma_3^2)] = \frac{\sigma_1^2}{2} + \sigma_3^2$$

Substitute J_1 and J_2 into (2.58):

$$J_{2D} = J_2 - \frac{J_1^2}{6} = \frac{\sigma_1^2}{2} + \sigma_3^2 - \frac{(\sigma_1 + 2\sigma_3)^2}{6} = \frac{1}{3}(\sigma_1 - \sigma_3)^2$$

2.7 STRAIN INVARIANTS

The strain characteristics equation is

$$\varepsilon^3 - I_{1\text{-strain}}\varepsilon^2 + I_{2\text{-strain}}\varepsilon - I_{3\text{-strain}} = 0 \tag{2.61}$$

where $I_{1\text{-strain}}$, $I_{2\text{-strain}}$, and $I_{3\text{-strain}}$ are the *strain invariants*, defined as

$$I_{1\text{-strain}} = \varepsilon_{11} + \varepsilon_{22} + \varepsilon_{33} \tag{2.62}$$

$$I_{2\text{-strain}} = \begin{vmatrix} \varepsilon_{11} & \varepsilon_{12} \\ \varepsilon_{21} & \varepsilon_{22} \end{vmatrix} + \begin{vmatrix} \varepsilon_{22} & \varepsilon_{23} \\ \varepsilon_{32} & \varepsilon_{33} \end{vmatrix} + \begin{vmatrix} \varepsilon_{11} & \varepsilon_{13} \\ \varepsilon_{31} & \varepsilon_{33} \end{vmatrix} \tag{2.63}$$

$$I_{3\text{-strain}} = \begin{vmatrix} \varepsilon_{11} & \varepsilon_{12} & \varepsilon_{13} \\ \varepsilon_{21} & \varepsilon_{22} & \varepsilon_{23} \\ \varepsilon_{31} & \varepsilon_{32} & \varepsilon_{33} \end{vmatrix} \tag{2.64}$$

A set of invariants related to the strain invariants, (2.62)–(2.64), are called *invariants of the strain tensor* and defined as

$$I_1 = I_{1\text{-strain}} = \varepsilon_{11} + \varepsilon_{22} + \varepsilon_{33} \tag{2.65}$$

$$I_2 = \tfrac{1}{2}(I_{1\text{-strain}}^2 - 2I_{2\text{-strain}}) \tag{2.66}$$

$$I_3 = \tfrac{1}{3}(I_{1\text{-strain}}^3 - 3I_{1\text{-strain}}I_{2\text{-strain}} + 3I_{3\text{-strain}}) \tag{2.67}$$

2.7.1 Decomposition of Strains

The strain matrix can be decomposed into a shear (distortion) strain matrix and a volumetric strain matrix:

$$\begin{pmatrix} \varepsilon_{11} & \varepsilon_{12} & \varepsilon_{13} \\ \varepsilon_{21} & \varepsilon_{22} & \varepsilon_{23} \\ \varepsilon_{31} & \varepsilon_{32} & \varepsilon_{33} \end{pmatrix} = \begin{pmatrix} E_{11} & \varepsilon_{12} & \varepsilon_{13} \\ \varepsilon_{21} & E_{22} & \varepsilon_{23} \\ \varepsilon_{31} & \varepsilon_{32} & E_{33} \end{pmatrix} + \begin{pmatrix} \dfrac{\varepsilon_v}{3} & 0 & 0 \\ 0 & \dfrac{\varepsilon_v}{3} & 0 \\ 0 & 0 & \dfrac{\varepsilon_v}{3} \end{pmatrix} \tag{2.68}$$

where

$$\varepsilon_v = I_1 = \varepsilon_{11} + \varepsilon_{22} + \varepsilon_{33} \tag{2.69}$$

On the right-hand side of (2.68), the first matrix represents the shear (distortion) strain in a soil element, and the second matrix represents the volumetric strains. From (2.68) the distortion strain matrix is calculated as

$$
\begin{pmatrix} E_{11} & \varepsilon_{12} & \varepsilon_{13} \\ \varepsilon_{21} & E_{22} & \varepsilon_{23} \\ \varepsilon_{31} & \varepsilon_{32} & E_{33} \end{pmatrix} = \begin{pmatrix} \varepsilon_{11} & \varepsilon_{12} & \varepsilon_{13} \\ \varepsilon_{21} & \varepsilon_{22} & \varepsilon_{23} \\ \varepsilon_{31} & \varepsilon_{32} & \varepsilon_{33} \end{pmatrix} - \begin{pmatrix} \dfrac{\varepsilon_v}{3} & 0 & 0 \\ 0 & \dfrac{\varepsilon_v}{3} & 0 \\ 0 & 0 & \dfrac{\varepsilon_v}{3} \end{pmatrix}
\tag{2.70}
$$

Using the distortion strains calculated above, define the *invariants of the distortion strains* as follows:

$$
I_{1D} = 0
\tag{2.71}
$$

$$
I_{2D} = I_2 - \frac{I_1^2}{6}
\tag{2.72}
$$

$$
I_{3D} = I_3 - \frac{2}{3} I_1 I_2 + \frac{2}{27} I_1^3
\tag{2.73}
$$

2.8   EXTENDED CAM CLAY MODEL

Next, we present a brief description of the extended Cam clay model. The model is described in greater detail in ABAQUS(2002). The modified Cam clay model, described in some detail earlier, is a special case of the extended Cam clay model described here.

The basic concepts of the extended Cam clay model are shown in Figure 2.19. An elasticity model in which the bulk elastic stiffness increases as the material undergoes compression, (2.26), is used to calculate the elastic strains. The plastic strains are calculated using the theory of plasticity: A yield surface with associated flow and a hardening rule that allows the yield surface to grow or shrink in the three-dimensional stress space is used. In the extended model, the critical-state surface is assumed to be a cone in the space of principal effective stress (Figure 2.19). The vertex of the cone is coincident with the origin (zero effective stress), and its axis is coincident with the hydrostatic pressure axis ($\sigma_1 = \sigma_2 = \sigma_3$).

The projection of the three-dimensional elliptic yield surface on the Π-plane (the plane in the principal stress space orthogonal to the hydrostatic pressure axis) has the general shape shown in Figure 2.20. The projection of the conical critical-state surface on the p–t plane is a straight line passing through the origin with slope M, as shown in Figure 2.21. Here the parameter t is a measure of shear stress as defined below. The yield surface in the p–t plane consists of two elliptic arcs. The first arc passes through the origin with its tangent perpendicular to the p-axis and intersects the critical-state line where its tangent is parallel to the p-axis. The second arc is a smooth continuation of the first arc through the critical-state line and

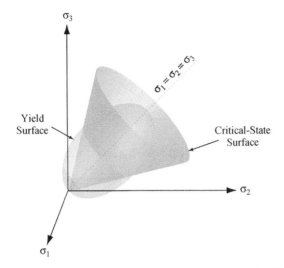

FIGURE 2.19   Elements of the extended Cam clay model: yield and critical-state surfaces in the principal stress space.

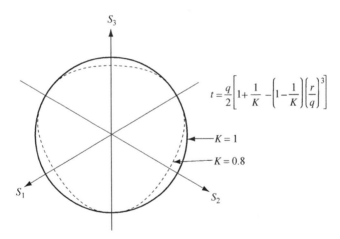

FIGURE 2.20   Projection of the extended Cam clay yield surface on the Π-plane. (Adapted from ABAQUS, 2002.)

intersects the p-axis with its tangent at a $90°$ angle to that axis (see Figure 2.21). Plastic flow is assumed to be normal to this surface.

The size of the yield surface is controlled by the hardening rule, which depends only on the volumetric plastic strain component. Thus, when the volumetric plastic strain is compressive, the yield surface grows in size. But when there is a dilative plastic strain, the yield surface contracts. The three-dimensional yield surface is defined as

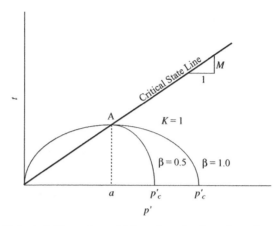

FIGURE 2.21 Extended Cam clay yield surface in the $p'-t$ plane. (Adapted from ABAQUS, 2002.)

$$f(p, q, r) = \frac{1}{\beta^2}\left(\frac{p}{a} - 1\right)^2 + \left(\frac{t}{Ma}\right)^2 - 1 = 0 \qquad (2.74)$$

where

$$p = \frac{J_1}{3} = \frac{\sigma_1 + \sigma_2 + \sigma_3}{3} \qquad (2.75)$$

$$q = \sqrt{3J_{2D}} = \sqrt{3\left(J_2 - \frac{J_1^2}{6}\right)} = \sqrt{\frac{1}{2}[(\sigma_1 - \sigma_2)^2 + (\sigma_2 - \sigma_3)^2 + (\sigma_1 - \sigma_3)^2]}$$

$$(2.76)$$

$$r = \left(\frac{27}{2}J_{3D}\right)^{1/3} = \left(\frac{27}{2}J_3 - 9J_1J_2 + J_1^3\right)^{1/3} \qquad (2.77)$$

in which β is a constant used to modify the shape of the yield surface on the "wet" side of the critical state; $\beta = 1$ can be used on the "dry" side of the critical state, and $\beta < 1$ can be used in most cases on the wet side (Figure 2.21) to make the curvature of the elliptic arc on the wet side different from that on the dry side. a is a hardening parameter defined as the point on the p-axis at which the evolving elliptic arcs of the yield surface intersect the critical-state line as indicated in Figure 2.21. M is the slope of the critical-state line in the $p-t$ plane (the ratio of t to p at the critical state). t is a measure of shear stress calculated as $t = q/g$, where g is a function used to control the shape of the yield surface in the Π-plane and is defined as

$$g = \frac{2K}{1 + K + (1 - K)(r/q)^3} \qquad (2.78)$$

where K is a constant. Setting $K = 1$ causes the yield surface to be independent of the third stress invariant, and the projection of the yield surface on the Π-plane becomes a circle, reduced to a modified Cam clay yield surface ($K = 1 \to g = 1 \to t = q$). The effect of different values of K on the shape of the yield surface in the Π-plane is shown in Figure 2.20. To ensure convexity of the yield surface, the range $0.778 \leq K \leq 1.0$ should not be violated.

Associated flow is used in the extended Cam clay model (i.e., the plastic potential is the same as the yield surface). The size of the yield surface is defined by the parameter a; the evolution of this variable therefore characterizes the hardening or softening of the material. The evolution of the parameter a is defined as

$$a = a_0 \exp \left[(1 + e_0) \frac{1 - J^{\text{pl}}}{\lambda - \kappa J^{\text{pl}}} \right] \tag{2.79}$$

where J^{pl} is the plastic part of the volume change J. The volume change is defined as the ratio of current volume to initial volume: $J = J^{\text{pl}} + J^e = (1 + e)/(1 + e_0)$. a_0 is a constant parameter that defines the position of a at the beginning of the analysis (preconsolidation pressure). The value of a_0 can be specified directly or can be computed as

$$a_0 = \frac{1}{2} \exp \left(\frac{e_N - e_0 - \kappa \ln p_0}{\lambda - \kappa} \right) \tag{2.80}$$

in which p_0 is the initial value of the mean effective stress and e_N is the intercept of the normal consolidation line (NCL) with the void ratio axis in the e–$\ln p'$ plane, as shown in Figure 2.9.

2.9 MODIFIED DRUCKER–PRAGER/CAP MODEL

The Drucker–Prager/cap plasticity model has been widely used in finite element analysis programs for a variety of geotechnical engineering applications. The cap model is appropriate to soil behavior because it is capable of considering the effect of stress history, stress path, dilatancy, and the effect of the intermediate principal stress. The yield surface of the modified Drucker–Prager/cap plasticity model consists of three parts: a Drucker–Prager shear failure surface, an elliptical *cap*, which intersects the mean effective stress axis at a right angle, and a smooth transition region between the shear failure surface and the cap, as shown in Figure 2.22.

Elastic behavior is modeled as linear elastic using the generalized Hooke's law. Alternatively, an elasticity model in which the bulk elastic stiffness increases as the material undergoes compression can be used to calculate the elastic strains [equation (2.26)]. The onset of plastic behavior is determined by the Drucker–Prager failure surface and the cap yield surface. The Drucker–Prager failure surface is given by

$$F_s = t - p \tan \beta - d = 0 \tag{2.81}$$

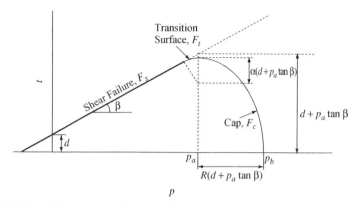

FIGURE 2.22 Yield surfaces of the modified cap model in the $p-t$ plane. (Adapted from ABAQUS, 2002.)

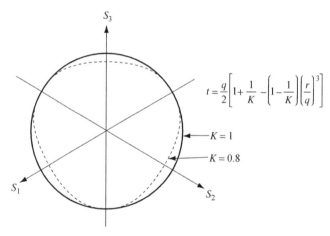

$$t = \frac{q}{2}\left[1 + \frac{1}{K} - \left(1 - \frac{1}{K}\right)\left(\frac{r}{q}\right)^3\right]$$

$K = 1$

$K = 0.8$

FIGURE 2.23 Projection of the modified cap yield/flow surfaces on the Π-plane. (Adapted from ABAQUS, 2002.)

where β is the soil's angle of friction and d is its cohesion in the $p-t$ plane, as indicated in Figure 2.22.

As shown in the figure, the cap yield surface is an ellipse with eccentricity $= R$ in the $p-t$ plane. The cap yield surface is dependent on the third stress invariant, r, in the deviatoric plane as shown in Figure 2.23 [equations (2.76) and (2.77)]. The cap surface hardens (expands) or softens (shrinks) as a function of the volumetric plastic strain. When the stress state causes yielding on the cap, volumetric plastic strain (compaction) results, causing the cap to expand (hardening). But when the stress state causes yielding on the Drucker–Prager shear failure surface, volumetric plastic dilation results, causing the cap to shrink (softening). The cap yield surface is given as

$$F_c = \sqrt{(p - p_a)^2 + \left(\frac{Rt}{1 + \alpha - \alpha/\cos\beta}\right)^2} - R(d + p_a \tan\beta) = 0 \qquad (2.82)$$

where R is a material parameter that controls the shape of the cap and α is a small number (typically, 0.01 to 0.05) used to define a smooth transition surface between the Drucker–Prager shear failure surface and the cap:

$$F_t = \sqrt{(p - p_a)^2 + \left[t - \left(1 - \frac{\alpha}{\cos\beta}\right)(d + p_a \tan\beta)\right]^2} - \alpha(d + p_a \tan\beta) = 0$$

$$(2.83)$$

p_a is an evolution parameter that controls the hardening–softening behavior as a function of the volumetric plastic strain. The hardening–softening behavior is simply described by a piecewise linear function relating the mean effective (yield) stress p_b and the volumetric plastic strain $p_b = p_b(\varepsilon_{vol}^{pl})$, as shown in Figure 2.24. This function can easily be obtained from the results of one isotropic consolidation test with several unloading–reloading cycles. Consequently, the evolution parameter, p_a, can be calculated as

$$p_a = \frac{p_b - Rd}{1 + R \tan\beta} \qquad (2.84)$$

2.9.1 Flow Rule

In this model the flow potential surface in the p–t plane consists of two parts, as shown in Figure 2.25. In the cap region the plastic flow is defined by a

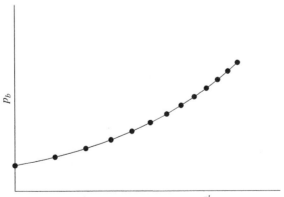

Plastic Volumetric Strain, ε_{vol}^{pl}

FIGURE 2.24 Typical cap hardening behavior. (Adapted from ABAQUS, 2002.)

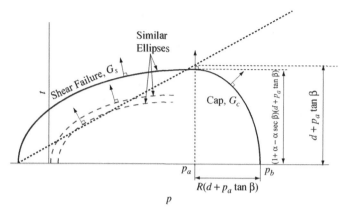

FIGURE 2.25   Flow potential of the modified cap model in the $p-t$ plane. (Adapted from ABAQUS, 2002.)

flow potential that is identical to the yield surface (i.e., associated flow). For the Drucker–Prager failure surface and the transition yield surface, a nonassociated flow is assumed: The shape of the flow potential in the $p-t$ plane is different from the yield surface as shown in Figure 2.25. In the cap region the elliptical flow potential surface is given as

$$G_c = \sqrt{(p - p_a)^2 + \left(\frac{Rt}{1 + \alpha - \alpha/\cos\beta}\right)^2}$$ (2.85)

The elliptical flow potential surface portion in the Drucker–Prager failure and transition regions is given as

$$G_s = \sqrt{[(p_a - p)\tan\beta]^2 + \left(\frac{t}{1 + \alpha - \alpha/\cos\beta}\right)^2}$$ (2.86)

As shown in Figure 2.25, the two elliptical portions, G_c and G_s, provide a continuous potential surface. Because of the nonassociated flow used in this model, the material stiffness matrix is not symmetric. Thus, an unsymmetric solver should be used in association with the cap model.

2.9.2   Model Parameters

We need the results of at least three triaxial compression tests to determine the parameters d and β. The at-failure conditions taken from the tests results can be plotted in the $p-t$ plane. A straight line is then best fitted to the three (or more) data points. The intersection of the line with the t-axis is d and the slope of the line is β. We also need the results of one isotropic consolidation test with several unloading–reloading cycles. This can be used to evaluate the hardening–softening

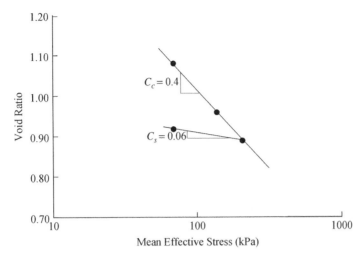

FIGURE 2.26 Isotropic consolidation test results for Example 2.2.

law as a piecewise linear function relating the hydrostatic compression yield stress p_b and the corresponding volumetric plastic strain $p_b = p_b(\varepsilon_{vol}^{pl})$(Figure 2.24). The unloading–reloading slope can be used to calculate the volumetric elastic strain that should be subtracted from the volumetric total strain to calculate the volumetric plastic strain.

Example 2.2   Three CD and three CU triaxial test results of a normally consolidated clay at three confining pressures are shown in Figures 2.15 and 2.18, respectively. (a) Calculate the soil's angle of friction β and its cohesion d in the p–t plane. (b) Using the results of an isotropic consolidation test performed on the same soil (Figure 2.26), calculate the hardening curve assuming the initial conditions $p_0' =210$ kPa and $e_0 =0.889$. Note that the compression index of the soil is 0.4 and the swelling index is 0.06.

SOLUTION: (a) The two tables below summarize the at-failure test results taken from Figures 2.15 and 2.18 for CD and CU triaxial test conditions, respectively. The shear stress measure t and the mean effective stress p' are calculated for all tests.

CD triaxial test (units: kPa):

Test	σ_3'	$t = q = \sigma_1' - \sigma_3'$	σ_1'	$p' = (\sigma_1' + 2\sigma_3')/3$
1	68.9	101.283	170.183	102.661
2	137.8	202.566	340.366	205.322
3	206.7	303.849	510.549	307.983

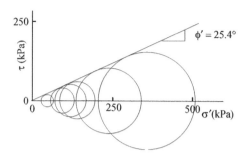

FIGURE 2.27   Mohr–Coulomb failure criterion.

CU triaxial test (units: kPa):

Test	σ_3	$\sigma_1 - \sigma_3$	u	σ_1	σ_1'	σ_3'	$t = q = \sigma_1' - \sigma_3'$	$p' = (\sigma_1' + 2\sigma_3')/3$
1	68.9	37.895	41.34	106.795	65.455	27.56	37.895	40.19167
2	137.8	75.101	81.302	212.901	131.599	56.498	75.101	81.53167
3	206.7	112.307	121.953	319.007	197.054	84.747	112.307	122.1827

Six effective-stress Mohr's circles corresponding to failure stresses obtained from the triaxial test results (Figures 2.15 and 2.18) are plotted in Figure 2.27. Subsequently, the effective-stress Mohr–Coulomb failure criterion is plotted as a straight line that is tangential to the six circles. The soil strength parameters $\phi' = 25.4°$ and $c' = 0\,\text{kPa}$ are obtained from the slope and intercept of the Mohr–Coulomb failure criterion.

For triaxial stress conditions, the Mohr–Coulomb parameters ($\phi' = 25.4°$ and $c' = 0$ kPa) can be converted to Drucker–Prager parameters as follows:

$$\tan \beta = \frac{6 \sin \phi'}{3 - \sin \phi'} \quad \text{for } \phi' = 25.4° \rightarrow \beta = 45°$$

$$d = \frac{18c \cos \phi'}{3 - \sin \phi'} \quad \text{for } c' = 0 \rightarrow d = 0$$

An alternative procedure for determining β and d is to plot the at-failure stresses of all six triaxial tests in the $p' = (\sigma_{1f}' + 2\sigma_{3f}')/3$ versus $t = q = \sigma_{1f}' - \sigma_{3f}'$ plane as shown in Figure 2.28. The data points are best fitted with a straight line whose slope is equal to $\tan \beta = 1$; thus, $\beta = 45°$. The line intersects with the vertical axis at $d = 0$.

(b) The cap hardening curve is obtained from the isotropic consolidation test results shown in Figure 2.26. From the figure we can calculate the plastic volumetric strain as

$$\varepsilon_v^p = \frac{\lambda - \kappa}{1 + e_0} \ln \frac{p'}{p_0'} = \frac{C_c - C_s}{2.3(1 + e_0)} \ln \frac{p'}{p_0'}$$

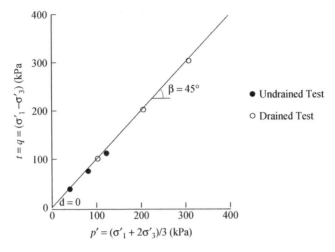

FIGURE 2.28   Evaluating the modified cap model parameters d and β.

For this clayey soil we have $\lambda = 0.174$, $\kappa = 0.026$, $p'_0 = 210\,\text{kPa}$, and $e_0 = 0.889$; therefore,

$$\varepsilon_v^p = \frac{0.174 - 0.026}{1 + 0.889}\ln\frac{p'}{210} = 0.07834\ln\frac{p'}{210}$$

which describes the evolution of plastic volumetric strain (the hardening parameter) with the mean effective stress. A graphic representation of the equation (cap hardening curve) is shown in Figure 2.29.

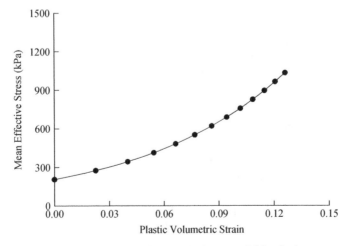

FIGURE 2.29   Evaluating the modified cap model hardening curve.

2.10 LADE'S SINGLE HARDENING MODEL

Lade's model is an elastoplastic model with a single yield surface expressed in terms of stress invariants. The hardening parameter in this model is assumed to be the total plastic work (i.e., the plastic work done by shear strains and volumetric strains), which is used to define the evolution of the yield surface. The model involves 11 parameters that can be determined from three CD triaxial compression tests and one isotropic compression test.

The total strain increments are divided into elastic and plastic strain components:

$$d\varepsilon = d\varepsilon^e + d\varepsilon^p \tag{2.87}$$

For a given *effective* stress increment, the elastic and plastic strain components are calculated separately, the elastic strains by a nonlinear form of Hooke's law and the plastic strains by a plastic stress–strain law.

2.10.1 Elastic Behavior

The elastic strain increments are calculated using (2.88), which accounts for the nonlinear variation of Young's modulus with a stress state (Lade and Nelson, 1987):

$$E = Mp_a \left[\left(\frac{I_1}{p_a} \right)^2 + 6 \frac{1+v}{1-2v} \left(\frac{J_2'}{p_a^2} \right) \right]^{\lambda} \tag{2.88}$$

where v is Poisson's ratio, I_1 is the first invariant of the stress tensor,

$$I_1 = \sigma_x + \sigma_y + \sigma_z \tag{2.89}$$

J_2' is the second invariant of the deviatoric stress tensor,

$$J_2' = \frac{1}{6}[(\sigma_x - \sigma_y)^2 + (\sigma_y - \sigma_z)^2 + (\sigma_z - \sigma_x)^2] + \tau_{xy}^2 + \tau_{yz}^2 + \tau_{zx}^2 \tag{2.90}$$

and p_a is the atmospheric pressure expressed in the same units as E, I_1, and J_2'. M is the modulus number and λ is the exponent, both are dimensionless constants. The parameters M, λ, and v can be determined from the unloading–reloading cycles of triaxial compression tests.

2.10.2 Failure Criterion

In Lade's model, the relationship of stresses at failure is expressed in terms of the first and third stress invariants, I_1 and I_3:

$$f_n = \begin{cases} \left(\dfrac{I_1^3}{I_3} - 27 \right) \left(\dfrac{I_1}{p_a} \right)^m & (2.91a) \\ & (2.91b) \\ \eta_1 \text{ at failure} \end{cases}$$

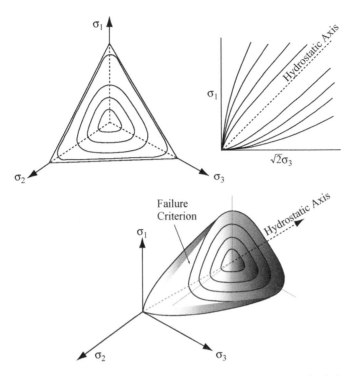

FIGURE 2.30   Lade's model failure criterion. (Adapted from Lade and Jakobsen, 2002.)

where

$$I_3 = \sigma_x \sigma_y \sigma_z + \tau_{xy} \tau_{yz} \tau_{zx} + \tau_{yx} \tau_{zy} \tau_{xz} - (\sigma_x \tau_{yz} \tau_{zy} + \sigma_y \tau_{zx} \tau_{xz} + \sigma_z \tau_{xy} \tau_{yx}) \quad (2.92)$$

The parameters η_1 and m are dimensionless constants that can be determined from triaxial compression test results. The shape of the failure criterion, (2.91), in principal stress space is shown in Figure 2.30. The projection of the failure criterion on the Π-plane is triangular with smoothly rounded edges, as shown in the figure. Lade and Kim (1988) indicated that the apex angle of the failure criterion increases with the value of η_1. The failure surface is always concave toward the hydrostatic pressure axis, and its curvature increases with the value of m.

2.10.3   Plastic Potential and Flow Rule

Plastic flow occurs when the state of stress touches the yield criterion f_n, causing the material to undergo plastic deformations. The plastic strain increments are calculated from the flow rule:

$$d(\varepsilon)^p = d\lambda_p \frac{dg_p}{d(\sigma)} \quad (2.93)$$

where: (σ) is the stress matrix, defined as

$$(\sigma) = \begin{pmatrix} \sigma_{xx} & \tau_{xy} & \tau_{xz} \\ \tau_{yx} & \sigma_{yy} & \tau_{yz} \\ \tau_{zx} & \tau_{zy} & \sigma_{zz} \end{pmatrix}$$

and $(\varepsilon)^p$ is the plastic component of the total strain matrix (ε), defined as

$$(\varepsilon) = \begin{pmatrix} \varepsilon_{xx} & \varepsilon_{xy} & \varepsilon_{xz} \\ \varepsilon_{yx} & \varepsilon_{yy} & \varepsilon_{yz} \\ \varepsilon_{zx} & \varepsilon_{zy} & \varepsilon_{zz} \end{pmatrix}$$

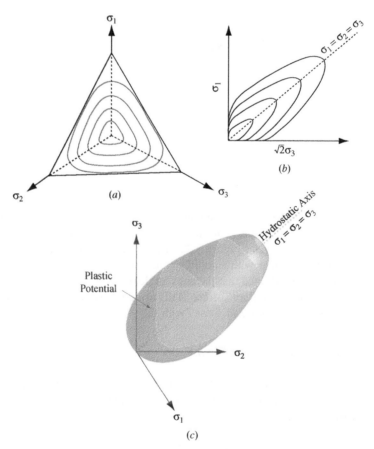

FIGURE 2.31 Lade's model plastic potential. (Adapted from Lade and Jakobsen, 2002.)

λ_p is the proportionality factor (a positive scalar) and g_p is the plastic potential function, given as

$$g_p = \left(\psi_1 \frac{I_1^3}{I_3} - \frac{I_1^2}{I_2} + \psi_2 \right) \left(\frac{I_1}{p_a} \right)^{\mu} \tag{2.94}$$

where I_1 and I_3 are as given in (2.89) and (2.92). I_2 is the second stress invariant, defined as

$$I_2 = \tau_{xy}\tau_{yx} + \tau_{yz}\tau_{zy} + \tau_{zx}\tau_{xz} - (\sigma_x\sigma_y + \sigma_y\sigma_z + \sigma_x\sigma_z) \tag{2.95}$$

ψ_2 and μ are material parameters that can be determined from the results of triaxial compression tests, and ψ_1 is a parameter related to the curvature parameter m of the failure criterion as suggested by Kim and Lade (1988):

$$\psi_1 = 0.00155m^{-1.27} \tag{2.96}$$

The parameter ψ_1 is a weighting factor between the triangular shape and the circular shape shown in Figure 2.31a. The parameter ψ_2 controls the intersection of the plastic potential with the hydrostatic pressure axis, and the exponent μ determines the curvature of the plastic potential in the principal stress space, as shown in Figure 2.31b.

The shape of the plastic potential function is shown in Figure 2.31b. The plastic potential is similar in shape to the failure surface (described earlier). The derivatives of g_p with respect to stress components [used in (2.93)] are defined as follows:

$$\left\{ \begin{array}{c} \dfrac{dg_p}{d\sigma_x} \\[2mm] \dfrac{dg_p}{d\sigma_y} \\[2mm] \dfrac{dg_p}{d\sigma_z} \\[2mm] \dfrac{dg_p}{d\sigma_{yz}} \\[2mm] \dfrac{dg_p}{d\sigma_{xz}} \\[2mm] \dfrac{dg_p}{d\sigma_{xy}} \end{array} \right\} = \left(\frac{I_1}{p_a} \right)^{\mu} \left\{ \begin{array}{c} G - (\sigma_y + \sigma_z)\dfrac{I_1^2}{I_2^2} - \psi_1(\sigma_y\sigma_z - \tau_{yz}^2)\dfrac{I_1^3}{I_3^2} \\[3mm] G - (\sigma_z + \sigma_x)\dfrac{I_1^2}{I_2^2} - \psi_1(\sigma_z\sigma_x - \tau_{zx}^2)\dfrac{I_1^3}{I_3^2} \\[3mm] G - (\sigma_x + \sigma_y)\dfrac{I_1^2}{I_2^2} - \psi_1(\sigma_x\sigma_y - \tau_{xy}^2)\dfrac{I_1^3}{I_3^2} \\[3mm] 2\dfrac{I_1^2}{I_2^2}\tau_{yz} - 2\psi_1(\tau_{xy}\tau_{zx} + \sigma_x\tau_{yz})\dfrac{I_1^3}{I_3^2} \\[3mm] 2\dfrac{I_1^2}{I_2^2}\tau_{zx} - 2\psi_1(\tau_{xy}\tau_{yz} + \sigma_y\tau_{zx})\dfrac{I_1^3}{I_3^2} \\[3mm] 2\dfrac{I_1^2}{I_2^2}\tau_{xy} - 2\psi_1(\tau_{yz}\tau_{zx} + \sigma_z\tau_{xy})\dfrac{I_1^3}{I_3^2} \end{array} \right\} \tag{2.97}$$

where G is the shear modulus, which is function of I_1, I_2, and I_3 as follows:

$$G = \psi_1(\mu + 3)\frac{I_1^2}{I_3} - (\mu + 2)\frac{I_1}{I_2} + \frac{\mu}{I_1}\psi_2 \tag{2.98}$$

2.10.4 Yield Criterion

Most materials behave elastically within their loading limits (i.e., within their initial yield surfaces); once the stresses or strains reach the limit, yield and plastic deformation occur. Lade and Kim (1988) employed an isotropic yield function given as

$$f_p = f_p'(I_1, I_2, I_3) - f_p''(W_p) = 0 \tag{2.99}$$

where

$$f_p' = \left(\psi_1\frac{I_1^3}{I_3} - \frac{I_1^2}{I_2}\right)\left(\frac{I_1}{p_a}\right)^h e^q \tag{2.100}$$

The parameter h in (2.100) is determined based on the assumption that the plastic work is constant along a yield surface. The parameter q varies with the stress level S. Let's define the stress level as

$$S = \frac{f_n}{\eta_1} = \frac{1}{\eta_1}\left(\frac{I_1^3}{I_3} - 27\right)\left(\frac{I_1}{p_a}\right)^m \tag{2.101}$$

in which f_n is given by (2.91a), and η_1 is the value of f_n at failure, (2.91b). The stress level S varies from zero at the hydrostatic pressure axis to unity at the failure surface (i.e., $f_n = \eta_1$). The parameter q [equation (2.100)] varies with S as follows:

$$q = \frac{\alpha S}{1 - (1 - \alpha)S} \tag{2.102}$$

where α is a constant that can be determined by fitting (2.102) to the results of triaxial compression tests, as we show later in an example.

Work hardening occurs when the yield surface expands isotropically as the plastic work increases:

$$f_p'' = \left(\frac{1}{D}\right)^{1/\rho}\left(\frac{W_p}{p_a}\right)^{1/\rho} \tag{2.103}$$

where ρ and D are constants. This means that f_p'' increases only if the plastic work increases. D and ρ are given as

$$D = \frac{C}{(27\psi_1 + 3)^\rho} \tag{2.104}$$

and

$$\rho = \frac{p}{h} \tag{2.105}$$

The parameters C and p in (2.104) and (2.105) can be estimated from the results of an isotropic compression test by best-fitting (2.106) with the test results:

$$W_p = C p_a \left(\frac{I_1}{p_a} \right)^p \tag{2.106}$$

The shape of the yield surface is shown in Figure 2.32. When a stress increment is applied, the plastic work increases and the isotropic yield surface expands until the current stress state hits the failure surface. The relationship between f_p'' and W_p is shown in Figure 2.33. Note that f_p'' increases with increasing W_p and that the slope of the curve decreases with increasing plastic work.

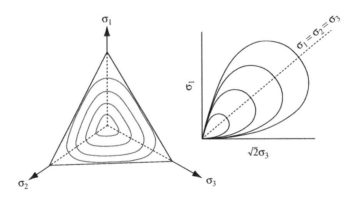

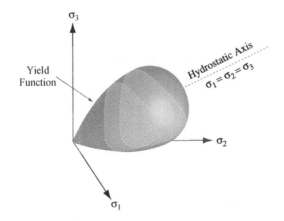

FIGURE 2.32   Lade's model yield function. (Adapted from Lade and Jakobsen, 2002.)

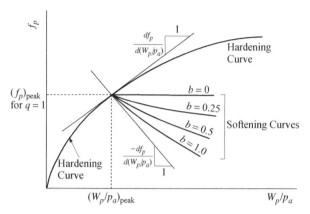

FIGURE 2.33   Hardening and softening definition in Lade's model. (Adapted from Lade and Jakobsen, 2002.)

For work softening, the yield surface contracts isotropically as

$$f_p'' = Ae^{-B(W_p/p_a)} \tag{2.107}$$

With the help of Figure 2.33, the constants A and B can be calculated from the value of f_p'' and the slope of the hardening curve at the point of peak failure at which $S = 1$; therefore,

$$A = [f_p'' e^{B(W_p/p_a)}]_{S=1} \tag{2.108}$$

and

$$B = \left[b \frac{df_p''}{d(W_p/p_a)} \frac{1}{f_p''} \right]_{S=1} \tag{2.109}$$

Note that df_p'' is negative during softening. Also, parameter b in (2.109) ranges from zero to unity. If $b = 0$ is used, the material will behave in a perfectly plastic manner.

The relationship between the plastic work increment and the proportionality factor $d\lambda_p$ [see (2.93)] can be expressed in terms of the plastic potential g_p [see (2.94)] as

$$d\lambda_p = \frac{dW_p}{\mu g_p} \tag{2.110}$$

The increment of plastic work in (2.110) can be calculated by differentiation of the hardening equation (2.103) and the softening equation (2.107). The main features of the single hardening model are summarized in Table 2.3, which also describes

TABLE 2.3   Components and Their Physical Significance in the Single Hardening Model (adapted from Lade, 2005)

	Component	Function
Elastic behavior	Hooke's law	Produces elastic strains whenever the stresses change
Plastic behavior	Failure criterion	Imposes limits on stress states that can be reached
	Plastic potential function	Produces relative magnitudes of plastic strain increments (function similar to Poisson's ratio for elastic strains)
	Yield criterion	Determines when plastic strain increments occur
	Hardening–softening region	Determines magnitudes of plastic strain increments

TABLE 2.4   Components and Number of Parameters in the Single Hardening Model

Component	Parameters
Hooke's law	ν, M, λ
Failure criterion	η_1, m
Plastic potential function	μ, ψ_2
Yield criterion	h, α
Hardening–softening region	C, p

the function of each feature and its physical significance. The single hardening model parameters are summarized in Table 2.4.

Example 2.3: *Obtaining Lade's Model Parameters*   In this example we explain the procedures for estimating Lade's model parameters for a dense silty sand (SP-SM of the USC system). The results of three conventional CD triaxial compression tests conducted on reconstituted soil specimens are shown in Figure 2.34. Also, the results of a hydrostatic compression test conducted on a reconstituted soil specimen are shown in Figure 2.35. Other soil characteristics are listed below.
    Gradation:

$$\text{Percent passing 19-mm sieve} = 100\%$$
$$\text{Percent passing No. 40 sieve} = 59\%$$
$$\text{Percent passing No. 200 sieve} = 8.5\%$$

Elastic behavior   Poisson's ratio ν may be determined from the initial slope, $\Delta\varepsilon_v/\Delta\varepsilon_1$, of the unloading–reloading line of the volume change curve in the triaxial

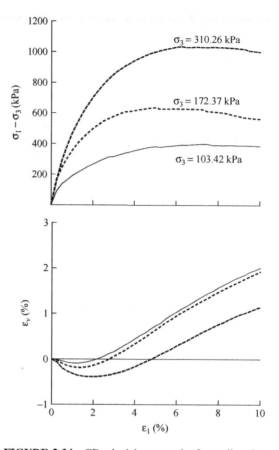

FIGURE 2.34   CD triaxial test results for a silty clay.

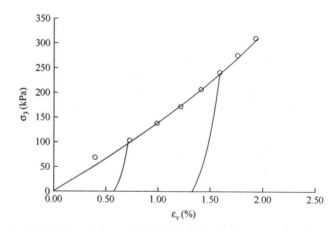

FIGURE 2.35   Isotropic consolidation test results for a silty clay.

compression test. The strains on the unloading–reloading line are purely elastic, and Poisson's ratio is determined as

$$\nu = -\frac{\Delta\varepsilon_3}{\Delta\varepsilon_1} = \frac{1}{2}\left(1 - \frac{\Delta\varepsilon_v}{\Delta\varepsilon_1}\right) \tag{2.111}$$

For the present material we assume that $\nu = 0.3$ because of lack of unloading–reloading cycles in the triaxial tests.

To determine the values of M and λ, let us rewrite (2.88) as

$$\log\left(\frac{E}{p_a}\right) = \log M + \lambda \log\left[\left(\frac{I_1}{p_a}\right)^2 + 6\frac{1+\nu}{1-2\nu}\frac{J_2'}{p_a^2}\right] \tag{2.112}$$

The initial slopes of the stress–strain curves (Figure 2.36) represent the initial elastic moduli (E's) of the soil under different confining pressures. By plotting E/p_a versus the stress function on the right-hand side of (2.112), using a log-log scale (Figure 2.37), the value of $M \approx 250$ is determined as the intercept of the best-fitting line with the vertical line

$$\log\left[\left(\frac{I_1}{p_a}\right)^2 + 6\frac{1+\nu}{1-2\nu}\frac{J_2'}{p_a^2}\right] = 1$$

The slope of the straight line is the exponent λ (≈ 0.21).

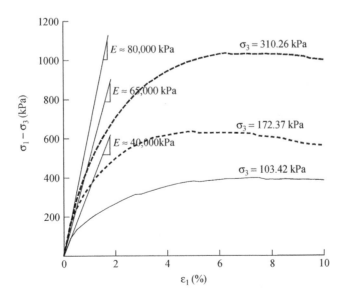

FIGURE 2.36 Determination of the initial elastic modulus for different confining pressures.

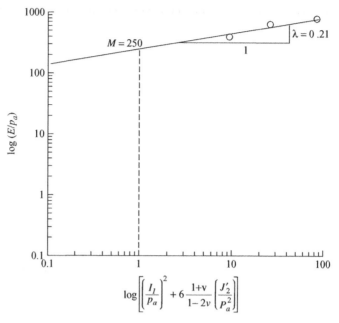

FIGURE 2.37   Determination of Lade's model parameters M and λ.

Failure criterion   The expression for the failure criterion in (2.91) is rewritten as

$$\log\left(\frac{I_1^3}{I_3} - 27\right) = \log \eta_1 + m \log \frac{p_a}{I_1} \tag{2.113}$$

By plotting $\log[(I_1^3/I_3) - 27]$ versus $\log(p_a/I_1)$, as shown in Figure 2.38, the value of η_1 (≈ 52) is determined as the intercept between the best-fitting line and the vertical line emanating from $\log(p_a/I_1) = 1$. The slope of the line is the exponent m (≈ 0.154).

Plastic Potential Parameters   For the plastic potential we need to determine the parameter ψ_1 from (2.96) and the parameters ψ_2 and μ from triaxial compression test data. To determine the parameters ψ_2 and μ we use the definition of the incremental plastic strain ratio given as

$$\nu_p = -\frac{d\varepsilon_3^p}{d\varepsilon_1^p} \tag{2.114}$$

The plastic strain increments in (2.114) are calculated from the results of the triaxial compression tests by subtracting the elastic strain increments from the total strain increments. To obtain the plastic strain increments under triaxial compression

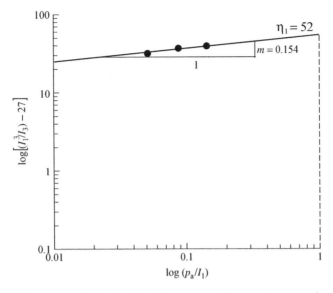

FIGURE 2.38   Determination of Lade's model parameters m and η_1.

conditions ($\sigma_2 = \sigma_3$), one can substitute (2.93) and (2.97) into (2.114) to derive

$$\xi_y = \frac{1}{\mu}\xi_x - \psi_2 \tag{2.115}$$

where

$$\xi_x = \frac{1}{1+v_p}\left[\frac{I_1^3}{I_2^2}(\sigma_1 + \sigma_3 + 2v_p\sigma_3) + \psi_1\frac{I_1^4}{I_3^2}(\sigma_1\sigma_3 + v_p\sigma_3^2)\right] - 3\psi_1\frac{I_1^3}{I_3} + 2\frac{I_1^2}{I_2} \tag{2.116}$$

and

$$\xi_y = \psi_1\frac{I_1^3}{I_3} - \frac{I_1^2}{I_2} \tag{2.117}$$

Now, if we plot (2.116) versus (2.117) for the three triaxial tests (i.e., ξ_x versus ξ_y, as shown in Figure 2.39), we can best fit the data with a straight line whose slope is $1/\mu$ and whose intercept with the vertical is $-\psi_2$ [see (2.115)]. As shown in Figure 2.39, the value of $1/\mu$ is approximately 0.44, and the value of $-\psi_2$ is approximately 3.41; thus, $\psi_2 \approx -3.41$.

Yield criterion and work hardening–softening relation   The work hardening relation along the hydrostatic axis, expressed by $W_p = Cp_a(I_1/p_a)^p$, (2.106), can be used to determine the parameters C and p that are required for the determination of q in the yield criterion, (2.100). The plastic work along the hydrostatic axis is calculated from

$$W_p = \int (\sigma)^T d(\varepsilon)^p \tag{2.118}$$

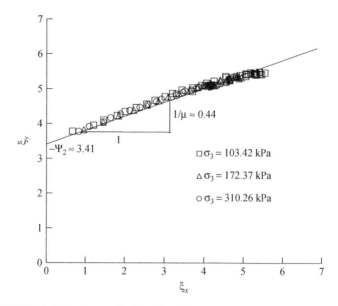

FIGURE 2.39   Determination of Lade's model parameters μ and ψ_2.

For isotropic compression test conditions, (2.118) reduces to

$$W_p = \int \sigma_3 \, d\varepsilon_v^p \tag{2.119}$$

The results of the isotropic compression test (Figure 2.35) can be plotted in the $\log(W_p/p_a)$ versus $\log(I_1/p_a)$ plane, as shown in Figure 2.40. The plastic, volumetric strains to be used in calculation of the plastic work [equation (2.119)] are calculated by subtracting the elastic volumetric strains from the total volumetric strains taken from the results of the isotropic compression test. Note that for the isotropic compression stress condition, I_1/p_a is equal to $3\sigma_3/p_a$. The data points in Figure 2.40 are best fitted with a line whose intercept with $\log(I_1/p_a) = 1$ line is C (≈ 0.0002) and whose slope is p (≈ 1.6).

The yield criterion in (2.100) requires two parameters, h and q. The value of h is determined on the basis that the plastic work is constant along a yield surface. Thus, for two stress points, A on the hydrostatic axis and B on the failure surface, the following expression is obtained for h:

$$h = \frac{\ln\{[\psi_1(I_{1B}^3/I_{3B}) - (I_{1B}^2/I_{2B})]e/(27\psi_1 + 3)\}}{\ln(I_{1A}/I_{1B})} \tag{2.120}$$

in which e is the base of the natural logarithm.

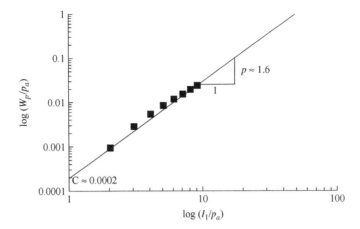

FIGURE 2.40   Determination of Lade's model parameters C and p.

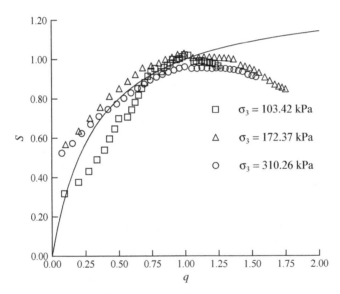

FIGURE 2.41   Determination of Lade's model parameter α.

Substituting (2.100) and (2.103) into (2.99) and solving for q, we get

$$q = \ln \frac{(W_p/Dp_a)^{\frac{1}{\rho}}}{[\psi_1(I_1^3/I_3) - (I_1^2/I_2)](I_1/p_a)^h} \qquad (2.121)$$

The variation of q from (2.121) with S from (2.101) is shown in Figure 2.41. This variation may be expressed by (2.102), in which α is constant. The best fit of (2.102) to the data shown in Figure 2.41 is obtained when $\alpha = 0.25$. Note that

TABLE 2.5   Lade's Model Parameters for Dense Silty Sand

	Parameter	Value
Failure criterion	m	0.154
	η_1	52
Elastic parameters	M	250
	λ	0.21
	υ	0.3
Plastic potential function	ψ_2	−3.42
	μ	2.3
Work hardening law	C	0.0002
	p	1.6
Yield function	α	0.25
	h	0.83

(2.102) is applicable only for $S < 1$ and $q < 1$. Table 2.5 summarizes Lade's model soil parameters.

2.10.5   Predicting Soil's Behavior Using Lade's Model: CD Triaxial Test Conditions

In the following, Lade's elastoplastic constitutive model with a single yield surface is implemented in a spreadsheet that can be used to predict soil behavior in consolidated drained triaxial test conditions. The model parameters can be obtained from a set of triaxial tests and one isotropic compression test following the procedures presented in the preceding section. To make certain that the model parameters are correct, one can use the spreadsheet (file: Example 2.4. spread sheet) developed here to back-calculate the experimental stress–strain and volumetric strain curves under different confining pressures. The reader can gain great insight into Lade's model by developing his or her own spreadsheet, or at least by studying the steps involved in the spreadsheet presented in this section. Once the model parameters are verified, they can be used for analysis of geotechnical problems using Lade's model embodied in a finite element program, for example.

The prediction of stress–strain behavior can be made using (2.87) to (2.121). For convenience, in the following step-by-step procedure, we present the equations needed again instead of referring to them.

Step 1: After obtaining all 11 material parameters, let us calculate the subparameters (ψ_1, D, and ρ):

$$(1.1) \qquad \psi_1 = 0.00155 m^{-1.27}$$

$$(1.2) \qquad D = \frac{C}{(27\psi_1 + 3)^\rho}$$

$$(1.3) \qquad \rho = \frac{p}{h}$$

Step 2: Set the initial value of σ_1 (axial stress) equal to σ_3 (confining pressure) to establish the initial stress conditions, then specify a small stress increment $d\sigma_1$ that can be added to the initial stress repeatedly until reaching the peak when $f_n = \eta_1$. After the peak, reduce the axial stress using the same stress increment until reaching the stress level required for which the strains will be calculated. Note that σ_2 and σ_3 are kept constant throughout the analysis ($\sigma_2 = \sigma_3$). f_n is calculated for each stress level using the equation

$$f_n = \left(\frac{I_1^3}{I_3} - 27 \right) \left(\frac{I_1}{p_a} \right)^m$$

in which

$$I_1 = \sigma_x + \sigma_y + \sigma_z$$

and

$$I_3 = \sigma_x \sigma_y \sigma_z + \tau_{xy} \tau_{yz} \tau_{zx} + \tau_{yx} \tau_{zy} \tau_{xz} - (\sigma_x \tau_{yz} \tau_{zy} + \sigma_y \tau_{zx} \tau_{xz} + \sigma_z \tau_{xy} \tau_{yx})$$

Step 3: Compute the stress invariants I_1, I_2, I_3, and J_2'; stress-level parameters, S and q; and the yield criterion, f_p'.

$$
\begin{align}
(3.1) \quad I_1 &= \sigma_x + \sigma_y + \sigma_z \\
(3.2) \quad I_2 &= \tau_{xy} \tau_{yx} + \tau_{yz} \tau_{zy} + \tau_{zx} \tau_{xz} - (\sigma_x \sigma_y + \sigma_y \sigma_z + \sigma_x \sigma_z) \\
(3.3) \quad I_3 &= \sigma_x \sigma_y \sigma_z + \tau_{xy} \tau_{yz} \tau_{zx} + \tau_{yx} \tau_{zy} \tau_{xz} \\
&\quad - (\sigma_x \tau_{yz} \tau_{zy} + \sigma_y \tau_{zx} \tau_{xz} + \sigma_z \tau_{xy} \tau_{yx})
\end{align}
$$

$$(3.4) \quad J_2' = \frac{1}{6}[(\sigma_x - \sigma_y)^2 + (\sigma_y - \sigma_z)^2 + (\sigma_z - \sigma_x)^2] + \tau_{xy}^2 + \tau_{yz}^2 + \tau_{zx}^2$$

$$(3.5) \quad S = \frac{f_n}{\eta_1} = \frac{1}{\eta_1} \left(\frac{I_1^3}{I_3} - 27 \right) \left(\frac{I_1}{p_a} \right)^m$$

$$(3.6) \quad q = \frac{\alpha S}{1 - (1 - \alpha)S}$$

$$(3.7) \quad f_p' = \left(\psi_1 \frac{I_1^3}{I_3} - \frac{I_1^2}{I_2} \right) \left(\frac{I_1}{p_a} \right)^h e^q$$

Step 4: Compute the elastic strain increments.

(4.1) Calculate E:

$$E = M p_a \left[\left(\frac{I_1}{p_a} \right)^2 + 6 \frac{1 + \nu}{1 - 2\nu} \left(\frac{J_2'}{p_a^2} \right) \right]^\lambda$$

(4.2) Calculate the stress increments $d\sigma_1$, $d\sigma_2$, and $d\sigma_3$:

$$d\sigma_1 = \sigma_1^{(i)} - \sigma_1^{(i-1)}$$

$$d\sigma_2 = \sigma_2^{(i)} - \sigma_2^{(i-1)}$$

$$d\sigma_3 = \sigma_3^{(i)} - \sigma_3^{(i-1)},$$

where (i) is the current step and $(i-1)$ is the preceding step.
(4.3) Calculate $d\varepsilon_1$, $d\varepsilon_2$, and $d\varepsilon_3$ using Hooke's law:

$$d\varepsilon_1^e = \frac{d\sigma_1 - v^e(d\sigma_2 - d\sigma_3)}{E} \times 100$$

$$d\varepsilon_2^e = \frac{d\sigma_2 - v^e(d\sigma_1 - d\sigma_3)}{E} \times 100$$

$$d\varepsilon_3^e = \frac{d\sigma_3 - v^e(d\sigma_1 - d\sigma_2)}{E} \times 100$$

Note that $d\sigma_2 = d\sigma_3 = 0$ in a triaxial test (σ_2 and σ_3 are kept constant!).

Step 5: Compute the plastic strain increments in the hardening regime.

(5.1) $\quad df_p' = f_p'^{(i)} - f_p'^{(i-1)}$ where (i) is the current step and $(i-1)$ is the preceding step.

(5.2) $\quad W_p = D \times (f_p')^\rho$

(5.3) $\quad dW_p = W_p^{(i)} - W_p^{(i-1)}$,   where (i) is the current step and $(i-1)$ is the preceding step.

(5.4) $\quad g_p = \left(\psi_1 \dfrac{I_1^3}{I_3} - \dfrac{I_1^2}{I_2} + \psi_2 \right) \left(\dfrac{I_1}{p_a} \right)^\mu$

(5.5) $\quad d\lambda_p = \dfrac{dW_p}{\mu g_p}$

(5.6) $\quad G = \psi_1(\mu+3)\dfrac{I_1^2}{I_3} - (\mu+2)\dfrac{I_1}{I_2} + \dfrac{\mu}{I_1}\psi_2$

(5.7)

$$\begin{Bmatrix} \dfrac{dg_p}{d\sigma_x} \\[2mm] \dfrac{dg_p}{d\sigma_y} \\[2mm] \dfrac{dg_p}{d\sigma_z} \\[2mm] \dfrac{dg_p}{d\sigma_{yz}} \\[2mm] \dfrac{dg_p}{d\sigma_{xz}} \\[2mm] \dfrac{dg_p}{d\sigma_{xy}} \end{Bmatrix} = \left(\dfrac{I_1}{p_a} \right)^\mu \begin{Bmatrix} G - (\sigma_y + \sigma_z)\dfrac{I_1^2}{I_2^2} - \psi_1(\sigma_y\sigma_z - \tau_{yz}^2)\dfrac{I_1^3}{I_3^2} \\[2mm] G - (\sigma_z + \sigma_x)\dfrac{I_1^2}{I_2^2} - \psi_1(\sigma_z\sigma_x - \tau_{zx}^2)\dfrac{I_1^3}{I_3^2} \\[2mm] G - (\sigma_x + \sigma_y)\dfrac{I_1^2}{I_2^2} - \psi_1(\sigma_x\sigma_y - \tau_{xy}^2)\dfrac{I_1^3}{I_3^2} \\[2mm] 2\dfrac{I_1^2}{I_2^2}\tau_{yz} - 2\psi_1(\tau_{xy}\tau_{zx} + \sigma_x\tau_{yz})\dfrac{I_1^3}{I_2^2} \\[2mm] 2\dfrac{I_1^2}{I_2^2}\tau_{zx} - 2\psi_1(\tau_{xy}\tau_{yz} + \sigma_y\tau_{zx})\dfrac{I_1^3}{I_3^2} \\[2mm] 2\dfrac{I_1^2}{I_2^2}\tau_{xy} - 2\psi_1(\tau_{yz}\tau_{zx} + \sigma_z\tau_{xy})\dfrac{I_1^3}{I_3^2} \end{Bmatrix}$$

(5.8) $$d(\varepsilon)^P = d\lambda_p \frac{dg_p}{d(\sigma)}$$

Step 6: Calculate plastic strain increments in the softening regime:

(6.1) $$\text{slope} = \left(\frac{1}{D}\right)^{\frac{1}{\rho}} \left(\frac{1}{\rho}\right) (W_{p(\max)})^{(1/\rho)-1}$$

(6.2) $$B = \frac{\text{slope}}{f_{p(\max)}}$$

(6.3) $$A = f_{p(\max)} \exp(B W_{p(\max)})$$

(6.4) $$W_p = \frac{\ln(A) - \ln(f_p')}{B}$$

After calculating W_p in the softening regime, repeat steps (5.3) through (5.8) to finish computing the plastic strain increments.

Step 7: Calculate the total strains (cumulative).

(7.1) $$\varepsilon_1 = \sum_{i=1}^{n}(d\varepsilon_1^e)i + \sum_{i=1}^{n}(d\varepsilon_1^p)i$$

where n is the total number of strain increments

(7.2) $$\varepsilon_2 = \sum_{i=1}^{n}(d\varepsilon_2^e)i + \sum_{i=1}^{n}(d\varepsilon_2^p)i$$

(7.3) $$\varepsilon_3 = \sum_{i=1}^{n}(d\varepsilon_3^e)i + \sum_{i=1}^{n}(d\varepsilon_3^p)i$$

(7.4) $$\varepsilon_v = \varepsilon_1 + \varepsilon_2 + \varepsilon_3$$

Example 2.4: *Back-Calculation of CD Triaxial Test Results Using Lade's Model*
Use Lade's model parameters for the dense silty sand obtained in Example 2.3 (Table 2.5) along with the spreadsheet described above to back-calculate the stress–strain behavior of the soil under CD triaxial conditions.

SOLUTION: You can establish your own spreadsheet (recommended) or use the one provided (file: Example2.4.spreadsheet) to calculate the stress–strain behavior of the soil under a given confining pressure. The spreadsheet provided automatically increases the stress level in small steps until reaching the peak. Thereafter, the stress level is decreased automatically to calculate the postpeak response. The calculation procedure is repeated for each confining pressure. Figure 2.42 presents the soil's triaxial behavior under three confining pressures. The calculated stress–strain behavior is also compared with the experimental data. Good agreement is noted in the figure. Of particular interest is the dilative behavior in the volumetric strain

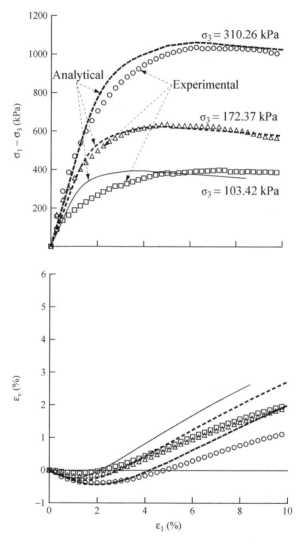

FIGURE 2.42 Predicted (Lade's model) and measured CD triaxial test results of a silty clay.

versus axial strain plane where the results calculated (Lade's model) captured the important dilation phenomenon.

PROBLEMS

2.1 The at-failure results of several CD and CU triaxial tests of a silty sand are presented in the $p'-q$ plane shown in Figure 2.43. The results of an isotropic

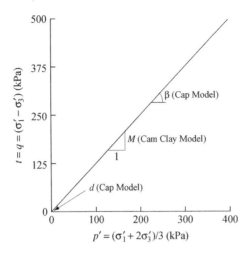

FIGURE 2.43 p' versus q curve for a silty sand.

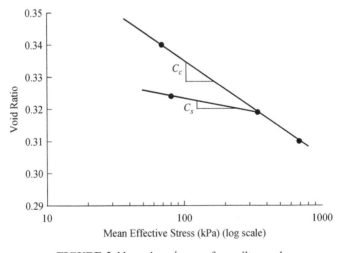

FIGURE 2.44 e–log p' curve for a silty sand.

compression test performed on the same soil are shown in Figure 2.44. Estimate the Cam clay model parameters M, λ, and κ. Note that you can obtain the compression and swelling indexes from Figure 2.44 (not λ and κ).

2.2 Using the Cam clay spreadsheet and model parameters for the silty sand (Problem 2.1), predict the consolidated drained triaxial behavior of this soil when subjected to a confining pressure of 70 kPa. Note that the initial void ratio corresponding to this confining pressure is 0.34 (Figure 2.44).

2.3 Repeat Problem 2.2 for consolidated undrained triaxial test conditions.

2.4   The at-failure results of several CD and CU triaxial tests of a silty sand are presented in the $p'-t$ plane shown in Figure 2.43. The results of an isotropic compression test performed on the same soil are shown in Figure 2.44. (*a*) Calculate the soil's angle of friction β and cohesion d in the $p'-t$ plane for the cap model. (*b*) Using the results of the isotropic compression test performed on the same soil (Figure 2.44), calculate the hardening curve assuming the initial conditions $p_0' = 70$ kPa and $e_0 = 0.34$.

2.5   The results of four consolidated drained triaxial tests and one isotropic compression test on loose Sacramento River sand are shown in Figures 2.45

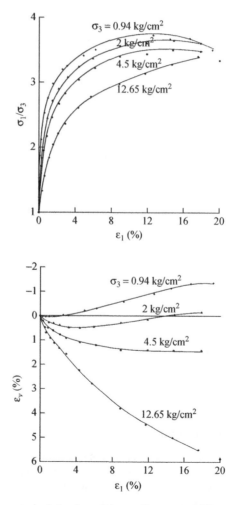

FIGURE 2.45   Stress–strain behavior of loose Sacramento River sand. (Adapted from Lade, 1977.)

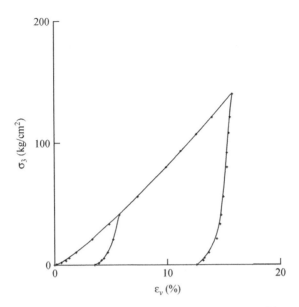

FIGURE 2.46   Isotropic compression behavior of loose Sacramento River sand. (Adapted from Lade, 1977.)

and 2.46, respectively. Estimate Lade's model parameters following the procedure discussed in Chapter 2.

2.6   Using Lade's model spreadsheet and Lade's model parameters obtained in Problem 2.5, predict the consolidated drained triaxial behavior of a loose Sacramento River sand specimen subjected to a confining pressure of 2 kg/cm^2.

2.7   From the results of the consolidated drained triaxial tests and the isotropic compression test on loose Sacramento River sand shown in Figures 2.45 and 2.46, respectively, estimate the Cam clay model parameters.

2.8   Using the Cam clay model spreadsheet and model parameters obtained in Problem 2.7, predict the consolidated drained triaxial behavior of a loose Sacramento River sand specimen subjected to a confining pressure of 2 kg/cm^2.

2.9   Obtain the cap model parameters using the results of the consolidated drained triaxial tests and the isotropic compression test on loose Sacramento River sand shown in Figures 2.45 and 2.46, respectively.

CHAPTER 3

STRESSES IN SOIL

3.1 INTRODUCTION

Accurate estimate of stress distribution in a soil mass is essential for calculations of elastic and consolidation settlements, of the bearing capacity of soil for shallow and deep foundations design, of lateral earth pressures for the design of earth-retaining structures, and of slope stability. In this chapter we show how to calculate in situ soil stresses and the additional soil stresses caused by external loads.

The in situ vertical stresses are the existing stresses in soil strata due to self-weight. The vertical stress at a point located at a depth z below the ground surface is equal to the weight of the soil above that point. When water is present within the soil strata, we need to distinguish between the total vertical stress and the effective vertical stress (Figure 3.1a). The concept of effective stress is presented in this chapter.

The stress increase within a soil mass caused by various types of external loading can be calculated based on the theory of elasticity. This stress increase is in excess of the in situ stress and has to be calculated separately (Figure 3.1b). Solutions for various types of loading are also presented in this chapter.

3.2 IN SITU SOIL STRESSES

In this chapter we assume that a soil located under the groundwater table is fully saturated [i.e., all voids between the soil grains are filled with water (no air)]. Also, we assume that the soil above the groundwater table is dry. When a total stress

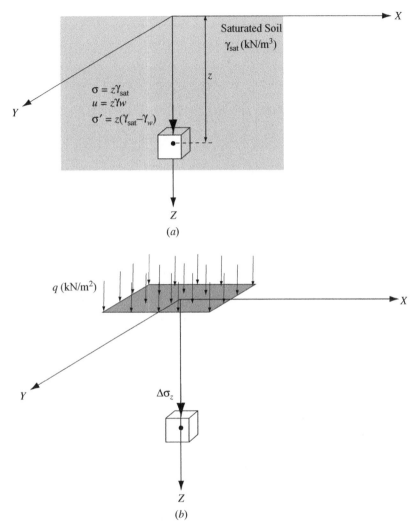

FIGURE 3.1   Stresses in a semi-infinite soil mass: (*a*) in situ vertical stresses; (*b*) stress increase due to external loads.

(σ) is applied to a saturated soil, it is carried by the water in the pores as well as the soil grains, as indicated in Figure 3.2. The stress carried by soil grains is called *effective stress* and given the symbol σ'. The stress carried by the water in the pores is termed *pore water pressure* and given the symbol u.

Define the effective stress (Figure 3.2) as:

$$\sigma' = \frac{F_1 + F_2 + F_3 + F_4}{A}$$

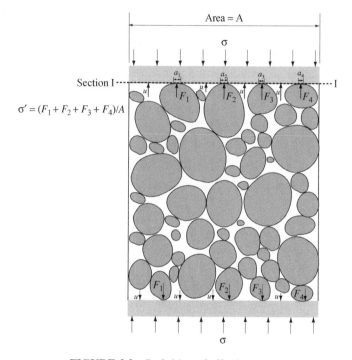

FIGURE 3.2 Definition of effective stress.

where F_1, F_2, F_3, and F_4 are the vertical components of the contact forces between the soil grains and section I–I, and A is the total cross-sectional area of the soil sample. From force equilibrium at Section I–I we can write

$$\sigma A = \sigma' A + u[A - (a_1 + a_2 + a_3 + a_4)]$$

where, the sum of a_1, a_2, a_3, and a_4 is the contact area between the soil grains and section I–I. This contact area is very small compared to the total area A:

$$a_1 + a_2 + a_3 + a_4 \approx 0$$

Therefore,

$$\sigma A = \sigma' A + uA$$

or

$$\sigma = \sigma' + u \tag{3.1}$$

This means that the effective stress can be calculated at any point below the ground surface by subtracting the pore water pressure from the total pressure at that point (i.e., $\sigma' = \sigma - u$). Note that the strength and compressibility of the soil depend on the effective stresses that exist within the soil grains—this is the essence of the effective stress principle that was formulated by Terzaghi (1936). The principle of effective stress is of fundamental importance in soil mechanics because soil behavior is governed by it. In this chapter we discuss only vertical in situ stresses. Horizontal stresses can be calculated as a fraction of the vertical stresses, as discussed in Chapter 7.

3.2.1 No-Seepage Condition

Let's consider the case of a homogeneous soil layer in a container, as shown in Figure 3.3. The thickness of the soil layer is H_2. Above the soil there is a layer of water H_1 thick. There is another reservoir that can be used to create an upward flow (upward seepage) through the soil sample. In this present discussion we assume that the valve leading to the upper reservoir is closed; thus, there is no water flowing through the soil sample. This is the case of no seepage.

First we calculate the total stress at various depths: $z = 0$ (top of the water layer), $z = H_1$ (top of the soil layer), and $z = H_1 + H_2$ (bottom of the soil layer). Then we calculate the pore water pressures at the same depths. Finally, we subtract the pore water pressure from the total stress to calculate the effective stress. The details of the calculations are shown. The stresses calculated are plotted in Figure 3.3 and connected with straight lines, since the stress distribution is a linear function of the depth z.

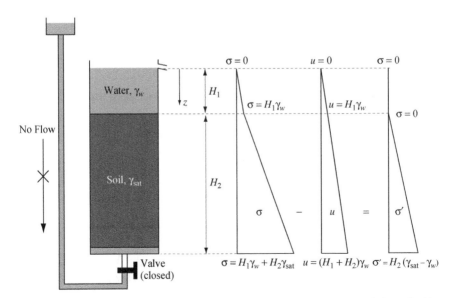

FIGURE 3.3 Calculation of effective stress distribution in a saturated soil layer inside a container (no-flow condition).

z	Total Stress, σ	Pore Water Pressure, u	Effective Stress, σ'
0	0	0	0
H_1	$H_1\gamma_w$	$H_1\gamma_w$	0
$H_1 + H_2$	$H_1\gamma_w + H_2\gamma_{sat}$	$(H_1 + H_2)\gamma_w$	$H_2(\gamma_{sat} - \gamma_w) = H_2\gamma'$

Note: γ' is the effective unit weight of the saturated soil ($\gamma' = \gamma_{sat} - \gamma_w$).

When the soil is dry, the pore water pressure is nonexistent and the effective stress is equal to the total stress. Figure 3.4*a* shows a homogeneous soil layer that has a dry unit weight γ_d and a thickness H. The groundwater table is deep and has no effect on the stress distribution in the soil; therefore, the soil layer can be assumed dry. The total and effective stress distributions are identical. They both start with a zero stress at $z = 0$ and increase linearly to a stress value of $H\gamma_d$ at $z = H$, as shown in Figure 3.4*b*.

Let us consider a homogeneous soil layer with a saturated unit weight γ_{sat} and a thickness H as shown in Figure 3.5*a*. The groundwater table is coincident with the ground surface. In this case the total stresses increases from zero at the ground

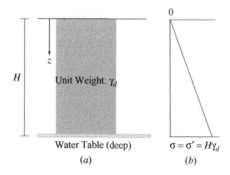

FIGURE 3.4   Effective stress distribution in a dry soil (deep water table).

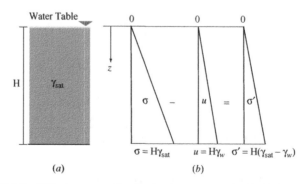

FIGURE 3.5   Effective stress distribution in a saturated soil (high water table).

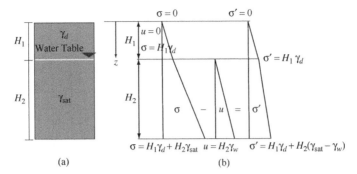

FIGURE 3.6 Effective stress distribution in a partially submerged soil layer.

surface to $H\gamma_{sat}$ at the bottom of the soil layer, as shown in Figure 3.5b. The pore water pressure also increases linearly with depth: from $u = 0$ at $z = 0$ to $u = H\gamma_w$ at $z = H$. The effective stress is calculated by subtracting the pore water pressure from the total stress: $\sigma' = 0$ at $z = 0$ to $\sigma' = H\gamma'$ at $z = H$ with a linear variation in between as indicated in Figure 3.5b.

When the water table is located below the surface of a homogeneous soil layer, as shown in Figure 3.6a, we can assume that the soil above the water table is dry and the soil below the water table is fully saturated. The effective stress calculations are summarized below. The stress profiles are shown in Figure 3.6b.

z	Total Stress, σ	Pore Water Pressure, u	Effective Stress, σ'
0	0	0	0
H_1	$H_1\gamma_d$	0	$H_1\gamma_d$
$H_1 + H_2$	$H_1\gamma_d + H_2\gamma_{sat}$	$H_2\gamma_w$	$H_1\gamma_d + H_2\gamma'$

Note: γ' is the effective unit weight of the saturated soil ($\gamma' = \gamma_{sat} - \gamma_w$).

Example 3.1 Plot the total stress, pore water pressure, and the effective stress distributions for the 4.5-m-thick soil layer shown in Figure 3.7a. The water table is located 1.5 m below the ground surface. The soil has a dry unit weight of 17 kN/m³ and a saturated unit weight of 19 kN/m³.

SOLUTION: The following is a detailed calculation of the stresses at $z = 0$, 1.5, and 4.5 m. The profiles calculated are shown in Figure 3.7b.

Z (m)	σ (kPa)	u (kPa)	σ' (kPa)
0	0	0	0
1.5	$(1.5)(17) = 25.5$	0	$25.5 - 0 = 25.5$
4.5	$25.5 + (3)(19) = 82.5$	$(3)(9.81) = 29.43$	$82.5 - 29.43 = 53.07$

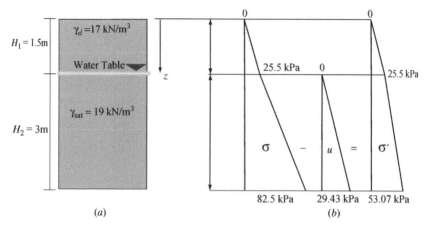

FIGURE 3.7   Soil profile for Example 3.1.

Example 3.2   Consider the homogeneous soil layer shown in Figure 3.8. In the winter, the water table is located at a distance H_1 below the ground surface, as shown in Figure 3.8a. In the spring, the water table rises a distance h above the winter level due to snow thaw (Figure 3.8b). Calculate the change in effective stress at the bottom of the soil layer from winter to spring. The soil has a dry unit weight of γ_d and a saturated unit weight of γ_{sat}.

SOLUTION: For winter conditions (Figure 3.8a), the effective stress at the bottom of the soil layer is

$$\sigma'_{\text{winter}} = H_1\gamma_d + H_2\gamma'$$

For spring conditions (Figure 3.8b), the effective stress at the bottom of the soil layer is

$$\sigma'_{\text{spring}} = (H_1 - h)\gamma_d + (H_2 + h)\gamma' = H_1\gamma_d + H_2\gamma' - h(\gamma_d - \gamma')$$

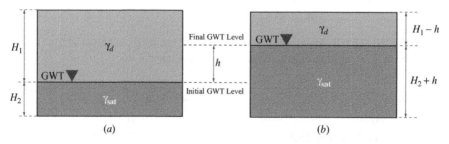

FIGURE 3.8   Effects of water table variation.

or

$$\sigma'_{\text{spring}} = \sigma'_{\text{winter}} - h(\gamma_d - \gamma')$$

Therefore,

$$\Delta\sigma' = \sigma'_{\text{winter}} - \sigma'_{\text{spring}} = h(\gamma_d - \gamma')$$

3.2.2 Upward-Seepage Conditions

Upward-seepage conditions can be induced in the laboratory using constant-head permeability test apparatus, as shown in Figure 3.9a. The upper reservoir causes the water to flow upward through the soil sample. If the hydraulic gradient is large enough ($i = i_{\text{cr}}$), the upward-seepage force will cause the effective stress within the soil to become zero, thus causing a sudden loss of soil strength in accordance with the effective-stress principle. This condition resembles that of the exit soil element on the downstream side of the sheet pile shown in Figure 3.10. If the hydraulic gradient in the exit element is large ($i_{\text{exit}} = i_{\text{cr}}$), the exit element becomes unstable—the upward-seepage force is large enough to cause the exit element to "float."

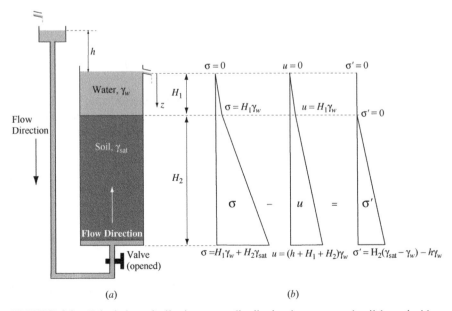

FIGURE 3.9 Calculation of effective stress distribution in a saturated soil layer inside a container (upward-flow condition).

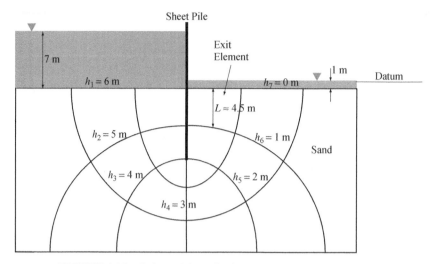

FIGURE 3.10   Safety of the exit element in a sheet pile problem.

Let's calculate the effective-stress profile of the soil sample shown in Figure 3.9a. This is done by calculating the total stresses and the pore water pressures at various depths and then subtracting the pore water pressures from the total stresses as follows:

z	Total Stress, σ	Pore Water Pressure u	Effective Stress, σ'
0	0	0	0
H_1	$H_1\gamma_w$	$H_1\gamma_w$	0
$H_1 + H_2$	$H_1\gamma_w + H_2\gamma_{\text{sat}}$	$(h + H_1 + H_2)\gamma_w$	$H_2\gamma' - h\gamma_w$

The triangular effective-stress profile along the depth of the soil specimen is shown in Figure 3.9b. The soil specimen will be totally destabilized when the effective-stress distribution becomes zero. We can obtain this condition if we set the effective stress at the bottom of the soil layer equal to zero:

$$H_2\gamma' - h\gamma_w = 0 \tag{3.2}$$

or

$$\frac{h}{H_2} = \frac{\gamma'}{\gamma_w} \tag{3.3}$$

As you recall, the hydraulic gradient through the soil specimen is given by

$$i = i_{\text{cr}} = \frac{h}{H_2} \tag{3.4}$$

where the hydraulic gradient, i, is equal to the critical hydraulic gradient, i_{cr}, because it causes the soil specimen to be destabilized. Substituting (3.3) into (3.4) yields

$$i_{cr} = \frac{\gamma'}{\gamma_w} \tag{3.5}$$

Now we can discuss the exit element on the downstream side of the sheet pile shown in Figure 3.10. The flow net indicates that the exit element is subject to a total head loss of 1 m ($= h_6 - h_7$). As the water flows from the bottom of the exit element toward the top, a distance L of approximately 4.5 m, it encounters a head loss of 1 m. Therefore, the exit hydraulic gradient can be calculated as

$$i_{exit} = \frac{h_6 - h_7}{L} = \frac{\Delta h}{L} \tag{3.6}$$

Let us define a hydraulic gradient safety factor for the exit element:

$$FS = \frac{i_{cr}}{i_{exit}} \tag{3.7}$$

When this safety factor is 1, the exit hydraulic gradient is equal to the critical hydraulic gradient, and the exit element is in the state of incipient failure. To prevent that, this safety factor should be equal to or greater than 1.5.

Is the exit element in Figure 3.10 safe? To answer that we need to calculate its safety factor as follows:

$$FS = \frac{i_{cr}}{i_{exit}} = \frac{\gamma'/\gamma_w}{\Delta h/L} = \frac{\gamma'}{\gamma_w} \frac{L}{\Delta h} \tag{3.8}$$

Assuming that the sand has $\gamma_{sat} = 19$ kN/m^3, then

$$FS = \left(\frac{19 - 9.81}{9.81} \right) \left(\frac{4.5}{1} \right) = 4.2 > 1.5$$

and the exit element is safe.

3.2.3 Capillary Rise

Soil pores are interconnected and they form a net of irregular tiny tubes. Due to the capillary phenomenon, water will rise above the water table through these tubes, forming a partially saturated zone of capillary rise. The height, h_w, above the water table to which the soil is partially saturated is called the *capillary rise*. This height is dependent on the grain size and soil type. In coarse soils capillary rise is very small, but in clays it can be over 10 m.

The pore pressure below the water table is considered positive and increases linearly with depth as discussed earlier. Above the water table, however, pore water pressure is negative (suction) and increases linearly, in absolute value, with the height above the water table, thus, at the water table level, $u = 0$; and at a distance h_w above the water table, $u \approx -(S\%/100)\gamma_w h_w$, where S is the degree of saturation of the soil in the zone of capillary rise.

The soil within the zone of capillary rise becomes substantially stronger because of the negative pore water pressure. Negative pore water pressure causes an increase in the effective stress: $\sigma' = \sigma - (-u) = \sigma + u$, hence the increase in strength. This is a direct consequence of the principle of effective stress. This capillary rise is the reason why we can build a sandcastle using moist sand (try building one with dry or very wet sand!).

Example 3.3   A 3.5-m-thick silt layer underlain by a 3-m-thick clay layer is shown in Figure 3.11a. Calculate the total stress, pore water pressure, and effective stress at points A, B, C, D, and E. The water table is located 2.5 m below the ground

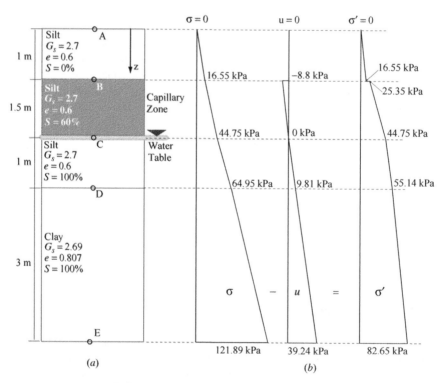

FIGURE 3.11   Soil profile for Example 3.3.

surface. The capillary rise in the silt layer is 1.5 m. Assume that the silt layer has a degree of saturation $S = 60\%$ in the zone of capillary rise.

SOLUTION: To calculate the in situ stresses, we need to estimate the unit weight of each soil layer. The dry unit weight of the top silt layer is

$$\gamma_d = \frac{G_s\gamma_w}{1+e} = \frac{(2.7)(9.81)}{1+0.6} = 16.55 \text{ kN/m}^3$$

The moist unit weight of the silt in the zone of capillary rise is

$$\gamma = \frac{(G_s + Se)\gamma_w}{1+e} = \frac{[2.7 + (0.6)(0.6)](9.81)}{1+0.6} = 18.8 \text{ kN/m}^3$$

The saturated unit weight of the silt layer below the water table is

$$\gamma_{sat} = \frac{(G_s + e)\gamma_w}{1+e} = \frac{(2.7 + 0.6)(9.81)}{1+0.6} = 20.2 \text{ kN/m}^3$$

The saturated unit weight of the clay layer is

$$\gamma_{sat} = \frac{(G_s + e)\gamma_w}{1+e} = \frac{(2.69 + 0.807)(9.81)}{1+0.807} = 18.98 \text{ kN/m}^3$$

Table 3.1 is a detailed calculation of the stresses at points A, B, C, D, and E (at $z = 0$, 1, 2.5, 3.5, and 6.5 m, respectively). The profiles calculated are shown in Figure 3.11b.

TABLE 3.1

Point	Z (m)	σ (kPa)	u, kPa	σ' (kPa)
A	0	0	0	0
B^{above}	0.999	(1)(16.55) = 16.55	0	16.55 − 0 = 16.55
B^{below}	1.001	(1)(16.55) = 16.55	(−0.6)(1.5)(9.81) = −8.8	16.55 − (−8.8) = 25.35
C	2.5	16.55+(1.5)(18.8) = 44.75	0	44.75 − 0 = 44.75
D	3.5	44.75+(1)(20.2) = 64.95	(1)(9.81) = 9.81	64.96 − 9.81 = 55.14
E	6.5	64.95+(3)(18.98) = 121.89	(4)(9.81) = 39.24	121.89 − 39.24 = 82.65

3.3 STRESS INCREASE IN A SEMI-INFINITE SOIL MASS CAUSED BY EXTERNAL LOADING

The stress increase within a soil mass caused by various types of external loading can be calculated based on the theory of elasticity. This stress increase is in excess of the in situ stress and has to be calculated separately. Solutions for various types of loading are presented next.

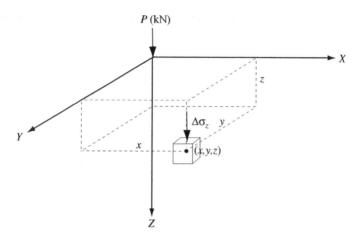

FIGURE 3.12 Vertical stresses caused by a point load.

3.3.1 Stresses Caused by a Point Load (Boussinesq Solution)

A *point load* is a concentrated load that can be applied at the surface of a semi-infinite soil mass as indicated in Figure 3.12. Boussinesq (1883) presented solutions for stresses within a semi-infinite soil mass subjected to a vertical point load applied at the surface. A *semi-infinite soil mass* is defined as an infinitely thick layer (in the z-direction) that is bounded by a horizontal plane at the top (x–y plane in Figure 3.12). A *Boussinesq solution* for a point load assumes that the soil mass is semi-infinite, homogeneous, linearly elastic, and isotropic. For the case of a vertical point load P applied at the origin of the coordinate system (Figure 3.12), the vertical stress increase at any point (x,y,z) within the semi-infinite soil mass is given by

$$\Delta\sigma_z = \frac{3P}{2\pi} \frac{z^3}{(x^2 + y^2 + z^2)^{5/2}} \tag{3.9}$$

where P is the intensity of the point load given in force units and x, y, and z are the coordinates of the point at which the increase of vertical stress is calculated.

Example 3.4 A vertical point load of 10 kN is applied at the surface of a semi-infinite soil mass. (a) Regarding the point of load application as the origin of the Cartesian coordinate system, calculate the increase of vertical stress directly under the applied load (i.e., at $x = 0$ and $y = 0$) for $z = 0$ to 1 m. Also calculate the increase in vertical stresses at $x = 0.1$ m, $y = 0$ m, for $z = 0$ to 1 m. (b) Repeat your solution using the finite element method and assuming that the soil is linear elastic with $E = 1 \times 10^7$ kPa and $v = 0.3$.

SOLUTION: (a) *Boussinesq solution*   To calculate the increase in vertical stress directly under the applied load for $z = 0$ to 1 m, we substitute $x = 0$ and $y = 0$ into (3.9):

$$\Delta\sigma_z = \frac{3P}{2\pi}\frac{1}{z^2}$$

Using this equation, we can calculate the increase in vertical stress as a function of z. The equation is plotted in Figure 3.13. According to this equation, $\Delta\sigma_z$ is linearly proportional to the intensity of the point load and inversely proportional to z^2. This means that $\Delta\sigma_z$ is very large near the point of load application but decreases very rapidly with depth as shown in Figure 3.13.

To calculate the increase in vertical stresses at $x = 0.1$ m, $y = 0$ m, for $z = 0$ to 1 m, we substitute these values in (3.9). The resulting increase in vertical stress is plotted as a function of depth in Figure 3.13. Note that near the surface the increase in vertical stress at $x = 0.1$ m and $y = 0$ m is much smaller than that immediately under the applied load (i.e., $x = 0$ m, $y = 0$ m). But at greater depths ($z > 0.4$ m) the increase in vertical stress is nearly identical at both locations.

(b) *Finite element solution* (filename: Chapter3_Example4.cae)   For simplicity, the semi-infinite soil mass is assumed to be a cylinder 2 m in diameter and 2 m in height, as shown in Figure 3.14. The reason of using a cylindrical shape in this simulation is to take advantage of axisymmetry, in which we can utilize axisymmetric two-dimensional analysis instead of three-dimensional analysis. The load is applied to the top surface at the center as shown in the figure. The purpose of the analysis is to calculate the increase in vertical stress within the soil mass due to the application of the point load, and to compare with Boussinesq analytical solution.

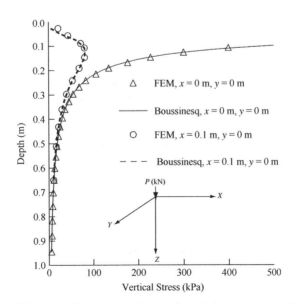

FIGURE 3.13   FEM versus Boussinesq solution of vertical stresses caused by a point load.

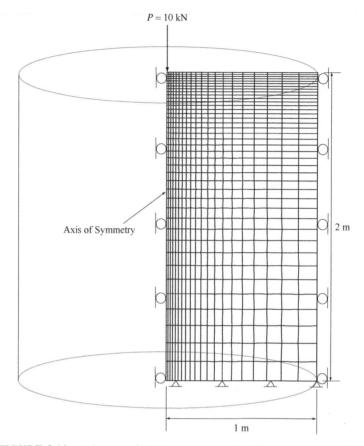

FIGURE 3.14   Axisymmetric finite element mesh of the point load problem.

The two-dimensional axisymmetric finite element mesh used has 20 elements in the x-direction and 40 elements in the z-direction, as shown in Figure 3.14. The finite element mesh is made finer in the zone around the point load where stress concentration is expected. The element chosen is a four-node bilinear axisymmetric quadrilateral element. The soil is assumed to be linear elastic with $E = 1 \times 10^7$ kPa and $\nu = 0.3$. The increase in vertical stress is plotted as a function of depth as shown in Figure 3.13 for $x = 0$ m, $y = 0$ m and $x = 0.1$ m, $y = 0$ m. The figure shows excellent agreement between the stresses calculated using the Boussinesq and finite element solutions.

3.3.2   Stresses Caused by a Line Load

A *line load* can be thought of as a point load that is applied repeatedly, in a uniform manner, along the y-axis as illustrated in Figure 3.15. The line load is applied infinitely along the y-axis. The units of a line load are given as force per

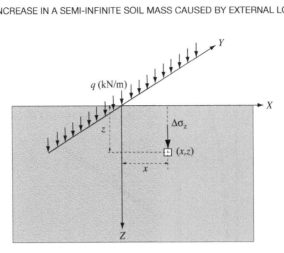

FIGURE 3.15   Stresses caused by a line load.

unit length, such as kN/m. Due to the nature of line load, the resulting stresses in the x–z plane are independent of y (i.e., we will get the same stresses in any x–z plane as we travel along the y-axis). This type of loading–geometry is termed *plane strain*. The vertical stress increase at any point (x,z) is given as

$$\Delta\sigma_z = \frac{2qz^3}{\pi(x^2 + z^2)^2} \tag{3.10}$$

where q is the line load (force/unit length) and x and z are the coordinates at which the stress increase is calculated.

Example 3.5   A vertical line load of 10 kN/m is applied at the surface of a semi-infinite soil mass. (a) Calculate the increase in vertical stress directly under the applied load for $z = 0$ to 0.3 m. (b) Repeat your solution using the finite element method and assuming that the soil is linear elastic with $E = 1 \times 10^7$ kPa and $v = 0.3$.

SOLUTION: (a) To calculate the increase in vertical stress directly under the applied load for $z = 0$ to 0.3 m, we substitute $x = 0$ into (3.10):

$$\Delta\sigma_z = \frac{2q}{\pi z}$$

Using this equation, we can calculate the increase in vertical stress as a function of z. The equation is plotted in Figure 3.16. According to this equation, $\Delta\sigma_z$ is linearly proportional to the intensity of the line load and inversely proportional to z. This means that $\Delta\sigma_z$ is very large near the point of load application but decreases rapidly with depth, as shown in Figure 3.16.

(b) *Finite element solution* (filename: Chapter3_Example5.cae)   A plane strain condition is assumed in which the semi-infinite soil mass is represented by a 1 m × 2 m (x–z) plane as shown in Figure 3.17. The purpose of the analysis is to calculate

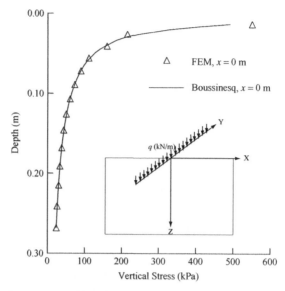

FIGURE 3.16   Stresses caused by a line load: FEM compared with the analytical solution.

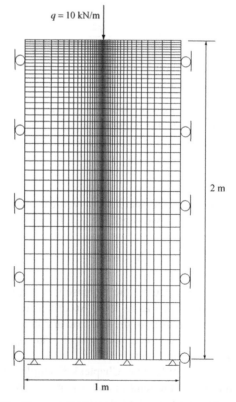

FIGURE 3.17   Plane strain finite element mesh of the line load problem.

the increase in vertical stress within the soil mass due to use of the line load and to compare with the analytical solution presented in part (a).

The two-dimensional plane strain finite element mesh used has 20 elements in the x-direction and 40 elements in the z-direction, as shown in Figure 3.17. The finite element mesh is made finer in the zone around the line load, where stress concentration is expected. The element chosen is a four–node bilinear plane strain quadrilateral element. The increase in vertical stress is plotted as a function of depth, as shown in Figure 3.16 for $x = 0$ m. The figure shows excellent agreement between the stresses calculated using the analytical solution (3.10) and the finite element solution.

Example 3.6 Two parallel line loads, 10 kN/m each, are applied at the surface of a semi-infinite soil mass as shown in Figure 3.18a. This type of loading can be thought as a load caused by the rails of a train while the train is standing still. (a) Calculate the increase in vertical stress directly above the crown of the underground tunnel located in the vicinity of the railroad, as shown in Figure 3.18a. (b) Repeat your solution using the finite element method and assuming that the soil is linear elastic with $E = 1 \times 10^7$ kPa and $\nu = 0.3$.

SOLUTION: (a) To calculate the total increase in vertical stress at the crown of the tunnel due to the line loads q_1 and q_2, we can calculate the stress increase caused by each line load separately and then combine the two. Called *superimposition*, this is permitted only when the loaded medium is linear elastic, which is the case in the present problem.

With the assistance of Figure 3.18b we can calculate $(\Delta\sigma_z)_1$, which is caused by q_1. For that we substitute $x = 0.51$ m and $z = 0.272$ m into (3.10):

$$(\Delta\sigma_z)_1 = \frac{(2)(10)(0.272)^3}{\pi(0.51^2 + 0.272^2)^2} = 1.148 \text{ kPa}$$

Also, with the help of Figure 3.18c we can calculate $(\Delta\sigma_z)_2$, which is caused by q_2. Substitute $x = 1.49$ m and $z = 0.272$ m into (3.10):

$$(\Delta\sigma_z)_2 = \frac{(2)(10)(0.272)^3}{\pi(1.49^2 + 0.272^2)^2} = 0.024 \text{ kPa}$$

Therefore,

$$\Delta\sigma_z = (\Delta\sigma_z)_1 + (\Delta\sigma_z)_2 = 1.148 + 0.024 = 1.172 \text{ kPa}$$

Note that the stress increase caused by q_1 at the crown of the tunnel is much greater than that caused by q_2. This is because q_1 is closer than q_2 to the tunnel.

(b) *Finite element solution* (filename: Chapter3_Example6.cae) A plane strain condition is assumed in which the semi-infinite soil mass is represented by a

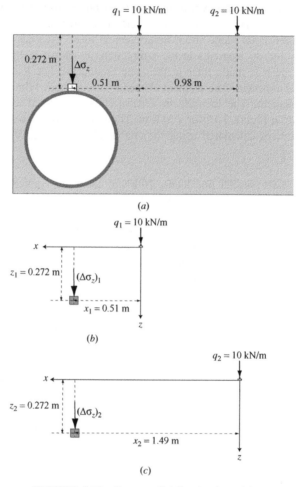

FIGURE 3.18 Two-parallel-line-loads problem.

2.8 m × 3.4 m (x–z) plane as shown in Figure 3.19. The purpose of the analysis is to calculate the increase of vertical stress at the crown of the tunnel due to the application of the two line loads and to compare with the analytical solution presented in part (a).

The two-dimensional plane strain finite element mesh used for this analysis has 22 elements in the x-direction and 50 elements in the z-direction, as shown in Figure 3.19. The element chosen is a four-node bilinear plane strain quadrilateral element. The underground tunnel is not modeled in this simplified analysis. For more elaborate analysis, the underground tunnel can be modeled as a cavity. The presence of an underground cavity will affect the stress distribution, especially around the tunnel.

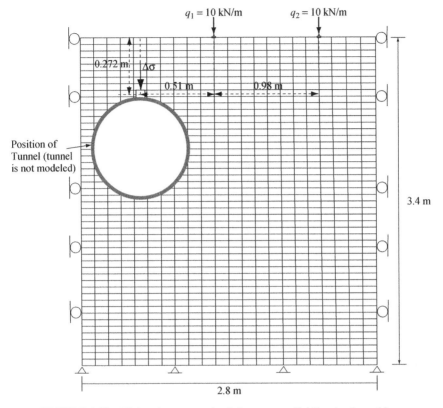

FIGURE 3.19 Finite element mesh of the two-parallel-line-loads problem.

The calculated increase in vertical stress at the crown of the tunnel is 1.188 kPa. This is in excellent agreement with the stress increase of 1.172 kPa calculated using the analytical solution (3.10).

3.3.3 Stresses Under the Center of a Uniformly Loaded Circular Area

For a uniformly loaded circular area (Figure 3.20), the stress increase under the center of the loaded area at any depth z is given by

$$\Delta\sigma_z = q\left\{1 - \frac{1}{[(R/z)^2 + 1]^{3/2}}\right\} \quad (3.11)$$

where q (force/unit area) is the applied pressure, R the radius of the loaded circle, and z the depth below the center of the loaded circle at which the stress increase is calculated. The elastic solution for stress increase elsewhere within the semi-infinite soil mass (not under the center) may be found in Ahlvin and Ulery (1962).

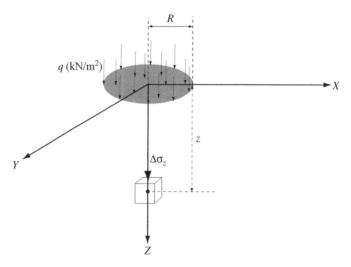

FIGURE 3.20   Stresses under the center of a uniformly loaded circular area.

Example 3.7   A pressure of 10 kPa is uniformly distributed on a circular area with $R = 0.5$ m. (a) Calculate the increase in vertical stress directly under the center of the applied load for $z = 0$ to 5 m. (b) Repeat your solution using the finite element method and assuming that the soil is linear elastic with $E = 1 \times 10^7$ kPa and $\nu = 0.3$.

SOLUTION: (a) For the increase in vertical stress directly under the center of the applied load for $z = 0$ to 5 m, we use (3.11):

$$\Delta\sigma_z = (10)\left\{1 - \frac{1}{[(0.5/z)^2 + 1]^{3/2}}\right\}$$

Using this equation, we can calculate the increase in vertical stress as a function of z. The equation is plotted in Figure 3.21. Note that $\Delta\sigma_z$ is large (10 kPa) near the surface but decreases very rapidly with depth.

(b) *Finite element solution* (filename: Chapter3_Example7.cae)   For simplicity, the semi-infinite soil mass is assumed to be a cylinder 100 m in diameter and 50 m in height, as shown in Figure 3.22. The reason for using a cylindrical shape in this simulation is to take advantage of axisymmetry, in which we can utilize axisymmetric two-dimensional analysis instead of three-dimensional analysis. The 10-kPa pressure is applied at the top surface on a circular area with 0.5-m radius. The purpose of the analysis is to calculate the increase in vertical stress within the soil mass due to the application of the 10-kPa pressure, and to compare with the analytical solution.

The two-dimensional axisymmetric finite element mesh used has 20 elements in the x-direction and 40 elements in the z-direction, as shown in Figure 3.22.

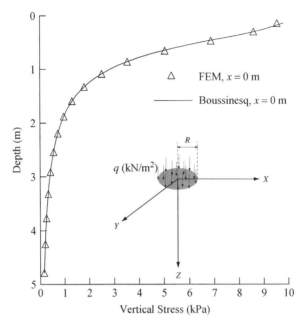

FIGURE 3.21 Comparison between FEM and analytical solution of the stresses under the center of a uniformly loaded circular area.

The finite element mesh is made finer in the zone around the pressurized circle where stress concentration is expected. The element chosen is a four-node bilinear axisymmetric quadrilateral element. The increase in vertical stress under the center of the pressurized circle is plotted as a function of depth as shown in Figure 3.21. The figure shows excellent agreement between the stresses calculated using the analytical elastic solution and the finite element solution. Note that the finite element solution is not limited to finding the stresses under the center of the loaded circle. It provides stresses, strains, and displacements at all nodal points within the loaded semi-infinite soil mass as well.

Vertical Stress Increase in a Layered Soil System   The equations presented above for point load (3.9), line load (3.10), and circularly loaded area (3.11) are based on the assumption that the underlying soil is homogeneous and infinitely thick. These equations are invalid for a soil system having several layers with varying elastic moduli (i.e., nonhomogeneous), such as the one shown in Figure 3.23. More complicated solutions based on the theory of elasticity are required for such cases. The following example is about a soil system with four different layers that resembles the structure of a highway pavement: an asphalt layer (top), a base layer, a subbase layer, and the existing soil (bottom). An analytical solution is not available for such a system. Thus, the finite element method can be extremely helpful for such a system.

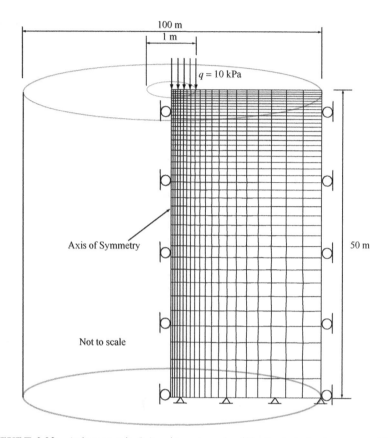

FIGURE 3.22 Axisymmetric finite element mesh of the loaded circular area problem.

Example 3.8 Consider a system with four layers of varying stiffness and thickness as shown in Figure 3.23. A pressure of 10 kPa is uniformly distributed on a circular area with $R = 0.5$ m. Using the finite element method, calculate the increase in vertical stress directly under the center of the circular area for $z = 0$ to 5 m. Compare this stress increase with that obtained in Example 3.7 for a single homogeneous soil layer.

SOLUTION: *Finite element solution* (filename: Chapter3_Example8.cae) Similar to what we did in Example 3.7, we assume that the semi-infinite soil mass is a cylinder 100 m in diameter and 50 m in height as shown in Figure 3.22. The 10-kPa pressure is applied at the top surface on a circular area with an 0.5 m radius. The purpose of the analysis is to calculate the increase in vertical stress within the stratified soil mass due to the application of a uniformly distributed load on a circular area, and to compare with the solution for a single homogeneous layer.

The two-dimensional axisymmetric finite element mesh used has 20 elements in the x-direction and 40 elements in the z-direction, as shown in Figure 3.22. The mesh includes four layers with the elastic moduli shown in Figure 3.23. The finite element mesh is made finer in the zone around the pressurized circle, where stress concentration is expected. The element chosen is a four-node bilinear axisymmetric quadrilateral element. The increase in vertical stress under the center of the pressurized circle is plotted as a function of depth as shown in Figure 3.24. For comparison, the increase in vertical stress in a single homogeneous layer is included

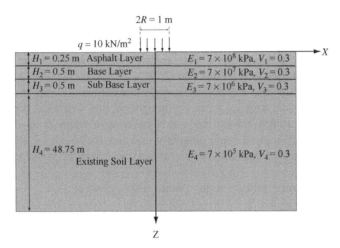

FIGURE 3.23 Stress increase in a layered soil system with a uniformly loaded circular area.

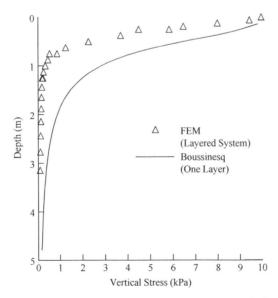

FIGURE 3.24 Comparison between FEM and analytical solution of a layered system with a uniformly loaded circular area.

in the figure (taken from Example 3.7). When the two are compared, the beneficial effects of the stiff asphalt layer and the base layer are seen clearly. These two layers absorbed most of the damaging vertical stress increase. Only a small fraction of stress increase is passed on to the subbase and existing softer soil layers. Thus, the asphalt layer acts as a shield that protects the underlying softer layers from excessive stress increases due to repeated traffic loads, which usually cause pavement rutting and cracking.

3.3.4 Stresses Caused by a Strip Load ($B/L \approx 0$)

Theoretically, a strip foundation is a rectangle of infinite length L and finite width B (i.e., $B/L \approx 0$). But foundations with $L/B > 10$ can be regarded as strip foundations. Examples of strip foundations include foundations for long structures such as retaining walls. A strip load can be thought of as a line load that is applied repeatedly and uniformly along the y-axis covering a width B as illustrated in Figure 3.25. This is a plane strain geometry in which the stresses in the x–z plane are independent of y. The units of a strip load are given as force per unit area, such as kN/m^2.

The vertical stress increase at any point (x,z) is given as:

$$\Delta\sigma_z = \frac{q}{\pi}\left\{ \tan^{-1}\left(\frac{x}{z}\right) - \tan^{-1}\left(\frac{x-B}{z}\right) \right.$$

$$\left. + \sin\left[\tan^{-1}\left(\frac{x}{z}\right) - \tan^{-1}\left(\frac{x-B}{z}\right) \right] \cos\left[\tan^{-1}\left(\frac{x}{z}\right) + \tan^{-1}\left(\frac{x-B}{z}\right) \right] \right\}$$

$$(3.12)$$

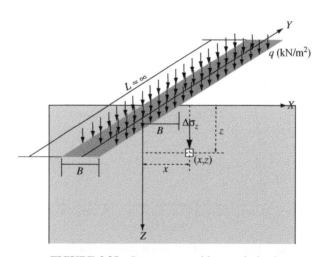

FIGURE 3.25 Stresses caused by a strip load.

where q is the applied pressure, B the width of the strip load, and x and z the coordinates of the point at which the stress increase is calculated. Usually, soil mechanics books and references tabulate $\Delta\sigma_z$ as a function of B, x, and z. With advancements in handheld preprogrammable calculators it became very easy to program and store equations such as (3.12). Thus, we do not include such tables in this chapter.

Example 3.9   A 1-m-wide strip load of 10 kN/m² is applied at the surface of a semi-infinite soil mass. (a) Using (3.12), calculate the increase in vertical stress directly under the center of the applied load for $z = 0$ to 8 m. (b) Repeat your solution using the finite element method and assuming that the soil is linear elastic with $E = 1 \times 10^7$ kPa and $\nu = 0.3$.

SOLUTION: (a) To calculate the increase in vertical stress directly under the strip load for $z = 0$ to 8 m, we substitute $x = 0.5$ m into (3.12), and then vary z from 0 to 8 m and use the equation to calculate the increase in vertical stress at various depths. The results are plotted in Figure 3.26. Note that the $\Delta\sigma_z$ calculated is equal to the applied pressure (= 10 kPa) at the surface where z is equal to zero. The stress decreases rapidly with depth, as shown in the figure.

(b) *Finite element solution* (filename: Chapter3_Example9.cae)   A plane strain condition is assumed. Because of symmetry, only one-half of the width of the strip load and one-half of the underlying soil are modeled, as shown in Figure 3.27. We calculate the increase in vertical stress within the soil mass due to the application of the strip load and compare it with the analytical solution presented above.

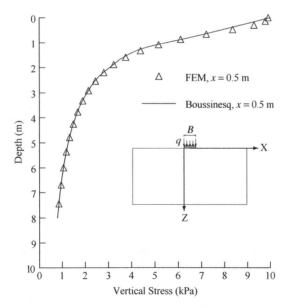

FIGURE 3.26   Stresses caused by a strip load: FEM versus analytical solution.

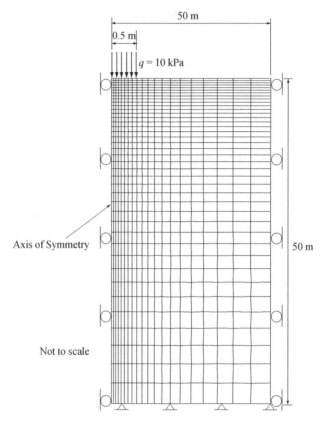

FIGURE 3.27 Plane strain finite element mesh of the strip load problem.

The two-dimensional plane strain finite element mesh used has 20 elements in the x-direction and 40 elements in the z-direction, as shown in Figure 3.27. The finite element mesh is made finer in the zone around the strip load, where stress concentration is expected. The element chosen is a four-node bilinear plane strain quadrilateral element. No mesh convergence studies have been performed. However, the dimensions of the soil layer are chosen such that the boundary effect on the solution is minimized. The increase in vertical stress is plotted as a function of depth as shown in Figure 3.26 for $x = 0.5$ m. The figure shows excellent agreement between the stresses calculated using the analytical solution (3.12) and the finite element solution.

3.3.5 Stresses Caused by a Uniformly Loaded Rectangular Area

Squares and rectangles are the most common shapes used in shallow foundations. The role of a shallow foundation is to spread the column load (from a superstructure) on a wider area in a uniform manner. Thus, instead of applying the concentrated column load directly to the "weak" soil, the shallow foundation will apply a much gentler uniform pressure to the soil.

Consider a uniformly loaded rectangular area with length L and width B as shown in Figure 3.28a. Note that L is always greater than B in a rectangle, and L is equal to B in a square. The uniform load q is expressed in force per unit area (pressure units). Equation (3.13) can be used to calculate the increase of vertical stress under the corner of a loaded rectangle as illustrated in Figure 3.28a.

$$\Delta\sigma_z = \frac{q}{4\pi}\left(\frac{2mn\sqrt{m^2+n^2+1}}{m^2+n^2+m^2n^2+1}\frac{m^2+n^2+2}{m^2+n^2+1} + \tan^{-1}\frac{2mn\sqrt{m^2+n^2+1}}{m^2+n^2-m^2n^2+1}\right)$$

(3.13)

where $m = B/z$ and $n = L/z$.

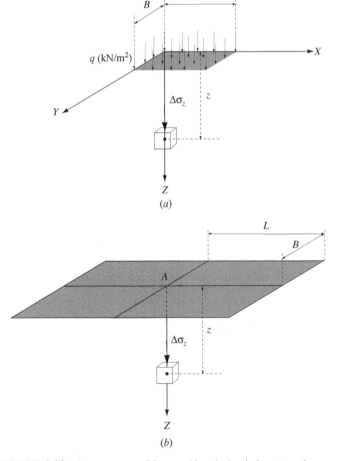

FIGURE 3.28 Stresses caused by a uniformly loaded rectangular area.

In most practical cases an increase in vertical stress under the center (point A in Figure 3.28b) of a uniformly loaded rectangular area is required. Remember that (3.13) is only for the stress increase below the corner of a rectangle. To use (3.13) to calculate the stress increase under the center of a loaded rectangle, we can divide the rectangle into four identical "small" rectangles each of which has point A as its corner. The increase in the vertical stress under the corner A of each small rectangle can be calculated using (3.13) and assuming that L and B are the length and width of a small rectangle (see Figure 3.28b). The total increase in vertical stress is then calculated by adding the four stress increases of the four identical small rectangles.

Example 3.10 (a) Using (3.13), calculate the increase in vertical stress under the center of a 4 m $\times$ 2 m rectangle that is loaded uniformly with $q = 10$ kPa. Assume that the soil layer underlying the loaded area is very thick and linear elastic with $E = 1 \times 10^7$ kPa and $\nu = 0.3$. (b) Calculate the increase in vertical stress under the center using the finite element method. Compare the two answers.

SOLUTION: (a) Since we need to calculate the stress increase under the center using (3.13), we will have to divide the 4 m $\times$ 2 m rectangle into four 2 m $\times$ 1 m small rectangles. Then for the small rectangles we have $m = B/z = 1/z$ and $n = L/z = 2/z$. Let's vary z from 0.01 to 20 m. For each z we calculate m and n and substitute those into (3.13) to get the stress increase $\Delta\sigma_z$ under the corner of the small rectangles. To calculate the stress increase under the center of the 4 m $\times$ 2 m rectangle, multiply $\Delta\sigma_z$ by 4 to account for the four identical small rectangles. The stress increase with depth calculated is listed in Table 3.2 and plotted in Figure 3.29. Note that near the surface (at $z = 0.01$ m), $\Delta\sigma_z$ is equal to $q (= 10$ kPa). The figure also shows that the stress declines very rapidly. At $z = 4B = 8$ m, $\Delta\sigma_z$ is only 0.56 kPa (i.e., less than 6% of the applied pressure q).

(b) *Finite element solution* (filename: Chapter3_Example10.cae)   This is a three-dimensional geometry that needs to be treated as such in a finite element analysis. The three-dimensional model analyzed is shown in Figure 3.30. The soil layer is 50 m deep and 100 m $\times$ 100 m in plan. The loaded area is 4 m $\times$ 2 m. The model considers only one-fourth of the soil layer and the loaded area, taking

TABLE 3.2

z (m)	$\Delta\sigma_z$ (kPa)	z (m)	$\Delta\sigma_z$ (kPa)
0.01	10.0	6	0.952
1	7.996	8	0.56
2	4.808	10	0.368
3	2.928	15	0.168
4	1.9	20	0.096
5	1.312		

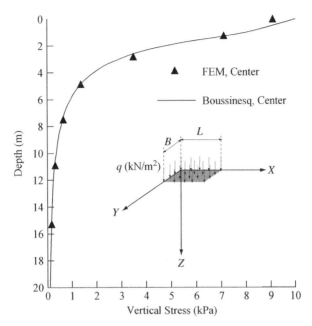

FIGURE 3.29 Stresses caused by a uniformly loaded rectangular area: FEM versus analytical solution.

advantage of symmetry as indicated in Figure 3.30a. The loaded area simulates a foundation with perfect contact with the soil. Reduced-integration eight-node linear brick elements are used for the soil layer. The base of the soil layer is fixed in all directions. All vertical boundaries are fixed in the horizontal direction but free in the vertical direction. The finite element mesh used in the analysis is shown in Figure 3.30b. It is noted that the mesh is finer in the vicinity of the foundation since that zone is a zone of stress concentration. No mesh convergence studies have been performed. However, the dimensions of the soil layer are chosen such that the boundary effect on foundation behavior is minimized. The increase in vertical stress under the center of the rectangular area is plotted as a function of depth in Figure 3.29. The figure shows excellent agreement between the stresses calculated using the analytical solution (3.13) and those calculated using the finite element method.

Note that the finite element analysis presented above provides stresses, strains, and displacements at all nodal points within the loaded semi-infinite soil mass—the finite element results are not limited to the stresses under the center of the loaded rectangle. Also note that this analysis assumes linear elastic soil behavior that is not suited for failure analysis. The present finite element analysis was carried out only to calculate the stress increase within the soil. In Chapter 6 we use elastoplastic soil models to predict the bearing capacity of various types of shallow foundations in a realistic manner.

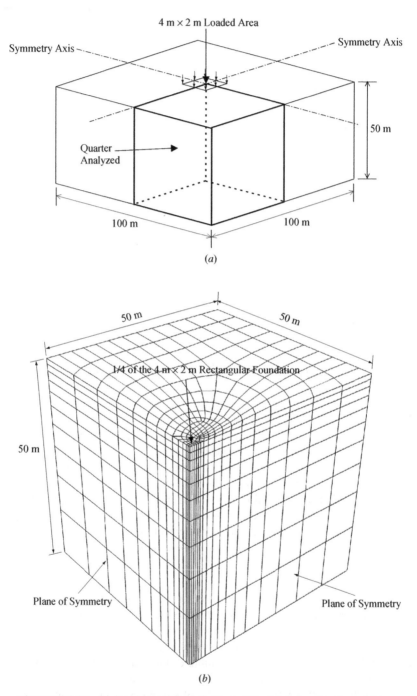

FIGURE 3.30 Finite element discretization of the loaded rectangle problem.

PROBLEMS

3.1 Plot the vertical stress, pore water pressure, and effective-stress distributions for the soil strata shown in Figure 3.31. The water table is located 5 m below the ground surface.

3.2 By the end of the summer, the water table in Problem 3.1 dropped 2 m. Plot the new distribution of the vertical stress, pore water pressure, and effective stress for the soil strata knowing that the bulk unit weight of the clay layer (above the water table) is 18 kN/m³ and its saturated unit weight is 19.7 kN/m³, as shown in Figure 3.31.

3.3 Refer to Figure 3.8. In the winter, the water table is located at a distance $H_1 = 10$ m below the ground surface, as shown in Figure 3.8*a*. In the spring, the water table rises a distance $h = 5$ m above the winter level, due to snow thaw (Figure 3.8*b*). Calculate the change in effective stress at the bottom of the soil layer from winter to spring. The soil has a dry unit weight of $\gamma_d = 17$ kN/m³ and a saturated unit weight of $\gamma_{sat} = 18.7$ kN/m³.

3.4 A 3.5-m-thick silt layer underlain by a 3-m-thick clay layer is shown in Figure 3.11*a* (refer to Example 3.3). Calculate the total stress, pore water pressure, and effective stress at points A, B, C, D, and E. The water table is located 2.5 m below the ground surface. Ignore the capillary rise in the silt layer and assume that the soil above the water table is dry. Compare the effective stresses at points A, B, C, D, and E with those calculated in Example 3.3.

3.5 Two vertical point loads of 100 kN each are applied at the surface of a semi-infinite soil mass. The horizontal distance between the two point loads is 0.5 m. Using a Boussinesq solution, calculate the increase in vertical stress directly under one of the applied loads at $z = 1$ m. If you were to solve this problem using the finite element method, what type of geometry would you use: axisymmetric, plane strain, or three-dimensional?

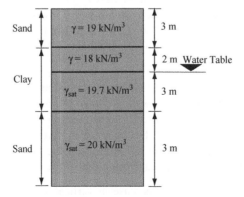

FIGURE 3.31

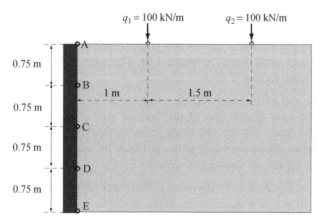

FIGURE 3.32

3.6 Two parallel line loads, 100 kN/m each, are applied at the surface of the backfill soil behind a rigid basement wall as shown in Figure 3.32. This type of loading can be thought of as the load caused by the rails of a train while the train is standing still or moving at a constant speed. **(a)** Calculate the increase in vertical stress at points A, B, C, D, and E. **(b)** Repeat your solution using the finite element method and assuming that the soil is linear elastic with $E = 1 \times 10^7$ kPa and $\nu = 0.3$. What is the horizontal stress increase at points A, B, C, D, and E?

3.7 A pressure of 100 kPa is distributed uniformly on a circular area with $R = 2$ m. **(a)** Calculate the increase in vertical stress directly under the center of the applied load for $z = 0$ to 12 m. **(b)** Repeat your solution using the finite element method and assuming that the soil is linear elastic with $E = 1 \times 10^7$ kPa and $\nu = 0.3$.

3.8 Consider the three-layer system shown in Figure 3.33. A pressure of 100 kPa is distributed uniformly on a circular area with $R = 2$ m. Using the finite element method, calculate the increase in vertical stress directly under the center of the circular area for $z = 0$ to 12 m. Compare the stress increase at points A, B, and C with that obtained in Problem 3.7 for a single homogeneous soil layer.

3.9 A 2-m-wide strip load of 100 kN/m² is applied at the surface of the backfill soil behind a rigid basement wall as shown in Figure 3.34. **(a)** Calculate the increase in vertical stress directly under the center of the applied load for $z = 0$ to 3 m, **(b)** Calculate the increase in vertical stress at points A, B, C, D, and E. **(c)** Repeat your solution using the finite element method and assuming that the soil is linear elastic with $E = 1 \times 10^7$ kPa and $\nu = 0.3$.

3.10 Using (3.13), calculate the increase in vertical stress under the center of a 4 m × 4 m foundation that is loaded uniformly with $q = 100$ kPa. Assume that the soil layer underlying the loaded area is very thick and linear elastic with $E = 1 \times 10^7$ kPa and $\nu = 0.3$. Also calculate the increase in vertical stress under the center using the finite element method. Compare the two answers.

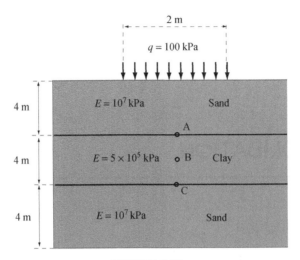

FIGURE 3.33

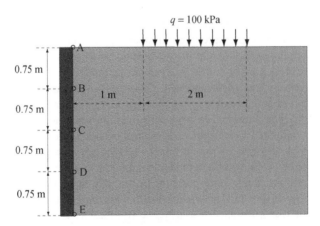

FIGURE 3.34

CHAPTER 4

CONSOLIDATION

4.1 INTRODUCTION

When a saturated soil is loaded, its pore pressure increases. This pore pressure increase, called *excess pore pressure*, *u*, dissipates from the boundaries of the soil layer as time goes by, resulting in consolidation settlement. This process is time dependent and is a function of the permeability of the soil, the length of the drainage path (defined later), and the compressibility of the soil.

When saturated sands and gravels are loaded slowly, volume changes occur, resulting in excess pore pressures that dissipate rapidly due to high permeability. This is called *drained loading*. On the other hand, when silts and clays are loaded, they generate excess pore pressures that remain entrapped inside the pores because these soils have very low permeabilities. This is called *undrained loading*. Consequently, the excess pore pressures generated by undrained loading dissipate slowly from the soil layer boundaries, causing consolidation settlement.

Consider a saturated clay layer sandwiched between two sand layers, with a groundwater table close to its top surface as shown in Figure 4.1. A uniform surcharge pressure of $\Delta\sigma = 10$ kPa is suddenly applied to the top surface. This loading is undrained and an excess pore pressure in the clay layer is generated instantaneously: $u = \Delta\sigma = 10$ kPa. The water level in a piezometer (standpipe) positioned at point A will rise a distance $h = u/\gamma_w = 10$ kPa/9.81 kN/m^3 ≈ 1 m above the level of the groundwater table. As time goes by, the excess pore pressure gradually dissipates from the top and bottom boundaries of the clay layer, and the water level in the piezometer drops accordingly. This loss of water from the boundaries is associated with consolidation settlements that take place gradually until the excess pore pressure is totally dissipated and the water level in the piezometer drops to the level of the groundwater table.

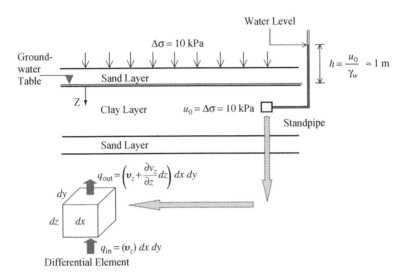

FIGURE 4.1   One-dimensional consolidation.

4.2 ONE-DIMENSIONAL CONSOLIDATION THEORY

To obtain approximate estimates of consolidation settlements in many geotechnical engineering problems, it is sufficient to consider that both water flow (due to excess pore pressure dissipation) and deformations take place in the vertical direction only. This is called *one-dimensional consolidation* and assumes that there is no lateral strain. Figure 4.1 illustrates a typical case of one-dimensional consolidation: a saturated clay layer that is loaded with a wide and uniform stress $\Delta\sigma$. An example of a wide and uniform stress is the stress resulting from depositing a wide even layer of sand on top of the existing sand layer in Figure 4.1.

Terzaghi (1925) considered the simple one-dimensional consolidation model shown in Figure 4.1, which consists of a cubic soil element subjected to vertical loading and through which only vertical water flow and deformation can take place. Several assumptions are used in the derivation of Terzaghi's one-dimensional consolidation equation:

1. The clay is fully saturated and homogeneous.
2. Water compressibility is negligible.
3. The compressibility of soil grains is also negligible, but soil grains can be rearranged during consolidation.
4. The flow of water obeys *Darcy's law* $(v = ki)$, where k is the soil permeability and i is the hydraulic gradient.
5. The total stress $(\Delta\sigma)$ applied to the element is assumed to remain constant.
6. The coefficient of volume compressibility, m_v, is assumed to be constant.
7. The coefficient of permeability, k, for vertical flow is assumed to be constant.

Using these assumptions and considering that the rate of volume change of the cubic element (Figure 4.1) is equal to the difference between the rate of outflow and the rate of inflow of water ($\partial V/\partial t = q_{out} - q_{in}$), one can derive the basic equation for one-dimensional consolidation (*Terzaghi's equation*):

$$c_v \frac{\partial^2 u}{\partial z^2} = \frac{\partial u}{\partial t} \tag{4.1}$$

where c_v is the coefficient of consolidation, given by

$$c_v = \frac{k}{m_v \gamma_w} \tag{4.2}$$

The solution of (4.1) must satisfy certain boundary and initial conditions (Figure 4.1):

Boundary conditions at the top of the clay layer: $z = 0$, $u = 0$ for $0 < t < \infty$

Boundary conditions at the bottom of the clay layer: $z = 2H_{dr}$, $u = 0$

for $0 < t < \infty$, (H_{dr} is defined in Section 4.2.1)

Initial conditions: $t = 0$, $u = u_0$ for $0 \leq z \leq 2H_{dr}$

For uniform (rectangular) initial excess pore pressure distribution with depth, and using a Fourier series, the exact solution of (4.1) is

$$u = \sum_{m=0}^{m=\infty} \left(\frac{2u_0}{M} \sin \frac{Mz}{H_{dr}} \right) \exp(-M^2 T_v) \tag{4.3}$$

In which $M = (\pi/2)(2m + 1)$; $m = 0, 1, 2, 3, \ldots, \infty$; u_0 is the initial excess pore pressure; and T_v is a nondimensional number called the *time factor*, defined as

$$T_v = \frac{c_v t}{H_{dr}^2} \tag{4.4}$$

Now define the degree of consolidation at depth z and time t as

$$U_z = \frac{u_0 - u}{u_0} = 1 - \frac{u}{u_0} = 1 - \sum_{m=0}^{m=\infty} \left(\frac{2}{M} \sin \frac{Mz}{H_{dr}} \right) \exp(-M^2 T_v) \tag{4.5}$$

The degree of consolidation at a point given by (4.5) is the ratio of the dissipated excess pore pressure ($= u_0 - u$) to the initial pore pressure at the same point ($= u_0$). For example, at $t = 0$, when stress is applied, the excess pore pressure (u) is equal to the initial excess pore pressure (u_0); therefore, $U_z = 0$ and no consolidation has occurred. But when $t \to \infty$, $u \to 0$ and $U_z \to 1$ (or 100%); that is, the consolidation is 100% complete.

Of more interest is the overall degree of consolidation of a clay layer rather than the degree of consolidation at a point within the clay layer. So let's define the average degree of consolidation for the entire thickness of the clay layer at time t as

$$U = 1 - \frac{(1/2H_{dr}) \int_0^{2H_{dr}} u \, dz}{u_0} = 1 - \sum_{m=0}^{m=\infty} \frac{2}{M^2} \exp(-M^2 T_v) \qquad (4.6)$$

It is to be noted that in this equation the initial pore pressure distribution is assumed to be uniform (rectangular) throughout the thickness of the clay layer. Also note that at $t = 0$, when stress is applied, the excess pore pressure is equal to the initial excess pore pressure; therefore, $U = 0$ for the entire layer and no consolidation has occurred. But when $t \to \infty$, $U \to 1$ (or 100%); that is, the consolidation is 100% complete for the entire layer.

4.2.1 Drainage Path Length

For the excess pore pressure to dissipate during consolidation, water must travel to the top boundary of the clay layer and sometimes to the bottom boundary as well, where there is a soil layer that is considerably more permeable than the clay layer itself. Logically, the rate of consolidation depends on the length of the longest path traveled by a drop of water. This length is called the *drainage path length*, H_{dr}. There are two possible drainage types:

1. Two-way drainage with a permeable layer both above and below the clay layer, as indicated in Figure 4.2*a*. In this case the longest path traveled by a drop of water located anywhere within the clay layer is $H_{dr} = H/2$, where H is the thickness of the clay layer.
2. One-way drainage with a permeable layer above the clay layer. In this case, $H_{dr} = H$, as indicated in Figure 4.2*b*.

4.2.2 One-Dimensional Consolidation Test

The consolidation characteristics of a soil can be measured in the laboratory using the one-dimensional consolidation test, shown schematically in Figure 4.3.

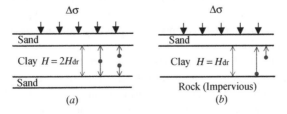

FIGURE 4.2 (*a*) Two- and (*b*) one-way drainage conditions.

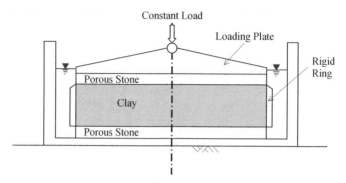

FIGURE 4.3   One-dimensional consolidation test apparatus.

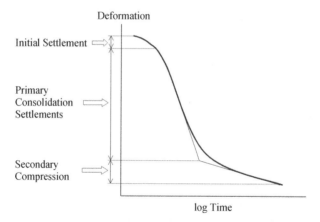

FIGURE 4.4   Deformation versus time curve (semilog).

A cylindrical specimen of soil measuring 75 mm in diameter and approximately 15 mm in thickness is enclosed in a metal ring and subjected to staged static loads. Each load stage lasts 24 hours, during which changes in thickness are recorded. The load is doubled with each stage up to the required maximum (e.g., 100, 200, 400, 800 kPa). At the end of the final loading stage, the loads are removed and the specimen is allowed to swell. Figure 4.4 shows an example of the settlement versus time curve obtained from one loading stage. Three types of deformations are noted in the figure: initial compression, primary consolidation settlement, and secondary compression or creep. Initial compression is caused by the soil's elastic response to applied loads. Primary consolidation settlement is caused by dissipation of the excess pore pressure generated by load application. Secondary compression is caused by the time-dependent deformation behavior of soil particles, which is not related to excess pore pressure dissipation. Primary consolidation settlement is our focus in this chapter.

Enclosing the soil specimen in a circular metal ring is done to suppress lateral strains. The specimen is sandwiched between two porous stones and kept submerged during all loading stages; thus, the specimen is allowed to drain from top and bottom. This is a two-way drainage condition in which the thickness of the specimen is $2H_{dr}$. The void ratio versus logarithm vertical effective-stress relationship (e–$\log \sigma'_v$) is obtained from the changes in thickness at the end of each load stage of a one-dimensional consolidation test. An example of an e–$\log \sigma'_v$ curve is shown in Figure 4.5.

Now define a *preconsolidation pressure* σ'_c as the maximum past pressure to which a clay layer has been subjected throughout time. A *normally consolidated* (NC) *clay* is defined as a clay that has a present (in situ) vertical effective stress σ'_0 equal to its preconsolidation pressure σ'_c. An *overconsolidated* (OC) *clay* is defined as a clay that has a present vertical effective stress of less than its preconsolidation pressure. Finally, define an *overconsolidation ratio* (OCR) as the ratio of the preconsolidation pressure to the present vertical effective stress (OCR $= \sigma'_c/\sigma'_0$).

To understand the physical meaning of OC clay, preconsolidation pressure, and OCR, imagine a 20-m-thick clay layer that was consolidated with a constant pressure of 1000 kPa caused by a glacier during the ice age. The glacier melted away and the pressure that had been exerted was gone totally. Thus, the preconsolidation pressure is 1000 kPa, the maximum past pressure exerted. Assuming that the groundwater table level is now at the top surface of the clay layer and that the saturated unit weight of the clay is 19.81 kN/m³, the present vertical effective stress in the middle of the clay layer is $\sigma'_0 = (\gamma_{sat} - \gamma_w)H/2 = (19.81 - 9.81)20/2 = 100$ kPa. The present vertical effective stress, 100 kPa, is less than the preconsolidation pressure of 1000 kPa. Therefore, the clay is overconsolidated. The overconsolidation ratio of this clay is $\sigma'_c/\sigma'_0 = 1000$ kPa/100 kPa $= 10$.

The preconsolidation pressure is a soil parameter that can be obtained from its e–$\log \sigma'_v$ curve deduced from the results of a one-dimensional consolidation test.

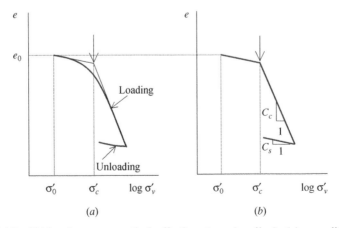

FIGURE 4.5 Void ratio versus vertical effective stress (semilog): (*a*) consolidation test results; (*b*) idealization.

The preconsolidation pressure is located near the point where the e–log σ'_v curve changes in slope, as shown in Figure 4.5a (Casagrande, 1936). Other consolidation parameters, such as the compression index (C_c) and swelling index (C_s), are also obtained from the e–log σ'_v curve. The compression index is the slope of the loading portion, in the e-log σ'_v curve, and the swelling index is the slope of the unloading portion, as indicated in Figure 4.5b.

The coefficient of consolidation, c_v, is an essential parameter for consolidation rate calculations [Terzaghi's one-dimensional consolidation equation (4.1)]. The coefficient of consolidation can be obtained from the changes in thickness recorded against time during one load stage of a one-dimensional consolidation test. This is done using the square-root-of-time method or the log-time method. The *square-root-of-time method* (Taylor, 1948) uses a plot of deformation versus the square root of time. Figure 4.6a shows a typical plot of deformation versus the square-root-of-time at a given applied load. The square-root-of-time method involves drawing a line AB through the early straight-line segment of the curve. Another line, AC, is drawn in such a way that $\overline{OC} = 1.15\overline{OB}$. The x-coordinate of point D, which is the intersection of AC with the curve, gives $\sqrt{t_{90}}$ (the square root of time corresponding to $U = 90\%$). One can use (4.6) to show that for a 90% degree of consolidation, the dimensionless time factor T_v is 0.848. Then (4.4) yields $c_v = 0.848 H_{dr}^2/t_{90}$. The value of t_{90} can be obtained from Figure 4.6a as explained above. H_{dr} is equal to the thickness of the soil specimen in the one-dimensional consolidation test if it is allowed to drain from one side only (one-way drainage). If the specimen is allowed to drain from both sides (top and bottom), H_{dr} is equal to half the thickness of the soil specimen.

The *log-time method* (Casagrande and Fadum, 1940) uses a plot of deformation versus the logarithm of time. Figure 4.6b shows a typical plot of deformation versus the logarithm of time at a given applied load. The log-time method involves extending the straight-line segments of the primary consolidation and the secondary compression. The extensions will intercept at point A. The y-coordinate of point

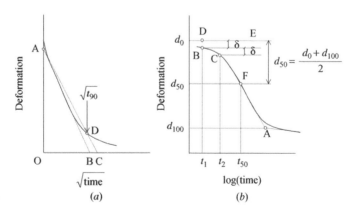

FIGURE 4.6 Graphical procedures for determining the coefficient of consolidation: (a) square-root-of-time method; (b) log-time method.

A is d_{100}, which is the deformation corresponding to 100% consolidation. On the initial portion of the deformation versus log-time plot, select two points, B and C, at times t_1 and t_2, respectively, such that $t_2 = 4t_1$. The vertical difference between points B and C is equal to δ. Draw a horizontal line DE at a vertical distance BD equal to δ. DE intercepts with the y-axis at d_0, which is the deformation corresponding to 0% consolidation. Now calculate d_{50} as the average of d_0 and d_{100}. Draw a horizontal line at d_{50} that will intercept with the curve at point F. The x-coordinate of point F is t_{50}, which is the time corresponding to 50% consolidation. Equation (4.6) can be used to calculate the dimensionless time factor $T_v (= 0.197)$, which corresponds to $U = 50\%$. Then (4.4) can be used to calculate the coefficient of consolidation: $c_v = 0.197 H_{dr}^2 / t_{50}$.

4.3 CALCULATION OF THE ULTIMATE CONSOLIDATION SETTLEMENT

Think of the e–log σ_v' curve as a strain–stress curve (inverse of stress–strain) that can be used to calculate the ultimate consolidation settlement of a clay layer subjected to a change of stress $= \Delta\sigma'$. The e–log σ_v' curve has two distinctive slopes, C_c and C_s, as shown in Figure 4.5. This curve is what we need to calculate the change in void ratio, Δe, caused by a change in stress, $\Delta\sigma'$, as shown in Figure 4.7. The consolidation settlement, S_c, can be calculated as

$$S_c = \frac{\Delta e}{1 + e_0} H \qquad (4.7)$$

where H is the thickness of the clay layer and e_0 is the initial (in situ) void ratio of the clay layer (i.e., the void ratio before $\Delta\sigma'$ was applied).

To calculate S_c, one must calculate Δe first. The calculation of Δe depends on the clay type (NC or OC) and on the stress condition ($\Delta\sigma' + \sigma_0'$). Figure 4.7 illustrates three possible cases.

Case 1: NC Clay   As you recall, NC clay has $\sigma_0' = \sigma_c'$. The e–log σ_v' curve for NC clay can be idealized as a straight line having a slope equal to C_c, as shown in Figure 4.7a. Adding a stress increment $\Delta\sigma'$ to the in situ stress σ_0' causes a change in void ratio equal to Δe that can be calculated as

$$\Delta e = C_c \log \frac{\sigma_0' + \Delta\sigma'}{\sigma_0'} \qquad (4.8)$$

Substituting (4.8) into (4.7) yields

$$S_c = C_c \frac{H}{1 + e_0} \log \frac{\sigma_0' + \Delta\sigma'}{\sigma_0'} \qquad (4.9)$$

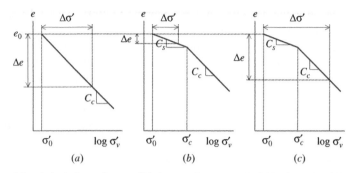

(a)   (b)   (c)

FIGURE 4.7 Calculation of consolidation settlements: (a) NC clay; (b) OC clay with $\Delta\sigma' + \sigma'_0 < \sigma'_c$; (c) OC clay with $\Delta\sigma' + \sigma'_0 > \sigma'_c$.

Case 2: OC Clay with $\Delta\sigma' + \sigma'_0 < \sigma'_c$ OC clay has $\sigma'_0 < \sigma'_c$. The e–log σ'_v curve for OC clay can be idealized as a bilinear curve having two distinctive slopes of C_s and C_c as shown in Figure 4.7b. Adding a stress increment $\Delta\sigma'$ to the in situ stress σ'_0 such that $\Delta\sigma' + \sigma'_0 < \sigma'_c$ causes a change in void ratio equal to Δe that can be calculated as

$$\Delta e = C_s \log \frac{\sigma'_0 + \Delta\sigma'}{\sigma'_0} \tag{4.10}$$

Substituting (4.10) into (4.7) yields

$$S_c = C_s \frac{H}{1 + e_0} \log \frac{\sigma'_0 + \Delta\sigma'}{\sigma'_0} \tag{4.11}$$

Case 3: OC Clay with $\Delta\sigma' + \sigma'_0 > \sigma'_c$ When $\Delta\sigma' + \sigma'_0$ is greater than σ'_c, the two segments of the bilinear curve must be considered for calculating Δe, as shown in Figure 4.7c. The following equation can easily be obtained for this condition:

$$S_c = \frac{H}{1 + e_0} \left(C_s \log \frac{\sigma'_c}{\sigma'_0} + C_c \log \frac{\sigma'_0 + \Delta\sigma'}{\sigma'_c} \right) \tag{4.12}$$

4.4   FINITE ELEMENT ANALYSIS OF CONSOLIDATION PROBLEMS

Consolidation analysis of saturated soils requires the solution of coupled stress–diffusion equations. In a coupled finite element analysis, the *effective-stress principle* is applied. Each point of the saturated soil mass is subject to a total stress σ, which is the sum of the effective stress σ' carried by the soil skeleton, and the pore pressure u. The pore pressure will increase by the addition of a load to the soil. Consequently, a hydraulic gradient of pore pressure will develop between two points within the soil mass. This hydraulic gradient between the two points will cause the water to flow. The flow velocity v is assumed to be proportional to the hydraulic gradient i according to Darcy's law, $v = ki$, where k is the soil permeability. As the external load is applied to the soil mass, the pore pressure rises initially; then as the soil skeleton absorbs the extra stress, the pore pressures decrease and the soil consolidates.

4.4.1 One-Dimensional Consolidation Problems

In many consolidation problems, it is sufficient to consider that both water flow (due to excess pore pressure dissipation) and deformations take place in the vertical direction only. This means that we can use the theory of one-dimensional consolidation for the analysis of such problems noting that in one-dimensional consolidation it is assumed that there is no lateral strain. The following five examples show how to solve simple one-dimensional consolidation problems using finite element. In each example, the finite element results are compared with the analytical results that are obtained using the consolidation concepts that we have learned in the preceding sections.

Example 4.1 *One-Dimensional Consolidation: Consolidation Analysis Assuming Linear Elastic Behavior of Soil* (a) Using the solution of Terzaghi's one-dimensional consolidation equation, plot the relationship between U and T_v for a cylindrical clay sample (Figure 4.8) that is 4 cm in diameter and 3.5 cm high confined by an impermeable, smooth, rigid cylindrical container and subjected to a sudden surcharge of 4 kg/cm². The top surface of the container is open and permeable. The clay has a constant permeability $k = 6 \times 10^{-6}$cm/min, a coefficient of consolidation $c_v = 0.16135$ cm²/min, and an initial void ratio $e_0 = 1.0$.

(b) Use the finite element method to solve the same problem assuming that the soil is linear elastic with Poisson's ratio $v = 0.33$.

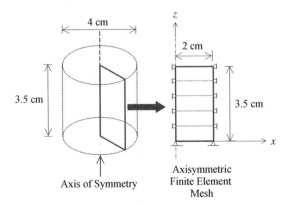

FIGURE 4.8 Axisymmetric finite element discretization of a one-dimensional consolidation test.

SOLUTION: (a) *Terzaghi's one-dimensional consolidation equation solution (exact solution)* First, let's calculate the dimensionless time factor as a function of time t. From (4.4),

$$T_v = \frac{c_v t}{H_{dr}^2} = \frac{0.16135t}{3.5^2} = 0.0132t$$

Next, let's calculate the average degree of consolidation of the 3.5-cm-thick clay layer using (4.6):

$$U = 1 - \sum_{m=0}^{m=\infty} \frac{2}{M^2} \exp(-M^2 T_v)$$

in which $M = (\pi/2)(2m + 1)$; $m = 0, 1, 2, 3, \ldots, \infty$; and $T_v = 0.0132t$. A spreadsheet was established to expand this expression for the evaluation of U as a function of t ranging from 0 to 100 minutes. The results are shown in Table 4.1 and plotted in Figure 4.9.

(b) *Finite element solution* (filename: Chapter4_Example1.cae) The problem is shown in Figure 4.8. A cylindrical soil specimen 4 cm in diameter and 3.5 cm in height is confined by an impermeable, smooth, rigid cylindrical container. The top surface is open and permeable. The load is applied suddenly to the top surface. The purpose of the analysis is to predict the time dependency of the effective stress and pore water pressure in the soil mass after load application and to compare those with the exact solution of the one-dimensional consolidation equation.

TABLE 4.1 T_v Versus U

t (min)	$T_v = 0.0132t$	U
0	0	0.017413
1	0.0133	0.1293
2	0.0266	0.1833
3	0.0399	0.2247
4	0.0533	0.2597
5	0.0666	0.2904
7	0.0933	0.3438
10	0.133	0.4111
15	0.199	0.5032
20	0.2667	0.5792
30	0.4	0.697
40	0.533	0.782
50	0.666	0.843
60	0.7998	0.887
70	0.9331	0.9186
80	1.0664	0.9414
90	1.1997	0.9578
100	1.333	0.9696

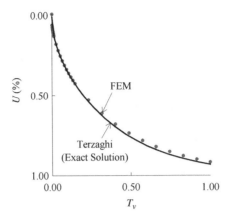

FIGURE 4.9   Average degree of consolidation versus time factor: comparison between Terzaghi's exact solution and FEM.

A two-dimensional axisymmetric mesh is used, with one element only in the x-direction (2 cm wide) and five elements in the z-direction (total height of 3.5 cm), as shown in Figure 4.8. The element chosen is a pore fluid/stress four-node axisymmetric quadrilateral element (with bilinear displacement and bilinear pore pressure) appropriate for finite-strain analysis. The soil is assumed to be linear elastic. Young's modulus, E, is not given explicitly in this problem. We attempt to calculate it from m_v knowing that $k = m_v c_v \gamma_w$, where m_v is the coefficient of volume compressibility that is equal to the inverse of the constrained modulus of elasticity, M. The constrained modulus is a function of Young's modulus and Poisson's ratio, v:

$$M = \frac{(1-v)E}{(1+v)(1-2v)}$$

The coefficient of volume compressibility is defined as

$$m_v = \frac{1}{M} = \frac{(1+v)(1-2v)}{(1-v)E}$$

But, $m_v = k/c_v \gamma_w$. Therefore,

$$\frac{(1+v)(1-2v)}{(1-v)E} = \frac{k}{c_v \gamma_w}$$

Finally,

$$E = \frac{c_v \gamma_w (1+v)(1-2v)}{k(1-v)} = \frac{(0.16135)(0.001)(1+0.33)[1-(2)(0.33)]}{(6 \times 10^{-6})(1-0.33)}$$

$$= 18.15 \text{ kg/cm}^2$$

where $v = 0.33$ and $\gamma_w = 0.001$ kg/cm^3.

When coupled analysis is performed on a saturated soil mass, two types of boundary conditions must be specified: displacement boundary conditions (or the equivalent) and hydraulic boundary conditions. Therefore, the boundary conditions of the finite element mesh shown in Figure 4.8 are as follows. On the bottom side, the vertical component of displacement is fixed ($u_z = 0$), and on the right-hand side, the horizontal component of displacement is fixed ($u_x = 0$) to simulate the frictionless interface between the soil and the rigid container. The left-hand side of the mesh is a symmetry line (no horizontal displacement). Note that the container is not modeled in the analysis. Flow of pore water through the walls of the container is not allowed—this is the natural boundary condition in the fluid mass conservation equation, so there is no need to specify it. On the top surface a uniform downward load of 4 kg/cm^2 is applied suddenly. During load application the top surface is made impervious, thereafter, a perfect drainage is assumed so that the excess pore pressure is always zero on this surface.

The problem is run in four steps. The first step is a single increment of analysis with a 0.01-minute time duration, with no drainage allowed across the top surface. When a sudden load is applied, a uniform pore pressure equal to the applied load is generated throughout the body. At that stage, the applied stress is totally carried by the pore water, u, and no stress is carried by soil skeleton ($\Delta u = \Delta \sigma = 4$kg/cm^2, $\Delta \sigma' = 0$). The hydraulic boundary condition of the top surface is then changed to a pervious boundary condition having $u = 0$, and the actual consolidation is done in three steps, each using fixed-time stepping. The durations of the three steps are 1, 10, and 100 minutes. Each step is divided into 10 equal substeps. The reason for choosing smaller time increments in the early stages is to capture the consolidation settlements, most of which occur following load application. Consolidation is a typical diffusion process; initially, the stress and pore pressure change rapidly with time, while more gradual changes in stress and pore pressure are seen at later times. This is also recognized through experiments. The one-dimensional consolidation test requires rapid readings of settlements following load application. Time increments between readings can be relaxed at later times.

From the results of the finite element analysis, the average degree of consolidation (U) is plotted with the dimensionless time factor (T_v) as shown in Figure 4.9. The average degree of consolidation is calculated as $U = 1 - u_{ave}/u_0$, where u_{ave} is the average excess pore pressure in the center of the five elements shown in Figure 4.8 and u_0 is the initial excess pore pressure ($= \Delta \sigma = 4$ kg/cm^2). For comparison, the exact solution [part (a) of this example] is also plotted in Figure 4.9. It is noted in the figure that the numerical solution agrees very well with the analytical solution.

The change in effective stress and pore pressure with respect to time is shown in Figure 4.10, as obtained from finite element analysis. Very good agreement with the exact solution (the solution of Terzaghi's equation) is noted in the figure.

In this example the permeability of the soil is assumed to be constant. However, it is possible in the finite element computer program used herein to consider the case in which the permeability decreases as the void ratio decreases from its initial value. If that assumption is used, the time required for the excess pore pressure

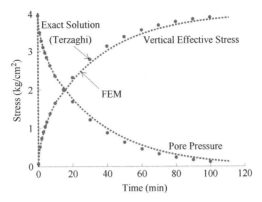

FIGURE 4.10   Variation of effective stress and excess pore pressure with time: comparison between Terzaghi's exact solution and FEM.

to dissipate will increase. Also, note that the final value of displacement (ultimate consolidation settlement) under the applied load is not a function of permeability as indicated by (4.12).

Example 4.2*   *One-Dimensional Consolidation Analysis Assuming Elastoplastic Behavior of Soil   Using the finite element method, calculate the consolidation settlement as a function of time for a cylindrical clay sample that is 20 cm in diameter and 10 cm high confined by an impermeable, smooth, rigid cylindrical container and subjected to a sudden surcharge of 20 kPa (Figure 4.11). Before applying the sudden surcharge, the clay specimen was subjected to a 9.19-kPa "seating" pressure. The top and bottom surfaces of the container are open and permeable. The clay is normally consolidated and has a constant permeability $k = 6 \times 10^{-9}$ m/s, and initial void ratio $e_0 = 1.5$, corresponding to its 9.19-kPa seating pressure. Assume that the clay is elastoplastic, obeying the extended Cam clay model (Chapter 2). The Cam clay model parameters for this clay are given in Table 4.2.

TABLE 4.2   Cam Clay Model Parameters

General		Plasticity	
ρ (kg/m^3)	1923	λ	0.1174
k (m/s)	6×10^{-9}	Stress ratio, M	1
γ_w(kN/m^3)	9.81	Initial yield surface	
e_0	1.5	size $= p_0'/2$ (kPa)	4.595
Elasticity		Wet yield surface size	1
κ	0.01957	Flow stress rate	1
ν	0.28		

SOLUTION(filename: Chapter4_Example2.cae): The parameter κ defines the elastic behavior of the soil in the Cam clay model. The parameter κ is related to the swelling index through the equation $\kappa = C_s/2.3$. The parameter λ is related to the compression index through $\lambda = C_c/2.3$. The strength parameter M is related to the internal friction angle of the soil, ϕ', as follows:

$$M = \frac{6 \sin \phi'}{3 - \sin \phi'}$$

A good practice is to solve the problem analytically, when possible, before seeking a numerical solution. Because this is a normally consolidated clay, we can use (4.9) to calculate its ultimate consolidation settlement. The initial condition of the clay layer is fully defined by its in situ vertical effective stress, $\sigma'_0 = 9.19$ kPa, and its in situ void ratio, $e_0 = 1.5$. The compression index C_c is equal to $2.3\lambda = 0.27$. Now we can use (4.9):

$$S_c = 0.27 \left(\frac{0.1}{1 + 1.5} \right) \log \frac{9.19 + 20}{9.19} = 0.00542 \text{ m} = 5.42 \text{ mm}$$

In the Cam clay model, the yield surface size is described fully by the parameter $p' = (\sigma'_1 + 2\sigma'_3)/3$. The evolution of the yield surface is governed by the volumetric plastic strain ε^p_{vol}, which is a function of p'. The relationship between ε^p_{vol} and p' can be deduced easily from an e–log σ'_v line. The consolidation curve (the e–log σ'_v line) is defined completely by its slope $C_c(= 2.3\lambda)$, and the initial conditions σ'_0 and e_0. Note that λ, σ'_0 and e_0 are part of the input parameters required in the finite element program used herein.

The preconsolidation pressure, σ'_c, is also a required parameter. It is specified by the size of the "initial" yield surface, as shown in Table 4.2. Consider the initial yield surface that corresponds to $\sigma'_0 = 9.19$ kPa (or $p'_0/2 = 4.595$ kPa), where $\sigma'_0 = 9.19$ kPa is the in situ vertical effective stress at the center of the clay specimen ($=$ seating pressure). In this example, σ'_c is assumed to be equal to the in-situ vertical effective stress, indicating that the clay is normally consolidated. If the clay were overconsolidated, its σ'_c can be set equal to its preconsolidation pressure, which is greater than its in situ vertical effective stress.

A two-dimensional axisymmetric mesh is used, with one element only in the x-direction (10 cm wide) and 12 elements in the z-direction (total height of 10 cm), as shown in Figure 4.11. The element chosen is a pore fluid/stress four-node axisymmetric quadrilateral element with bilinear displacement and bilinear pore pressure. The boundary conditions of the finite element mesh shown in Figure 4.11 are as follows. On the bottom side, the vertical and horizontal components of displacement are fixed ($u_x = u_z = 0$), and on the right-hand side, the horizontal component of displacement is fixed ($u_x = 0$) to simulate the frictionless interface between the soil and the rigid container. The left-hand side of the mesh is a symmetry line (no horizontal displacement). Note that the container is not modeled in the analysis. Flow of pore water through the walls of the container is not allowed. On the top surface a uniform downward pressure of 20 kPa is applied suddenly. During

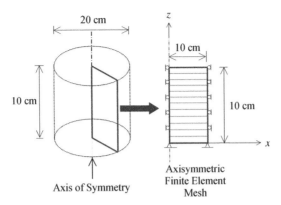

FIGURE 4.11   Axisymmetric finite element discretization of a one-dimensional consolidation test.

load application the top and bottom surfaces are made impervious; thereafter, perfect drainage is assumed so that the excess pore pressure is always zero on these surfaces.

The problem is run in three steps. The first step is a single increment of analysis with no drainage allowed across the top and bottom surfaces. In this step the seating pressure of 9.19 kPa is applied at the top surface. During step 1, the "geostatic" command is invoked to make sure that equilibrium is satisfied within the clay layer. The geostatic option makes sure that the initial stress condition in any element within the clay specimen falls within the initial yield surface of the Cam clay model. In step 2, a sudden 20-kPa pressure is applied, and a uniform pore pressure equal to the applied pressure is generated throughout the body. At that stage, the applied stress is carried totally by the pore water, u, and no stress is carried by the soil skeleton ($\Delta u = \Delta \sigma = 20$ kPa, $\Delta \sigma' = 0$). In step 3, the hydraulic boundary condition of the top and bottom surfaces is changed to a pervious boundary condition having $u = 0$, and the actual consolidation is performed using automatic time stepping. The durations of the three steps are 0.01, 10, and 100,000 seconds.

The time history calculated for vertical settlement is plotted in Figure 4.12, where it is compared with the solution obtained using (4.9). There is a relatively good agreement between the theoretical solution and the finite element solution. The difference between the results of the theoretical and numerical analyses is attributed primarily to stress anisotropy (K_0 condition). The Cam clay model used here is better suited for isotropic stress conditions. This shortcoming can be overlooked, however, because of the other many advantages of using the Cam clay model.

Example 4.3   *Ultimate Consolidation Settlement of an NC Clay Layer*   Consider a 1-m-thick NC clay layer sandwiched between two sand layers, with a groundwater table at the top ground surface as shown in Figure 4.13a. The initial void ratio of the clay layer is 0.8, its compression index is 0.27, and its swelling index is 0.045. A uniform surcharge pressure $q = 100$ kPa is applied to the top surface. Calculate

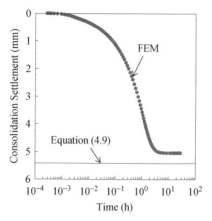

FIGURE 4.12 Consolidation settlement of the NC clay specimen: FEM versus analytical solution.

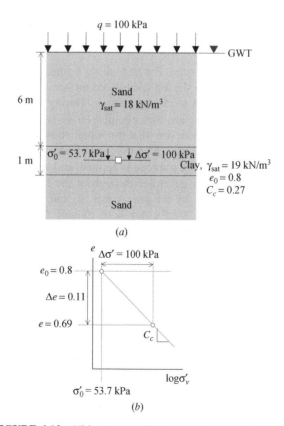

FIGURE 4.13 Ultimate consolidation settlement calculations.

TABLE 4.3   Cam Clay Model Parameters

General		Plasticity	
ρ (kg/m³)	1923	λ	0.1174
k (m/sec)	6×10^{-8}	Stress ratio, M	1
γ_w(kN/m³)	9.81	Initial yield surface size $= p_0'/2$ (kPa)	26.85
e_0	0.8	Wet yield surface size	1
Elasticity		Flow stress rate	1
κ	0.01957		
ν	0.28		

the ultimate consolidation settlement of the clay layer (a) analytically and (b) using finite element analysis. Assume that the clay layer is elastoplastic obeying the extended Cam clay model with the parameters given in Table 4.3.

SOLUTION: (a) *Analytical solution*   Because this is normally consolidated clay, we need to use (4.9) to calculate its ultimate consolidation settlement. Let's calculate the in situ vertical effective stress at the center of the clay layer:

$$\sigma_0' = (6)(18 - 9.91) + (\tfrac{1}{2})(19 - 9.81) = 53.7 \text{ kPa}$$

The initial condition of the clay layer is fully defined by its in situ vertical effective stress and in situ void ratio. Figure 4.13b shows this initial condition and illustrates how an increase in the vertical effective stress ($\Delta\sigma' = 100$ kPa) can cause a change in void ratio ($\Delta e = 0.11$). Now we can use (4.9) to calculate the ultimate consolidation settlement of the clay layer:

$$S_c = 0.27 \left(\frac{1}{1 + 0.8}\right) \log \frac{53.7 + 100}{53.7} = 0.06848 \text{ m} = 68.48 \text{ mm}$$

(b) *Finite element analysis* (filename: Chapter4_Example3.cae)   In this analysis, the clay layer is assumed elastoplastic, obeying the extended Cam clay model (Chapter 2). The Cam clay model parameters for the clay layer are given in Table 4.3. An important input parameter for the Cam clay model is the preconsolidation pressure, σ_c', which is specified by the size of the initial yield surface (see Table 4.3). Consider the initial yield surface that corresponds to $\sigma_0' = 53.7$ kPa in Figure 4.13b (or $p_0'/2 = 26.85$ kPa), where $\sigma_0' = 53.7$ kPa is the in situ vertical effective stress at the center of the clay layer. In this example, σ_c' is assumed to be equal to the in situ vertical effective stress, indicating that the clay is normally consolidated. If the clay were overconsolidated, its σ_c' can be set equal to its preconsolidation pressure, which is greater than its in situ vertical effective stress.

A two-dimensional plane strain mesh is used, with one element only in the x-direction (20 cm wide) and five elements in the z-direction (total height of 100 cm) for the clay layer, and six elements (total height of 6 m) for the sand layer, as shown

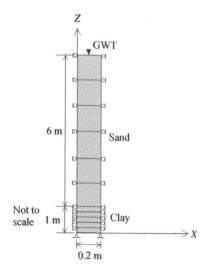

FIGURE 4.14   Finite element discretization.

in Figure 4.14. There is no need to consider the bottom sand layer in the finite element mesh, but its effect as a drainage layer is considered by assuming that the bottom surface of the clay layer is drained. The element chosen is a pore fluid/stress eight-node plane strain quadrilateral element with biquadratic displacement, bilinear pore pressure, and reduced integration.

On the bottom side of the finite element mesh, shown in Figure 4.14, the horizontal and vertical components of displacement are fixed ($u_x = u_z = 0$), and on the right- and left-hand sides the horizontal component of displacement is fixed ($u_x = 0$). On the top surface a uniform downward pressure of 100 kPa is applied suddenly. During load application the top and bottom surfaces are made impervious, thereafter, perfect drainage is assumed, so the excess pore pressure is always zero on these surfaces. The initial void ratio and vertical and horizontal effective stress profiles of the upper sand layer and the clay layer are part of the input data that must be supplied to the finite element program for this consolidation analysis. In this analysis, the initial horizontal effective stress is assumed to be 50% of the vertical effective stress.

The problem is run in eight steps. In step 1 the effective self-weights of the top sand layer and the clay layer are applied using the "body-force" option. As mentioned earlier, the clay layer is assumed to be elastoplastic obeying the extended Cam clay model. In general, using such a model is essential because we are concerned with the ability of the clay layer to withstand the stresses caused by various types of loading, and a model like the Cam clay model can detect failure within the clay layer. During step 1 the "geostatic" command is invoked to make sure that equilibrium is satisfied within the clay layer. The geostatic option makes sure that the initial stress condition in any element within the clay layer falls within the initial yield surface of the Cam clay model.

The second step is a single increment of analysis with a 1-second time duration, with no drainage allowed across the top and bottom surfaces of the clay layer. In this step, a sudden stress of 100 kPa is applied, and a uniform pore pressure equal to the applied stress is generated throughout the clay layer. At that stage, the applied stress is carried totally by the pore water u; none is carried by the soil skeleton ($\Delta u = \Delta\sigma = 100$ kPa, $\Delta\sigma' = 0$). The hydraulic boundary conditions of the top and bottom surfaces are then changed to pervious boundary conditions with $u = 0$, and the actual consolidation is done in six steps. The durations of the six steps are 1, 10, 100, 1000, 10,000, and 100,000 seconds. Each step is divided into 10 equal substeps. The reason for choosing smaller time increments in the early stages is to capture the consolidation settlements, most of which occur following load application. Consolidation is a typical diffusion process: Initially, the stress and pore pressure change rapidly with time; more gradual changes in stress and pore pressure are seen at later times.

From the results of the finite element analysis, the average pore pressure of the five elements of the clay layer is plotted along with the average vertical effective stress in the same elements as shown in Figure 4.15. The consolidation settlement versus time curve for the clay layer is shown in Figure 4.16. The ultimate consolidation settlement calculated using (4.9) is shown in the figure for comparison. It is noted from the figure that the numerical solution agrees with the analytical solution.

Example 4.4 *Ultimate Consolidation Settlement of an OC Clay Layer* ($\Delta\sigma' + \sigma'_0 < \sigma'_c$) Consider a 1-m-thick OC clay layer sandwiched between two sand layers, with a groundwater table at the top ground surface as shown in Figure 4.17a. The initial void ratio of the clay layer is 0.6715, its preconsolidation pressure is 200 kPa, its compression index is 0.27, and its swelling index is 0.045. A uniform surcharge pressure $q = 50$ kPa is applied to the top surface. Calculate the

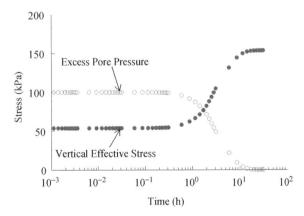

FIGURE 4.15 Variation of effective stress and excess pore pressure with time.

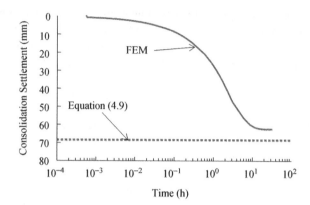

FIGURE 4.16   Consolidation settlement of the NC clay layer (Example 4.3).

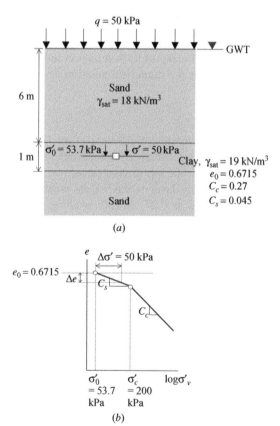

FIGURE 4.17   Ultimate consolidation settlement calculations: OC clay with $\Delta\sigma' + \Delta\sigma'_0 < \sigma'_c$.

ultimate consolidation settlement of the clay layer (a) analytically and (b) using finite element analysis.

SOLUTION: (a) *Analytical solution*   We need to calculate the in situ vertical effective stress at the center of the clay layer:

$$\sigma_0' = (6)(18 - 9.91) + (\tfrac{1}{2})(19 - 9.81) = 53.7 \text{ kPa}$$

The initial condition of the clay layer is fully defined by its in situ vertical effective stress $= 53.7$ kPa and in situ void ratio $= 0.6715$. Since $\Delta\sigma' + \sigma_0' = 50$ kPa $+ 53.7$ kPa $= 103.7$ kPa $< \sigma_c' = 200$ kPa, we need to use (4.11):

$$S_c = 0.045 \left(\frac{1}{1 + 0.6715} \right) \log \frac{53.7 + 50}{53.7} = 0.0077 \text{ m} = 7.7 \text{ mm}$$

(b) *Finite element analysis* (filename: Chapter4_Example4.cae)   This example is the same as Example 4.3 except that the clay layer in the present example is overconsolidated. The solution procedure is almost identical to that of Example 4.3. The OC clay layer is assumed elastoplastic, obeying the extended Cam clay model (Chapter 2). The Cam clay model parameters for the clay layer are given in Table 4.4.

In the present example, the initial yield surface size is taken as $p_0' = 200$ kPa, as indicated in Table 4.4. This initial yield surface corresponds to $\sigma_c' = 200$ kPa, which is the preconsolidation pressure of the clay. The in situ vertical effective stress at the center of the clay layer, $\sigma_0' = 53.7$ kPa (part of the input data), is less than the preconsolidation pressure, indicating that this clay layer is overconsolidated.

The finite element mesh and solution procedures are identical to those used in Example 4.3. Note that the applied pressure in this example is 50 kPa. The consolidation settlement versus time curve for the clay layer is shown in Figure 4.18. The ultimate consolidation settlement calculated using (4.11) is shown in the figure for comparison. It is noted from the figure that the numerical solution agrees with the analytical solution.

TABLE 4.4   Cam Clay Model Parameters

General		Plasticity	
ρ (kg/m^3)	1923	λ	0.1174
k (m/s)	6×10^{-8}	Stress ratio, M	1
γ_w (kN/m^3)	9.81	Initial yield surface size $= p_0'/2$ (kPa)	100
e_0	0.6715	Wet yield surface size	1
Elasticity		Flow stress rate	1
κ	0.01957		
ν	0.28		

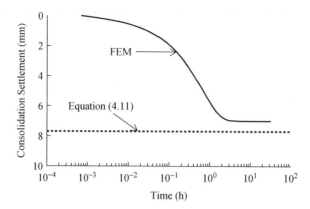

FIGURE 4.18   Consolidation settlement of the OC clay layer (Example 4.4).

Example 4.5   *Ultimate Consolidation Settlement of an OC Clay Layer* $(\Delta\sigma' + \sigma_0' > \sigma_c')$ Repeat Example 4.4 with an applied pressure of 400 kPa.

SOLUTION: (a) *Analytical solution*   The in situ vertical effective stress at the center of the clay layer is $\sigma_0' = 53.7$ kPa. The initial condition of the clay layer is fully defined by its in situ vertical effective stress $= 53.7$ kPa and in situ void ratio $= 0.6715$. Since $\Delta\sigma' + \sigma_0' = 400$ kPa $+ 53.7$ kPa $= 453.7$ kPa $> \sigma_c' = 200$ kPa, we need to use (4.12) (Figure 4.19):

$$S_c = \frac{1}{1 + 0.6715}\left(0.045\log\frac{200}{53.7} + 0.27\log\frac{53.7 + 400}{200}\right) = 0.0728 \text{ m}$$

$$= 72.8 \text{ mm}$$

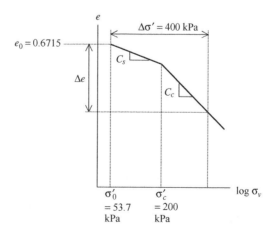

FIGURE 4.19   Calculation of consolidation settlements: OC clay with $\Delta\sigma' + \sigma_0' > \sigma_c'$.

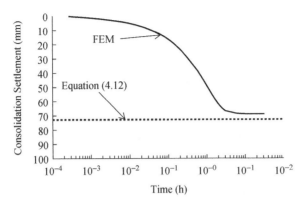

FIGURE 4.20   Consolidation settlement of the OC clay layer (Example 4.5).

(b) *Finite element analysis* (filename: Chapter4_Example5.cae)   The finite element mesh, material parameters, and solution procedures are identical to those used in Example 4.4. Note that the applied pressure in this example is 400 kPa. The consolidation settlement versus time curve for the clay layer is shown in Figure 4.20. The ultimate consolidation settlement calculated using (4.12) is shown in the figure for comparison. It is noted from the figure that the numerical solution agrees well with the analytical solution.

4.4.2   Two-Dimensional Consolidation Problems

Many practical consolidation problems are two- or three-dimensional. Examples include a strip foundation situated on a clay layer (two-dimensional), a rectangular or square foundation on a clay layer (three-dimensional), an embankment underlain by a clay layer (two-dimensional), and so on. Terzaghi's one-dimensional consolidation theory offers a rough estimate of consolidation settlements for such problems. The following three examples illustrate cases of special interest: (1) the settlement history of a partially loaded linear elastic soil layer—this particular case is chosen to illustrate two-dimensional consolidation because an exact solution is available (Gibson et al., 1970), thus providing verification of this capability in the finite element computer program used herein; (2) the two-dimensional consolidation of a clay layer subjected to sequential construction of an embankment; and (3) the two-dimensional consolidation of a thick elastoplastic clay layer subjected to strip loading.

Example 4.6   *Two-Dimensional Consolidation—Numerical Solution Versus Gibson's Solution*   A fully saturated soil layer with infinite horizontal extension in the x-direction and a finite height of 50 mm is subjected to a 50-mm-wide strip load of 3.45 MPa as shown in Figure 4.21. The soil layer is supported by a frictionless rigid surface (i.e., the bottom boundary of the layer is free to slide in the horizontal

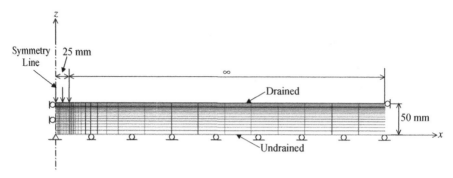

FIGURE 4.21 Finite element discretization of Gibson's two-dimensional consolidation problem.

direction). The bottom boundary is assumed to be impervious. The soil is linear elastic with $E = 690$ GPa, $v = 0$, and $k = 5.08 \times 10^{-7}$ m/day. The pore fluid has a specific weight of 272.9 kN/m^3. The theoretical treatment of this two-dimensional consolidation problem is available elsewhere (Gibson et al., 1970), and the consolidation settlement at the center of the loaded strip area with respect to time is readily available using the material parameters above (see Figure 4.22). Calculate the consolidation settlement at the center of the loaded strip area with respect to time using a finite element consolidation program and compare with the theoretical curve shown in Figure 4.22. (Note that the material parameters in this example are chosen only for illustration purposes and do not reflect the behavior of a real soil and pore fluid.)

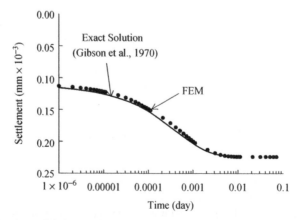

FIGURE 4.22 Consolidation settlement at the center of the loaded area with time: comparison between Gibson's exact solution and FEM.

SOLUTION(filename: Chapter4_Example6.cae): Clearly, this is a two-dimensional consolidation problem since the loaded strip region is as wide as the depth of the soil layer, which means that the soil will consolidate vertically and horizontally as well, especially in the vicinity of the loaded area. The geometry of the problem conforms to the plane strain condition since the soil layer and the applied strip load are extended infinitely in the y-direction (normal to Figure 4.21). Only one-half of the geometry is considered because of symmetry, as indicated by the finite element mesh shown in Figure 4.21. In the theoretical derivation of the solution of this two-dimensional problem, the soil layer was assumed to be infinitely long in the x-direction. Therefore, the finite element mesh is made very long in the horizontal direction (20 × foundation width)—this distance is very large compared to the width of the loaded strip and will cause the boundary effect to be diminished. An alternative solution may consider the use of infinite elements at the right-hand side of the finite element mesh.

A reduced-integration plane strain element with pore pressure is used in this analysis. The finite element mesh shown in Figure 4.21 is made finer (biased) near the zone where the load is applied. Coarser elements are used away from the loaded zone. The soil is assumed linear elastic with $E = 690$ GPa; $v = 0$; $k = 5.08 \times 10^{-7}$ m/day, and $\gamma_{\text{fluid}} = 272.9$ kN/m^3. The load was applied instantaneously with a magnitude of 3.45 MPa. The soil layer is assumed to lie on a smooth, impervious base, so the vertical component of displacement is set to zero on that surface ($u_z = 0$). The left-hand side of the mesh is a symmetry line (no horizontal displacement). The right-hand side of the mesh is assumed to slide freely in the vertical direction with $u_x = 0$.

The problem is run in six steps. In the first step, the load is applied and no drainage is allowed across the top surface of the mesh. This step establishes the initial distribution of excess pore pressures within the soil. These excess pore pressures will dissipate from the top surface of the soil layer during subsequent consolidation steps. To allow that, the hydraulic boundary condition of the top surface is changed to a pervious boundary condition with $u = 0$, and the actual consolidation is done with five time steps, each using fixed-time stepping. The durations of the five steps are 0.00001, 0.0001, 0.001, 0.01, and 0.1 day. Each step is divided into 10 equal substeps. The reason for choosing smaller time increments in the early stages is to capture the consolidation settlements, most of which occur following load application.

The time history calculated for vertical settlement of the central point under the strip load is plotted in Figure 4.22, where it is compared with the exact solution by Gibson et al. (1970). There is a good agreement between the theoretical solution and the finite element solution.

Example 4.7 *Sequential Construction of an Embankment on a Clay Layer*   Calculate the consolidation settlement under the center of a 1.8-m-high embankment founded on a 4.57-m-thick clay layer underlain by an impermeable layer of rock. The groundwater table is coincident with the top surface of the clay layer, as indicated in Figure 4.23. The embankment is constructed in three equal layers, each

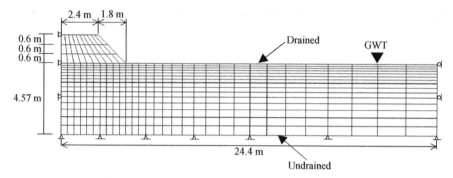

FIGURE 4.23   Finite element Discretization of an embankment on a soft foundation.

TABLE 4.5   Cam Clay Model Parameters

General			Plasticity	
ρ (kg/m^3)	1923		λ	0.174
k (m/s)	2.5×10^{-8}		Stress ratio, M	1.5
γ_w (kN/m^3)	9.81		Initial yield surface	
e_0	0.889		size $= p'_0/2$ (kPa)	103
Elasticity			Wet yield surface size	1
κ	0.026		Flow stress rate	1
ν	0.28			

0.6 m thick. Each embankment layer is constructed in a two-day period, during which consolidation of the clay layer takes place. The total construction time is six days. Assume that the embankment material is linear elastic with $\rho = 1923$ kg/m^3, $E = 478$ kPa, $\nu = 0.3$, $k = 0.1$ m/s, and $e_0 = 1.5$. Assume that the clay layer is elastoplastic, obeying the extended Cam clay model (Chapter 2). The Cam clay model parameters for the clay layer are given in Table 4.5.

SOLUTION(filename: Chapter4_Example7.cae): A finite element mesh (Figure 4.23), is constructed to simulate the sequential construction procedure of the embankment and to calculate the resulting consolidation settlements in the clay layer. The mesh includes half of the geometry because of symmetry. The mesh consists of one part that includes the clay layer and the three embankment layers. In the first calculation step, the embankment is removed from the finite element mesh. Then the embankment layers are added, layer by layer, in subsequent calculation steps. When an embankment layer is added, it is situated on the deformed layer that was added earlier. The new layer is assumed to be strain-free at the time of construction.

Initially, the clay layer is constructed in one calculation step (step 1) and its effective self-weight is applied using the "body-force" option. The top surface of the clay layer is assumed to be permeable. The clay layer is assumed to be elastoplastic, obeying the modified Cam clay model. Using such a model is essential for this class of analysis because we are concerned with the ability of the clay layer to withstand the stresses caused by the sequential construction of the embankment. A model like this one can detect failure within the clay layer during and after construction. During step 1, a "geostatic" command is invoked to make sure that equilibrium is satisfied within the clay layer. The geostatic option makes sure that the initial stress condition in any element within the clay layer falls within the initial yield surface of the Cam clay model.

Next, the first embankment layer is constructed on top of the clay layer in a separate calculation step (step 2). The self-weight of the first embankment layer is gradually applied, using the "body-force" option, in a two-day period. Now, it is important to change the hydraulic boundary condition at the interface between the clay layer and the first embankment layer. As you recall, in step 1 the top surface of the clay layer was assumed to be permeable. In the present step 2, we will need to make the permeable boundary condition inactive only at the interface between the clay layer and the first embankment layer. If we don't, the old boundary condition will act as a drain between the clay layer and the embankment. Further, we need to apply a permeable boundary condition on top of the first embankment layer and also on its sloping face. This will allow the excess pore pressure to drain during the two-day period. Note that the embankment soil is assumed to be more permeable than the underlying clay layer.

In step 3 we construct the second embankment layer in the manner described above. Again, we will need to make the permeable boundary condition at the interface between the first and second embankment layers inactive. Also, we need to apply a permeable boundary condition on top of the second embankment layer and on its sloping face. In step 4 we construct the third (last) embankment layer following the exact procedure for layer 2. Step 5 is a consolidation step with a duration of 200 days.

Figure 4.24 shows the time history of the settlements under the center of the embankment (top of the clay layer). Note that the consolidation settlements started immediately when the first embankment layer was constructed. The effect of

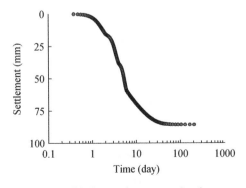

FIGURE 4.24   Calculated consolidation settlements under the center of an embankment.

sequential construction on consolidation settlements can be noted in the figure. No comparison with analytical solutions is possible in this case since such solutions are nonexistent due to the complexity of the problem (two-dimensional consolidation, sequential loading, with non-uniformly distributed embankment loading). Nevertheless, readers are encouraged to calculate the consolidation settlements at the center of the embankment using Terzaghi's one-dimensional consolidation theory. That should yield a rough estimate of consolidation settlements that can be compared with the finite element solution presented here. Figure 4.25 shows the time history of excess pore pressure at the midpoint of the clay layer under the center of the embankment. The figure shows how the excess pore pressure increases in steps as the embankment is constructed. Note how the excess pore pressure dissipates gradually, after the end of construction, until the consolidation process is completed.

Figures 4.26 shows the contours of excess pore water pressures as they develop and dissipate during and after construction for up to 15 days. Figure 4.27 shows the distribution of shear strains in the clay layer at the end of construction of the embankment. The maximum strains are concentrated under the edge of the embankment. These strains are within acceptable limits and failure did not occur anywhere within the clay layer. Note that failure is most likely to occur during construction or at the end of construction because loads are applied rapidly and the excess pore pressures substantially increase, due to this near-undrained type of loading, thus decreasing the effective stresses within the clay layer. As you know, according to the effective stress principle, the strength of soil decreases as the effective stresses decrease, so the clay layer will be most vulnerable to failure during construction or at the end of construction.

Example 4.8 *Consolidation of a Thick Clay Layer with Strip Loading* A 6-m-thick clay layer drained at the top and bottom is subjected to a 3-m-wide strip load of 100 kPa, as shown in Figure 4.28. The load is applied suddenly. The 1-m-thick sand layer can be assumed linear elastic with $E = 13{,}780$ kPa and $v = 0.3$. The

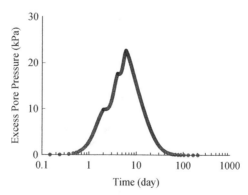

FIGURE 4.25 Calculated excess pore pressure in the middle of a clay layer under the center of an embankment.

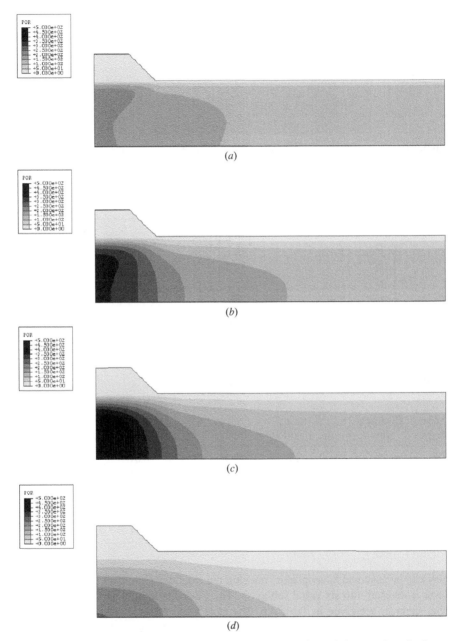

FIGURE 4.26 Distribution of pore pressure at different elapsed times: (*a*) end of construction of layer 1: elapsed time, 2 days; (*b*) end of construction of layer 2: elapsed time, 4 days; (*c*) end of construction of layer 3: elapsed time, 6 days; (*d*) elapsed time, 15 days.

FIGURE 4.27   Distribution of shear strains at the end of construction of layer 3: elapsed time, 6 days.

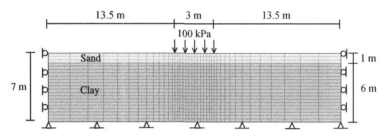

FIGURE 4.28   Finite element discretization: strip load on a clay foundation.

TABLE 4.6   Cam Clay Model Parameters

Sub-layer	Thickness (m)	e_0	σ'_0 (kPa)	σ'_c (kPa)	λ	κ	Stress Ratio, M	k (m/s)
1	1	1.42	13.785	36.76	0.1174	0.01	1.5	6×10^{-12}
2	1	1.37	22.975	55.14	0.1174	0.01	1.5	6×10^{-12}
3	1	1.34	32.165	73.52	0.1174	0.01	1.5	6×10^{-12}
4	1	1.31	41.355	91.9	0.1174	0.01	1.5	6×10^{-12}
5	1	1.29	50.545	110.28	0.1174	0.01	1.5	6×10^{-12}
6	1	1.27	59.735	128.66	0.1174	0.01	1.5	6×10^{-12}

clay can be assumed elastoplastic, obeying the extended Cam clay model. The clay layer is divided into six equal sublayers. The Cam clay model parameters for each clay sublayer are given in Table 4.6. The table also gives the in situ stress and the preconsolidation stress for each sublayer. Calculate the consolidation settlement at the center of the loaded strip area with respect to time using a finite element consolidation program.

SOLUTION(filename: Chapter4_Example8.cae): As you know, the parameter κ defines the elastic behavior of the soil in the Cam clay model, and it is related to the swelling index through the equation: $\kappa = C_s/2.3$. The parameter λ is related to the compression index through $\lambda = C_c/2.3$. The strength parameter M is related to the internal friction angle of the soil, ϕ', as follows:

$$M = \frac{6 \sin \phi'}{3 - \sin \phi'}$$

It is a good idea to solve the problem analytically, if possible, before seeking a numerical solution. As indicated in Table 4.6, the six clay sublayers are overconsolidated. We can use (4.11) or (4.12) to calculate the ultimate consolidation settlement of each sublayer. The choice of (4.11) or (4.12) depends on the stress condition in each sublayer as discussed in Section 4.3. The initial condition of a clay sublayer is fully defined by its in situ vertical effective stress σ'_0 and its in situ void ratio e_0. All clay sublayers have a compression index C_c equal to $2.3\lambda = 0.27$ and a swelling index C_s equal to $2.3\kappa = 0.023$. A spreadsheet was used to do the settlement calculations. A summary of the spreadsheet calculations is given in Table 4.7. The ultimate consolidation settlement is calculated to be 91 mm.

In the Cam clay model, the yield surface size is fully described by the parameter $p' = (\sigma'_1 + 2\sigma'_3)/3$. The evolution of the yield surface is governed by the volumetric plastic strain ε^p_{vol}, which is a function of p'. The relationship between ε^p_{vol} and p' can be deduced easily from an e–log σ'_v line. The consolidation curve (the e–log σ'_v line) is defined completely by its slope C_c ($= 2.3\lambda$) and the initial conditions σ'_0 and e_0. Note that λ, σ'_0 and e_0 are part of the input parameters required in the finite element program used herein. Also, the preconsolidation pressure, σ'_c, is a required parameter (Table 4.6). This parameter specifies the size of the initial yield surface of the Cam clay model.

A two-dimensional plane strain finite element mesh was established as shown in Figure 4.28. The clay layer is divided into six sublayers. Each sublayer has a different set of material parameters, as shown in Table 4.6. A reduced-integration plane strain element with pore pressure is used in this analysis.

TABLE 4.7   Ultimate Consolidation Settlement Calculations

Sub layer	H (m)	e_0	Depth (m)	σ'_0 (kPa)	σ'_c (kPa)	$\Delta\sigma'$ (kPa)	$\sigma'_0 + \Delta\sigma'$ (kPa)	Equation	S_c (m)	S_c (mm)
1	1	1.42	1.5	13.785	36.76	81.8	95.585	(4.12)	0.050352	50.35
2	1	1.37	2.5	22.975	55.14	62	84.975	(4.12)	0.025088	25.08
3	1	1.34	3.5	32.165	73.52	49.4	81.565	(4.12)	0.008733	8.73
4	1	1.31	4.5	41.355	91.9	39.6	80.955	(4.11)	0.002905	2.90
5	1	1.29	5.5	50.545	110.28	33	83.545	(4.11)	0.002192	2.19
6	1	1.27	6.5	59.735	128.66	28.6	88.335	(4.11)	0.001721	1.72
										90.97

The boundary conditions of the finite element mesh shown in Figure 4.28 are as follows. On the bottom side, the vertical and horizontal components of displacement are fixed ($u_x = u_z = 0$), and on the left- and right-hand sides, the horizontal component of displacement is fixed ($u_x = 0$). Note that the entire geometry is considered in this problem (not taking advantage of symmetry). On the top surface a uniform downward strip load of 100 kPa is applied suddenly. During load application the top and bottom surfaces are made impervious; thereafter, perfect drainage is assumed so that the excess pore pressure is always zero on these surfaces (two-way drainage).

The problem is run in three steps. The first step is a single increment of analysis with no drainage allowed across the top and bottom surfaces. In this step, the effective self-weight is applied. The initial stress conditions indicated in Table 4.6 are also applied in this step. During this step, the "geostatic" command is invoked to make sure that equilibrium is satisfied within the clay layer. The geostatic option makes sure that the initial stress condition in any element within the clay specimen falls within the initial yield surface of the Cam clay model. In step 2, the 100-kPa strip load is applied and a nonuniform pore pressure is generated throughout the clay layer, especially in the vicinity of the applied load. At that stage, the applied stress is carried totally by the pore water, u; no stress is carried by the soil skeleton. In step 3, the hydraulic boundary condition of the top and bottom surfaces is changed to a pervious boundary condition having $u = 0$, and the actual consolidation is performed using automatic time stepping. The durations of the three steps are 1, 10^{-5} (sudden), and 10^9 seconds (30 years).

The calculated time history of the vertical settlement is shown in Figure 4.29, where it is compared with the ultimate consolidation settlement obtained analytically. There is a very significant difference between the theoretical solution and the finite element solution. The difference is attributed primarily to the fact that this problem is a two-dimensional consolidation problem; the one-dimensional analytical solution [equations (4.11) and (4.12)] yields only a rough approximation of ultimate consolidation settlements. Another cause of this difference is the assumption of soil homogeneity and linear elasticity when calculating $\Delta\sigma'$ at the center of

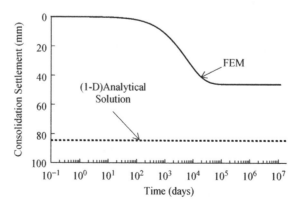

FIGURE 4.29 Calculated consolidation settlements under the center of a strip load.

each clay sublayer to be used in the analytical solution. In reality, the soil in this example is heterogeneous and inelastic: It consists of a layer of sand and a thick elastoplastic clay layer with parameters that vary with depth and that is considered in the finite element solution.

PROBLEMS

4.1 A soil profile is shown in Figure 4.30. A uniformly distributed load, $q = 200$ kPa, is applied at the ground surface. Calculate the primary consolidation settlement of the normally consolidated clay layer. Assume the following: The thickness of the top sand layer is 1 m. The thickness of the clay layer is 3 m. For sand, $\gamma_{dry} = 15$ kN/m^3. For clay, $\gamma_{sat} = 19$ kN/m^3, $C_c = 0.27$, $C_v = 3.0 \times 10^{-6}$ m^2/min, and $k = 4.37 \times 10^{-9}$ m/min.

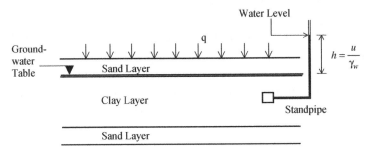

FIGURE 4.30

4.2 Refer to Figure 4.30. Calculate the height of the water level in the piezometer immediately after applying the load $q = 200$ kPa on top of the sand layer. How long will it take for h to become zero? Note that the piezometer tip is located at the center of the clay layer.

4.3 Refer to Figure 4.30. How long will it take for 70% primary consolidation to take place (for the entire clay layer)?

4.4 Use the finite element method to solve Problems 4.1, 4.2, and 4.3. Assume that both sand and clay are linear elastic. $E_{sand} = 7 \times 10^4$ kN/m^2, $\nu_{sand} = 0.3$, $E_{clay} = 5 \times 10^3$ kN/m^2, and $\nu_{clay} = 0.3$. Compare your finite element solutions with the analytical solutions. (You can prepare your own data file, or you can modify the data file of Example 4.2.)

4.5 A 2-m-thick clay layer with two different drainage conditions (case A and case B) is shown in Figure 4.31. A uniformly distributed load, $q = 500$ kPa, is applied at the ground surface. Calculate the primary consolidation settlement of the normally consolidated clay layer for the two cases. Which case has more settlement? Assume the following: For sand, $\gamma_{sat} = 18$ kN/m^3. For clay, $\gamma_{sat} = 19$ kN/m^3, $C_c = 0.27$, $C_v = 3 \times 10^{-6}$ m^2/min, and $k = 4.37 \times 10^{-9}$ m/min.

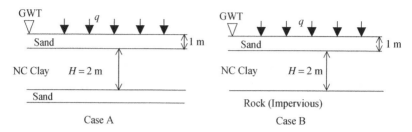

Case A   Case B

FIGURE 4.31

4.6 Refer to Figure 4.31. How long will it take for 50% primary consolidation to take place for cases A and B? Derive an expression relating t_{50} for case A to t_{50} for case B. (*Note:* t_{50} is the time required to achieve 50% consolidation.)

4.7 Use the finite element method to solve Problems 4.5 and 4.6. Assume that both sand and clay are linear elastic. $E_{sand} = 7 \times 10^4$ kN/m^2, $\nu_{sand} = 0.3$, $E_{clay} = 5 \times 10^3$ kN/m^2, and $\nu_{clay} = 0.28$. Compare your finite element solutions with the analytical solutions. (You can prepare your own data file, or you can modify the data file of Example 4.3.)

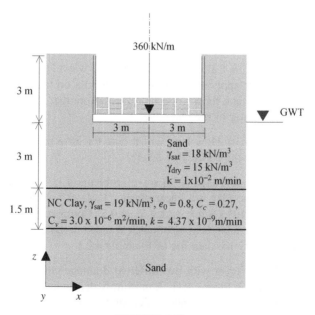

FIGURE 4.32

4.8 A uniformly distributed load, $q = 60$ kPa, is applied at the top of the basement floor as shown in Figure 4.32. The basement is 6 m wide and is very
long in the y-direction (plane strain condition). Calculate the primary consolidation settlement of the normally consolidated clay layer under the center
of the basement.

4.9 Use the finite element method to solve Problem 4.8. Assume that both
sand and clay are linear elastic. $E_{sand} = 7 \times 10^4$ kN/m^2, $\nu_{sand} = 0.3$, $E_{clay} = 5 \times 10^3$ kN/m^2, and $\nu_{clay} = 0.28$. Other parameter are given in Figure 4.32.
Compare your finite element solution with the analytical solution.

4.10 A nonuniformly distributed load is applied at the top of the basement floor
as shown in Figure 4.33. The basement is 6 m wide and is very long in the
y-direction (plane strain condition). Assuming a triangular load distribution,
calculate the primary consolidation settlement of the normally consolidated
clay layer. Compare the consolidation settlement under the center and under
the corner of the basement concrete slab. Use the finite element method to
solve this problem. Assume that both sand and clay are linear elastic. $E_{sand} = 7 \times 10^4$ kN/m^2, $\nu_{sand} = 0.3$, $E_{clay} = 5 \times 10^3$ kN/m^2, and $\nu_{clay} = 0.28$. Other
parameters are given in Figure 4.33.

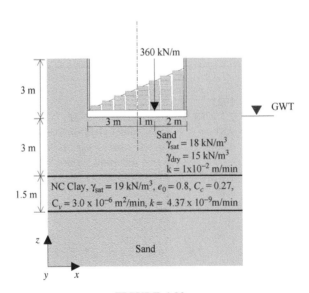

FIGURE 4.33

4.11 Calculate the consolidation settlement under the center of a 1.5-m-high
embankment founded on a 1-m-thick sand layer underlain by a 6-m-thick
overconsolidated clay layer. An impermeable layer of rock lies below the
clay layer. The groundwater table is coincident with the top surface of the
clay layer as indicated in Figure 4.34. The embankment is constructed in

three equal layers each 0.5 m thick. Each embankment layer is constructed in a 7-day period, during which some consolidation of the clay layer takes place. The total construction time is 21 days. Assume that the embankment material is linear elastic with $\rho = 1923$ kg/m^3, $E = 478$ kPa, $\nu = 0.3$, $k = 0.1$ m/s, and $e_0 = 1.5$. Assume that the clay layer is elastoplastic, obeying the extended Cam clay model. The Cam clay model parameters for the clay layer are given in Table 4.8. Also, assume that the clay layer has a preconsolidation pressure of 206 kPa (i.e., the initial yield surface size is 103 kPa).

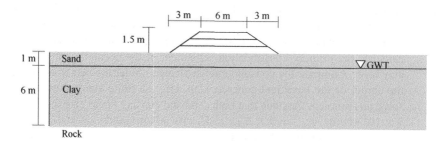

FIGURE 4.34

TABLE 4.8   Cam Clay Model Parameters

General		Plasticity	
ρ (kg/m^3)	1923	λ	0.1174
k (m/sec)	2.5×10^{-8}	Stress ratio, M	1
γ_w (kN/m^3)	9.81	Initial yield surface	
e_0	0.889	size $= p'_0/2$ (kPa)	103
Elasticity		Wet yield surface size	1
κ	0.026	Flow stress rate	1
ν	0.28		

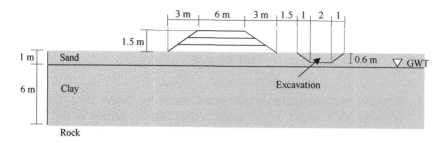

FIGURE 4.35

4.12 A year after the end of construction of the embankment (refer to Problem 4.11), the city of Milwaukee decided to make an excavation parallel to the embankment as shown in Figure 4.35. The depth of the excavation is 0.6 m. The excavation was dug in two stages, each involving the removal of 0.3 m of sand. Evaluate the effect of the excavation on the deformation performance of the embankment for one year following the excavation. Use the soil parameters given in Problem 4.11.

CHAPTER 5

SHEAR STRENGTH OF SOIL

5.1 INTRODUCTION

The *shear strength* of soil is the shear resistance offered by the soil to overcome applied shear stresses. Shear strength is to soil as tensile strength to steel. When you design a steel truss for a bridge, for example, you have to make sure that the tensile stress in any truss element is less than the tensile strength of steel, with some safety factor. Similarly, in soil mechanics one has to make sure that the shear stress in any soil element underlying a shallow foundation, for example, is less than the shear strength of that particular soil, with some safety factor.

In soils the shear strength, τ_f, is a function of the applied normal effective stress, σ'. The *Mohr–Coulomb failure criterion* (discussed in the next section) provides a relationship between the two:

$$\tau_f = c' + \sigma' \tan \phi' \tag{5.1}$$

where c' is the cohesion intercept of the soil and ϕ' is the internal friction angle of the soil. These two parameters are termed the *strength parameters* of a soil. They can be obtained from laboratory and field tests.

Consider the stability of the soil slope shown in Figure 5.1. Soil slopes usually fail in the manner shown in the figure—at failure, there exists a failure surface along which the applied shear stress is equal to the shear strength of the soil. Let's assume that this particular soil has $c' = 10\,\text{kPa}$ and $\phi' = 30°$. Also, let's assume that the applied normal effective stress and shear stress at point A are $\sigma' = 110$ kPa

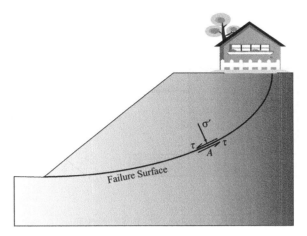

FIGURE 5.1   Shear strength concept.

and $\tau = 40$ kPa, respectively. The shear strength offered by the soil at point A can be calculated using (5.1):

$$\tau_f = c' + \sigma' \tan \phi' = 10 + 110 \tan 30° = 73.5 \text{ kPa}$$

Note that at point A the applied shear stress τ is 40 kPa, which is less than the shear strength $\tau_f = 73.5$ kPa at the same point. This means that point A is not on the verge of failure. The soil at point A will fail only when the applied shear stress is equal to the shear strength.

The shear strength of soil has to be determined accurately because it is crucial for the design of many geotechnical structures, such as natural and human-made slopes, retaining walls, and foundations (both shallow and deep). The shear strength parameters can be measured in the field using the vane shear test, for example. They can be obtained from correlations with the standard number N obtained from the standard penetration test, or from correlations with the cone resistance obtained from the cone penetration test. The shear strength parameters can also be measured in the laboratory using direct shear and/or triaxial compression testing methods on undisturbed or reconstituted soil samples. These testing methods are described next. Other laboratory shear tests are available primarily for research purposes, including simple shear tests and hollow cylinder triaxial tests.

5.2 DIRECT SHEAR TEST

To measure the frictional resistance between wood and steel, a wooden block and a steel block can be stacked vertically and placed on a rough surface as shown in Figure 5.2. A constant vertical load F is applied on top of the steel block. Then a gradually increasing lateral load is applied to the steel block until it starts sliding against the wooden block. Sliding indicates that shear failure has occurred between

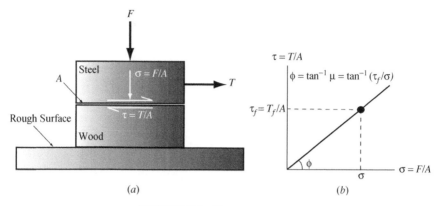

FIGURE 5.2 Shear test concept.

the two blocks. Thus, the applied shear stress at failure ($\tau_f = T_f/A$) is equal to the shear resistance (or shear strength) between the blocks, where T_f is the applied lateral force at failure and A is the cross-sectional area of the block. In reference to Figure 5.2b, the friction factor between the two materials can be calculated as $\mu = \tan\phi = \tau_f/\sigma$, where, ϕ is the friction angle between the two blocks and σ is the normal stress applied ($= F/A$). In this testing arrangement, termed *direct shear*, the shear strength at a predetermined shear plane is measured at a constant normal stress.

The direct shear test for soils uses the same concept as that illustrated above. The soil is placed in a "split" shear box that consists of two halves, as shown in Figure 5.3. The box has a square cross section and can accommodate a soil

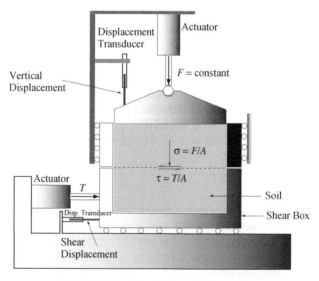

FIGURE 5.3 Direct shear test apparatus.

specimen 10 cm × 10 cm in plane and 2.5 cm in height. The upper half of the shear box is not allowed to move laterally, whereas the bottom half can slide laterally by the action of a horizontal actuator. After placing the soil inside the box, a loading plate is seated on top of the soil and a vertical load is applied using a vertical actuator. A gradually increasing lateral displacement (at a constant rate) is then applied to the bottom half of the box, via the horizontal actuator, to generate shear stresses within the soil. The shear force, T, is measured using a load cell attached to the piston of the horizontal actuator. The horizontal displacement (shear displacement) of the bottom half of the shear box is measured during shearing using a horizontal displacement transducer. Also, the vertical displacement of the loading plate is measured during shearing using a vertical displacement transducer. Figure 5.4 is a photo of a direct shear apparatus.

The results of a direct shear test are plotted in the shear displacement versus shear stress plane such as the one shown in Figure 5.5a. The vertical displacement of the loading plate is plotted against the shear displacement as shown in Figure 5.5b. The test results shown in Figure 5.5a and 5.5b are typical for loose sands. Note that the shear stress is calculated by dividing the measured shear force by the cross-sectional area of the soil specimen. Figure 5.5a shows that the shear stress increases in a nonlinear manner as the shear displacement increases. The shear

FIGURE 5.4   Direct shear test apparatus. (Courtesy of Geocomp)

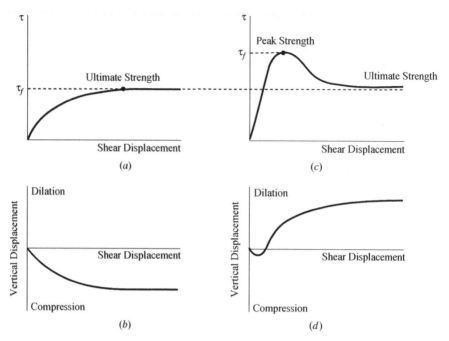

FIGURE 5.5 Typical direct shear test results on (a, b) loose sand and (c, d) dense sand.

stress approaches failure at $\tau = \tau_f$, where τ_f is the shear strength of the soil. Figure 5.5b shows a downward displacement of the loading plate, indicating soil compression. Loose sand consists of loosely packed grains with large voids in between. During shearing, some of the voids in the shear zone and its vicinity will collapse, causing the soil specimen to compress (downward plate displacement).

The behavior of dense sand during shearing is quite different from that of loose sand even though the two sands are assumed to be identical in terms of their gradation and specific gravity—they differ only in relative density. In the shear displacement versus shear stress plane (Figure 5.5c) the dense sand exhibits greater strength, or peak strength, at an early stage during shearing. After reaching peak strength, the shear stress decreases as the shear displacement increases until reaching an ultimate strength that is approximately the same as the ultimate strength of the loose sand (Figure 5.5a). Figure 5.5d shows the direct shear test results for dense sand in the shear displacement versus vertical displacement plane. In this figure, downward displacement of the loading plate indicates soil compression, and upward displacement of the plate indicates soil expansion (dilation). A peculiar characteristic is noted in the figure. At an early stage of shearing, the sand compresses slightly and then starts to dilate until a later stage of shearing, when it levels out as shown in the figure. Dense sand consists of densely packed grains with small voids in between. During shearing, some of the grains will slide and roll on top of other densely packed particles in the shear zone and its vicinity, causing the soil specimen to dilate (upward plate displacement).

The results of a direct shear test provide the shear strength (τ_f) of the soil at a specific normal stress (σ'). The direct shear test is repeated several times on identical soil specimens using different normal stresses. Typical direct shear test results for sand using three different normal stresses are shown in Figure 5.6a. The shear strength of the soil at different normal stresses can be determined from the figure as indicated by points 1, 2, and 3. The at-failure tests results (points 1, 2,

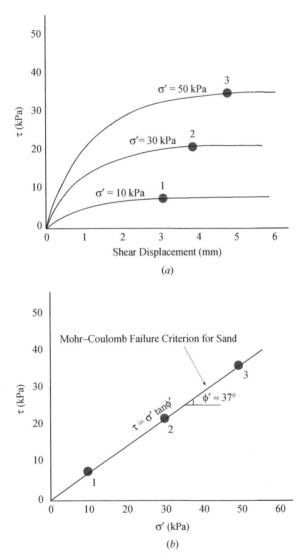

FIGURE 5.6   Determination of the Mohr–Coulomb failure criterion for sand (direct shear test).

and 3) are presented in the normal stress versus shear stress plane in Figure 5.6b. The three data points in Figure 5.6b are best fitted with a straight line. This straight line is the Mohr–Coulomb failure criterion. The slope of this line is the internal friction angle of the soil, ϕ', and its intercept with the shear stress axis is the cohesion intercept, c'. The parameters c' and ϕ' are the strength parameters of the soil. They are unique parameters for a given soil. For a sandy soil c' is zero, that is, the Mohr–Coulomb failure criterion passes through the origin, so $\tau_f = \sigma' \tan \phi'$. In Figure 5.6$b$ the internal friction angle can be calculated from the slope of the Mohr-Coulomb failure criterion: $\phi' = \tan^{-1}(\tau_f/\sigma') \approx 37°$.

Figure 5.7a presents typical direct shear test results for clay under three different normal stresses. The Mohr–Coulomb failure criterion for this clay is shown in Figure 5.7b. The figure shows that this soil has a cohesion intercept of approximately 9 kPa and an internal friction angle of approximately 26.5°. With the help of Figure 5.7b, one can define the *cohesion intercept* (or *cohesion*) as the shear strength of the soil at *zero normal stress* (or *zero confining pressure*). This means that clays have some shear strength even when they are not subjected to confining pressure. It also means that sands do not have any shear strength without confining pressure. That is why we cannot make shapes out of dry sand, although we can certainly make shapes out of clay. For this reason sands and gravels are called *cohesionless*, whereas clays are called *cohesive*.

Note that we used effective stresses in the discussion above when describing the Mohr–Coulomb failure criterion and the shear strength parameters of the soil. This is because the shear strength of soil is dependent on effective stresses rather than total stresses (recall the effective-stress principle). Also, this means that when we conduct a direct shear test on wet or saturated soils, we have to facilitate drainage while shearing the soil specimen to prevent the development of excess pore water pressures in the soil. When the soil is saturated, the shear stress must be applied very slowly to prevent the development of excess pore water pressure. That way the total stress is equal to the effective stress because the pore water pressure is kept equal to zero: $\sigma' = \sigma - u = \sigma - 0 = \sigma$.

When measuring the shear strength of a soil using a direct shear test, one needs to duplicate the field conditions of the soil being tested. Take, for example, the case of the slope shown in Figure 5.1. To measure the shear strength at point A, we can estimate the normal stress σ' at that point. This requires knowledge of the in situ unit weight of the soil and the location of the groundwater table. In the laboratory we can reconstitute a soil specimen in the direct shear box, aiming for the same in situ soil density. The soil sample can be submerged in a water basin (not shown in Figure 5.3) to simulate field conditions. A constant vertical stress equal to the normal stress σ' calculated at point A is then applied to the soil specimen. The shearing stage should not start until equilibrium within the soil specimen is achieved. This means that if the soil is a clayey soil, we have to wait until the excess pore water pressure generated as a result of stress application is dissipated. Shearing can then be applied until failure. The shear stress applied at failure should reflect the true shear strength of the soil at point A.

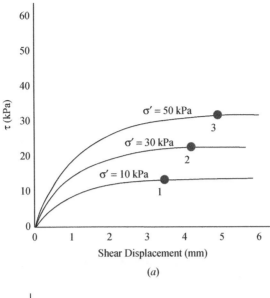

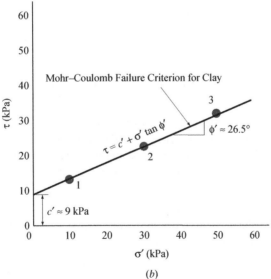

FIGURE 5.7   Determination of the Mohr–Coulomb failure criterion for clay (direct shear test).

Example 5.1   A dry sand sample is subjected to a normal stress $\sigma' = 20\,\text{kPa}$ in a direct shear test (Figure 5.3). Calculate the shear force at failure if the soil sample is 10 cm × 10 cm in plane and 2.5 in height. The strength parameters of the sand are $c' = 0$ and $\phi' = 38°$.

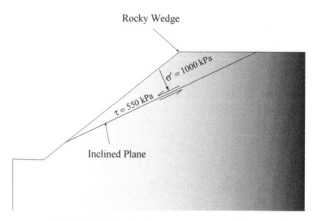

FIGURE 5.8 Is the wedge going to slide?

SOLUTION: From the Mohr–Coulomb failure criterion we can calculate the shear strength of the sand at $\sigma' = 20$ kPa:

$$\tau_f = c' + \sigma' \tan \phi' = 0 + 20 \times \tan 38° = 15.6 \text{ kPa}$$

At failure, the applied shear stress is equal to the shear strength of the soil (i.e., $\tau = \tau_f$); therefore, the shear force at failure is:

$$T = \tau \times A = 15.6 \times 0.1 \times 0.1 = 0.156 \text{ kN} = 156 \text{ N}.$$

Example 5.2 An intact rocky wedge is situated on an inclined plane as shown in Figure 5.8. Due to the self-weight of the wedge, a normal stress of 1000 kPa and a shear stress of 550 kPa are applied to the inclined plane. Determine the safety of the wedge against sliding given that the friction angle between the wedge and the inclined plane is 30°.

SOLUTION: The shear strength offered at the wedge–plane interface can be calculated using the Mohr–Coulomb failure criterion with $\sigma' = 1000$ kPa:

$$\tau_f = c' + \sigma' \tan \phi' = 0 + 1000 \times \tan 30° = 577.35 \text{ kPa}$$

To avoid sliding, the applied shear stress must be less than the shear strength offered by the wedge–plane interface (i.e., $\tau < \tau_f$). Luckily, τ is 550 kPa, which is smaller than $\tau_f = 577.35$ kPa. Therefore, the wedge is safe against sliding.

5.3 TRIAXIAL COMPRESSION TEST

The triaxial compression test is used to determine the shear strength of soil and to determine the stress–strain behavior of the soil under different confining pressures.

The test involves a cylindrical soil sample that is subjected to a uniform confining pressure from all sides and then subjected to an additional vertical load until failure.

Figure 5.9 shows the triaxial test apparatus schematically, and Figure 5.10 is a photo of the apparatus. The cylindrical soil sample can have different dimensions. A typical triaxial soil specimen is 5 cm in diameter and 15 cm in height. The sample is situated on top of the pedestal that is part of the base of the triaxial chamber, as shown in Figure 5.9. A loading plate is then placed on top of the specimen. The pedestal, the soil specimen, and the top loading plate are carefully enclosed in a thin rubber membrane. O-rings are used to prevent the confining fluid from entering the soil specimen. Finally, the triaxial chamber is positioned on top of the base, the loading ram is lowered to the position shown in Figure 5.9, and the triaxial chamber is filled with water.

As noted in Figure 5.9, two drainage tubes connect the top and bottom of the soil specimen to the outside of the triaxial chamber. These tubes have valves that are used to control drainage into the soil specimen. A third tube leads to the space

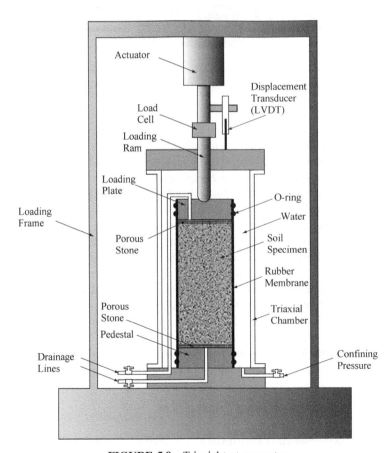

FIGURE 5.9   Triaxial test apparatus.

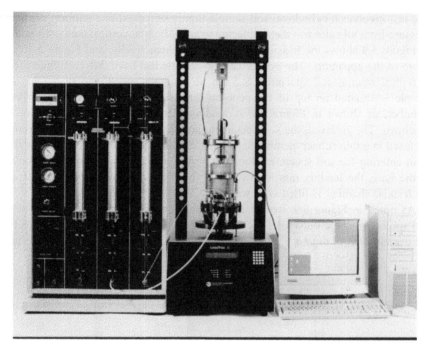

FIGURE 5.10 Triaxial test apparatus. (Courtesy of Geocomp)

inside the triaxial chamber, which is usually filled with water. This tube is used to pressurize the confining fluid (water) using pressurized air and a pressure regulator.

There are three types of triaxial compression tests: CD, CU, and UU. Each triaxial test consists of two stages: Stage I is the conditioning stage, during which the initial stress condition of the soil specimen is established; stage II is the shearing stage, during which a deviator stress is applied until failure. The designation of a triaxial test consists of two letters, the first letter describes stage I and the second describes stage II. Stage I can be either consolidated (C) or unconsolidated (U), and stage II can be either drained (D) or undrained (U). A triaxial CD test means that the soil specimen is allowed to consolidate in stage I of the triaxial test, and during stage II the specimen is allowed to drain while being sheared. On the other hand, a triaxial CU test means that the soil specimen is allowed to consolidate in stage I, and during stage II the specimen is not allowed to drain while being sheared. Finally, the UU test means that the specimen is not allowed to consolidate in stage I and is not allowed to drain during shearing.

5.3.1 Consolidated–Drained Triaxial Test

The consolidated–drained (CD) triaxial test is used to obtain the effective strength parameters of soils. First, a soil specimen is saturated by circulating deaired water through the specimen, from bottom to top, utilizing the two drainage tubes shown in Figure 5.9. After the specimen is fully saturated, the triaxial test is done in two stages: a stress initialization stage and a shearing stage. In the first stage a confining pressure is applied via the confining fluid. Because the soil specimen is fully saturated, excess pore water pressure will be generated ($\approx$ confining pressure). The soil specimen is allowed to consolidate by opening the two drainage valves throughout this stage. That will allow the excess pore water pressure to dissipate gradually and the specimen to consolidate. The volume of the dissipated water can be measured using a graduated flask. The volume of the dissipated water is equal to the volume change of the specimen because the specimen is fully saturated. The volumetric strain can be calculated by dividing the volume change by the initial volume of the specimen. The consolidation curve for this stage can be plotted in the time versus volumetric strain (ε_v) plane, as shown in Figure 5.11.

In the shearing stage of the CD test a deviator stress $\Delta\sigma_d = \sigma_1 - \sigma_3$ is applied very slowly while the drainage valves are opened, to ensure that no excess pore

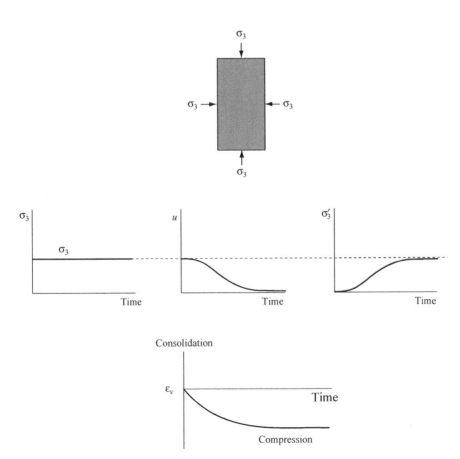

FIGURE 5.11   Stage I of a consolidated drained triaxial test: consolidation stage.

water pressure is generated. Consequently, the effective stresses are equal to the total stresses during this stage of the CD test. Because of its stringent loading requirements, the CD test may take days to carry out, making it an expensive test. Figure 5.12a shows stress–strain behavior typical of loose sand and normally consolidated clay. Note that ε_a is the axial strain and $\sigma_1 - \sigma_3$ is the deviator stress. The soil shows smooth nonlinear stress–stress behavior reaching an ultimate shear strength $(\sigma_1 - \sigma_3)_f$. Figure 5.12b shows the deformation behavior in the axial strain

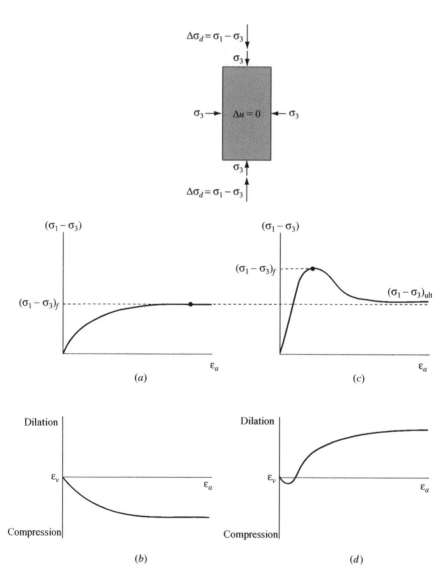

FIGURE 5.12 Stage II of a consolidated drained triaxial test: shearing stage (a, b) loose sand; (c, d) dense sand.

(ε_a) versus volumetric strain (ε_v) plane typical of loose sand and normally consolidated clay. In this figure the soil compresses as the shearing stress is increased. When the ultimate shear strength is approached, the curve levels out, indicating that the volumetric strain is constant (i.e., no further volume change). From the measured deviator stress at failure ($\sigma_1 - \sigma_3$)$_f$, one can plot Mohr's circle, which describes the stress state of the soil specimen at failure. Such a plot is shown in Figure 5.13. The Mohr–Coulomb failure criterion can be obtained by drawing a line that is tangent to Mohr's circle and passing through the origin. This is applicable to sands and normally consolidated clays since these soils have $c' = 0$. The slope of the Mohr–Coulomb failure criterion is the effective (or drained) friction angle ϕ' of the soil.

Figure 5.12c shows typical stress–strain behavior of dense sand and overconsolidated clay in the axial strain (ε_a) versus deviator stress plane. The soil shows nonlinear stress–stress behavior reaching a peak shear strength ($\sigma_1 - \sigma_3$)$_f$ at an early stage of shearing. After the peak strength is reached, a sharp decrease in strength is noted. Then the stress–strain curve levels out, approaching an ultimate strength ($\sigma_1 - \sigma_3$)$_{ult}$. Note that this ultimate strength is the same as the ultimate strength of the loose sand if the two sands were identical (not in terms of their relative density). Figure 5.12d shows typical deformation behavior of dense sand and overconsolidated clay in the axial strain (ε_a) versus volumetric strain (ε_v) plane. Initially, the soil compresses as the shearing stress is increased. But shortly

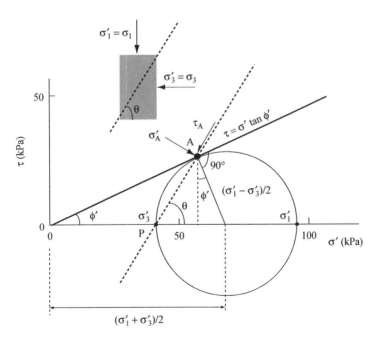

FIGURE 5.13 Determination of the Mohr–Coulomb failure criterion for sand (CD triaxial test).

after that the soil starts expanding (dilating) as the axial strain increases. When the ultimate shear strength is approached, the curve levels out, indicating that the volumetric strain is constant.

We need only one CD test to determine the strength parameters of dense sand (because $c' = 0$). From the peak deviator stress measured $(\sigma_1 - \sigma_3)_f$, we can plot Mohr's circle, which describes the stress state of the soil specimen at failure as shown in Figure 5.13. The Mohr–Coulomb failure criterion can be obtained by drawing a line that is tangent to Mohr's circle and passing through the origin. The slope of the Mohr–Coulomb failure criterion is the effective (or drained) friction angle ϕ' of the soil.

For overconsolidated clays the cohesion intercept, c', is not equal to zero. Therefore, we will need to have the results of at least two CD triaxial tests on two identical specimens subjected to two different confining pressures. From the measured peak deviator stress $(\sigma_1 - \sigma_3)_f$ of these tests, we can plot Mohr's circles that describe the stress states of the soil specimens at failure, as shown in Figure 5.14. The Mohr–Coulomb failure criterion in this case can be obtained by drawing a line that is tangent to the two Mohr's circles. The Mohr-Coulomb failure criterion will intersect with the shear stress axis at $\tau = c'$, as shown in Figure 5.14. This is the effective cohesion intercept of the overconsolidated clay. The slope of the Mohr–Coulomb failure criterion is the effective friction angle ϕ' of the overconsolidated clay.

You recall from the strength of materials laboratory that when a concrete cylindrical specimen was crushed between the jaws of a compression machine, there existed a failure plane making an angle θ with the horizontal. Soils exhibit similar behavior—they also have a failure plane that occurs at failure in a triaxial

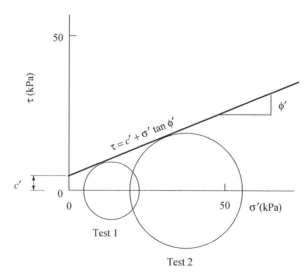

FIGURE 5.14 Determination of the Mohr–Coulomb failure criterion for overconsolidated clay (CD triaxial test).

compression test. Let us calculate, both graphically and analytically, the orientation of the failure plane. In reference to Figure 5.13, a sandy soil will have a zero cohesion intercept, and the Mohr–Coulomb failure criterion is tangent to Mohr's circle. The tangency point (point A in Figure 5.13) represents the stress state on the failure plane at failure. The x and y coordinates of point A give, respectively, the normal and shear stresses exerted on the failure plane. We can measure the x and y coordinates at point A directly from the figure provided that our graph is drawn to scale. We can determine the failure plane orientation using the *pole method*. For a triaxial compression test, the pole (the origin of planes) is located at point P in Figure 5.13. The orientation of the failure plane can be obtained by connecting point P with point A. The orientation of the failure plane is the angle between line PA and the horizontal. We can use a protractor to measure the angle θ. This is the graphical solution.

The analytical solution for the angle θ can be obtained easily from Figure 5.13. Using simple trigonometry, you can show that

$$\theta = 45° + \frac{\phi'}{2} \tag{5.2}$$

As mentioned above, the x and y coordinates of point A give, respectively, the normal stress (σ'_A) and shear stress (τ_A) exerted on the failure plane at failure. Let's calculate those analytically. In reference to Figure 5.13 we can write

$$\sigma'_A = \frac{\sigma'_1 + \sigma'_3}{2} - \frac{\sigma'_1 - \sigma'_3}{2} \sin \phi' \tag{5.3}$$

$$\tau_A = \frac{\sigma'_1 - \sigma'_3}{2} \cos \phi' \tag{5.4}$$

The friction angle ϕ' can be calculated from

$$\sin \phi' = \frac{\sigma'_1 - \sigma'_3}{\sigma'_1 + \sigma'_3} \tag{5.5}$$

Example 5.3   A CD triaxial compression test was conducted on a sand specimen using a confining pressure of 42 kPa. Failure occurred at a deviator stress of 53 kPa. Calculate the normal and shear stresses on the failure plane at failure. Also calculate the angle made by the failure plane with the horizontal. (a) Solve the problem graphically and (b) confirm your solution analytically.

SOLUTION: (a) *Graphical solution*   Given: $\sigma'_{3f} = 42$ kPa and $(\sigma'_1 - \sigma'_3)_f = 53$ kPa. We need to calculate the major principal stress in order to draw Mohr's circle:

$$\sigma'_{1f} = \sigma'_{3f} + 53 \text{ kPa} = 42 \text{ kPa} + 53 \text{ kPa} = 95 \text{ kPa}$$

The radius of Mohr's circle is

$$\frac{(\sigma'_1 - \sigma'_3)_f}{2} = \frac{53}{2} = 26.5 \text{ kPa}$$

The x-coordinate of the center of Mohr's circle is

$$\frac{(\sigma_1' + \sigma_3')_f}{2} = \frac{95+42}{2} = 68.5 \text{ kPa}$$

Now we can plot Mohr's circle with its center located at (68.5 kPa, 0) and radius $= 26.5\,kPa$, as shown in Figure 5.15.

Since the soil is sand, we have $c' = 0$. The Mohr–Coulomb failure criterion is drawn as a straight line passing through (0,0) and tangent to the circle as shown in the figure. Using the same scale as that used in the drawing, we can now measure the coordinates of the tangency point: The normal stress at point A is $\sigma_A' \approx 57\,\text{kPa}$, and the shear stress at point A is $\tau_A' \approx 26\,\text{kPa}$. The orientation of the failure plane can be obtained by connecting pole P with point A. The orientation of the failure plane is the angle between line PA and the horizontal. Using a protractor, we can measure the angle $\theta \approx 56°$.

(b) *Analytical solution* First, we calculate the friction angle of the sand using (5.5):

$$\phi' = \sin^{-1}\frac{\sigma_1' - \sigma_3'}{\sigma_1' + \sigma_3'} = \sin^{-1}\frac{53}{95 + 42} = 22.75°$$

Therefore,

$$\theta = 45° + \frac{\phi'}{2} = 45° + \frac{22.75°}{2} = 56.38°$$

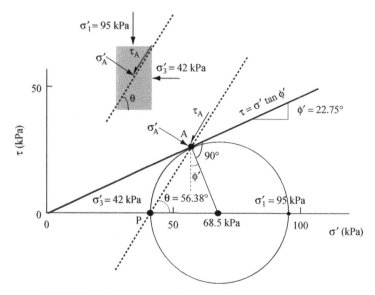

FIGURE 5.15 Mohr's circle for a CD triaxial test on sand.

In reference to Figure 5.15, we can write

$$\sigma'_A = \frac{\sigma'_1 + \sigma'_3}{2} - \frac{\sigma'_1 - \sigma'_3}{2} \sin \phi' = \frac{95 + 42}{2} - \frac{95 - 42}{2} \sin 22.75° = 58.25 \text{ kPa}$$

$$\tau_A = \frac{\sigma'_1 - \sigma'_3}{2} \cos \phi' = \frac{95 - 42}{2} \cos 22.75° = 24.44 \text{ kPa}$$

Note that there is a relatively good agreement between the graphical and analytical solutions. Solving problems graphically first can reveal better ways of solving them analytically. Also, sometimes it is more efficient to solve problems graphically.

Example 5.4   Three CD triaxial compression tests were conducted on three overconsolidated clay specimens using three confining pressures: 4, 20, and 35 kPa. Failure occurred at the deviator stresses of 19, 36 and 54 kPa, respectively. Determine the shear strength parameters of the soil.

SOLUTION: Let's calculate the major principal stress at failure for each test. Then let's determine the x-coordinate of the center of each Mohr's circle and its radius:

σ'_{3f} (kPa)	$(\sigma'_1 - \sigma'_3)_f$ (kPa)	σ'_{1f} (kPa)	$(\sigma'_1 + \sigma'_3)_f/2$ (kPa)	$(\sigma'_1 - \sigma'_3)_f/2$ (kPa)
4	19	23	13.5	9.5
20	36	56	38	18
35	54	89	62	27

Using the results from the fourth and fifth columns, we can draw three Mohr's circles as shown in Figure 5.16. The failure envelope (Mohr–Coulomb failure criterion) is then established. The friction angle and the cohesion intercept are measured from the figure as $\phi' \approx 23°$ and $c' \approx 5$ kPa, respectively.

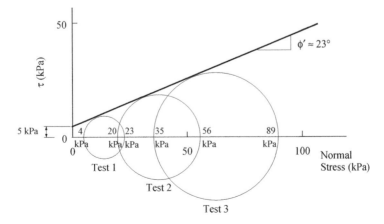

FIGURE 5.16   CD triaxial compression tests on three overconsolidated clays.

5.3.2   Consolidated–Undrained Triaxial Test

The consolidated–undrained (CU) triaxial test includes two stages: stage I, the consolidation stage, and stage II, the shearing stage. In stage I the specimen is consolidated under a constant confining pressure. The drainage valves are opened to facilitate consolidation. The volume change is measured and plotted against time as shown in Figure 5.17a. In stage II the soil specimen is sheared in an undrained condition. The undrained condition is realized by closing the drainage valves, thus preventing the water from flowing out of the sample or into the sample. The undrained condition makes it possible to apply the deviator stress in a much faster manner than in the consolidated–drained triaxial test. This makes the CU test more economical than the CD test. During stage II there will be no volume change since water is not allowed to leave the specimen. Therefore, the volumetric strain (ε_v) is always equal to zero during this stage. Because the soil has the tendency to compress (or dilate) during shearing, there will be increase (or decrease) in excess pore water pressure within the saturated specimen. The excess pore water pressure is measured using a pressure transducer connected to one of the drainage tubes. The presence of pore water pressure indicates that the total stress is different from the effective stress in a CU test.

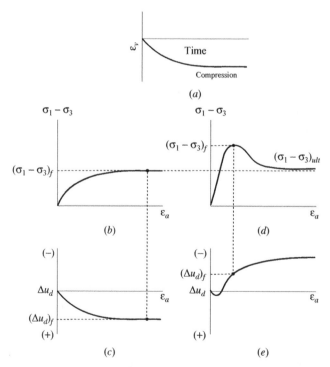

FIGURE 5.17   (a) Stage I (consolidation) and ($b - e$) stage II (shearing) of a consolidated–undrained triaxial test (b, c) loose sand; (d, e) dense sand.

Figure 5.17*b* shows stress–strain behavior typical of loose sand and normally consolidated clay in the axial strain (ε_a) versus deviator stress plane. The soil shows a smooth nonlinear stress–stress behavior reaching an ultimate shear strength $(\sigma_1 - \sigma_3)_f$. Figure 5.17*c* shows the corresponding behavior in the axial strain (ε_a) versus pore water pressure (Δu_d) plane. In this figure the soil has a tendency to compress as the shearing stress is increased. This causes a positive change in pore water pressure, as shown in the figure. When the ultimate shear strength is approached, the pore water pressure levels out, approaching its maximum value at failure $(\Delta u_d)_f$. From the deviator stress measured at failure $(\sigma_1 - \sigma_3)_f$, one can plot a Mohr's circle that describes the total stress state of the soil specimen at failure. Such a plot is shown in Figure 5.18. Knowing the value of the pore water pressure at failure $(\Delta u_d)_f$, one can calculate the effective principal stresses at failure: $\sigma'_{1f} = \sigma_{1f} - (\Delta u_d)_f$ and $\sigma'_{3f} = \sigma_{3f} - (\Delta u_d)_f$. The corresponding effective Mohr's circle is shown in Figure 5.18. Note that the two circles have the same size and that the effective-stress Mohr's circle can be obtained from the total stress Mohr's circle if the latter is shifted to the left a distance equal to $(\Delta u_d)_f$. It follows that there are two Mohr–Coulomb failure criteria: one for total stresses and one for effective stresses. The slope of the effective-stress Mohr–Coulomb failure criterion is the effective (or drained) friction angle ϕ' of the soil, whereas the slope of the total stress Mohr–Coulomb failure criterion is the consolidated–undrained friction angle ϕ (without a prime) of the soil.

Figure 5.17*d* shows stress–strain behavior typical of dense sand and overconsolidated clay in the axial strain (ε_a) versus deviator stress plane. The soil shows nonlinear stress–strain behavior, reaching a peak shear strength $(\sigma_1 - \sigma_3)_f$ at an early stage of shearing. After the peak strength is reached, a sharp decrease in strength is noted. Then the stress–strain curve levels out, approaching an ultimate strength $(\sigma_1 - \sigma_3)_{ult}$. Figure 5.17*e* shows the corresponding behavior in the axial

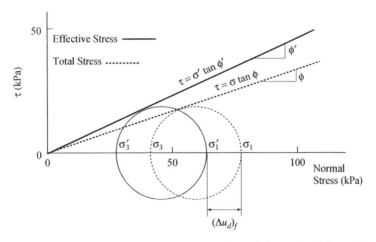

FIGURE 5.18 Total and effective-stress Mohr–Coulomb failure criteria from CU triaxial test results: loose sand (or NC clay).

strain (ε_a) versus pore water pressure (Δu_d) plane. Initially, the pore water pressure increases as the shearing stress is increased. But shortly after that the pore water pressure starts to decrease, and then becomes negative (suction), as the shearing stress increases. When the ultimate shear strength is approached, the curve levels out and the pore water pressure reach its maximum negative value.

We need only one CU test to determine the total and effective shear strength parameters of a dense sand (because $c = c' = 0$). From the peak deviator stress measured $(\sigma_1 - \sigma_3)_f$, we can plot the total stress Mohr's circle that describes the total stress state of the soil specimen at failure, as shown in Figure 5.19. Knowing that the pore water pressure at failure is $-(\Delta u_d)_f$, we can calculate the effective stresses at failure and plot the effective-stress Mohr's circle as shown in the same figure. Note that the effective-stress Mohr's circle has the same diameter as the total stress Mohr's circle. Also note that the effective-stress circle results from the total stress circle by a shift $=(\Delta u_d)_f$ from left to right. The total stress Mohr–Coulomb failure criterion can be obtained by drawing a line that is tangent to the total stress Mohr's circle and passing through the origin. The slope of the total stress Mohr–Coulomb failure criterion is the consolidated–undrained friction angle ϕ of the soil. The angle ϕ can be obtained as

$$\sin \phi = \frac{\sigma_1 - \sigma_3}{\sigma_1 + \sigma_3} \tag{5.6}$$

Also, the effective-stress Mohr–Coulomb failure criterion for dense sand can be obtained by drawing a line tangent to the effective-stress Mohr's circle and passing through the origin. The slope of the effective-stress Mohr–Coulomb failure criterion is the effective (or drained) friction angle ϕ' of the soil. Equation (5.5) can be used to calculate the angle ϕ'.

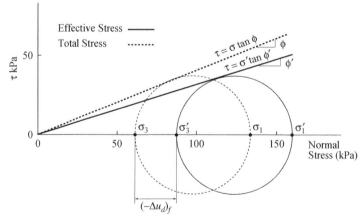

FIGURE 5.19   Total and effective-stress Mohr–Coulomb failure criteria from CU triaxial test results: dense sand.

For overconsolidated clays, the cohesion intercept, c', is not equal to zero. Therefore, we will need to have the results of at least two CU triaxial tests on two identical
specimens subjected to two different confining pressures. From the measured peak
deviator stress $(\sigma_1 - \sigma_3)_f$ of these tests we can plot Mohr's circles that describe
the total stress state of the soil specimens at failure as shown in Figure 5.20. The
total stress Mohr–Coulomb failure criterion in this case can be obtained by drawing
a line that is tangent to the two Mohr's circles. The total stress Mohr–Coulomb
failure criterion will intersect with the shear stress axis at $\tau = c$, as shown in
Figure 5.20. This is the cohesion intercept of the overconsolidated clay. The slope
of the Mohr–Coulomb failure criterion, is the consolidated-undrained friction angle
ϕ of the overconsolidated clay. Figure 5.20 also shows the effective-stress Mohr's
circles for the two tests and the corresponding effective-stress Mohr-Coulomb failure criterion, from which the effective shear strength parameters c' and ϕ', can be
obtained.

Example 5.5   A consolidated–undrained triaxial test was performed on a dense
sand specimen at a confining pressure $\sigma_3 = 23$ kPa. The consolidated undrained
friction angle of the sand is $\phi = 37.5°$, and the effective friction angle is $\phi' = 30°$.
Calculate (a) the major principal stress at failure, σ_{1f}; (b) the minor and the major
effective principal stresses at failure, σ'_{3f} and σ'_{1f}; and (c) the excess pore water
pressure at failure, $(\Delta u_d)_f$. Solve the problem graphically and analytically.

SOLUTION: *Graphical solution* (a) As shown in Figure 5.21a, we can draw the
Mohr–Coulomb failure criteria for both total stresses and effective stresses. To do
so we need the consolidated undrained friction angle $\phi = 37.5°$ and the effective
friction angle $\phi' = 30°$. Both criteria pass through the origin because c and c' for
sand are equal to zero. Next, we plot the confining pressure $\sigma_3 = 23$ kPa on the

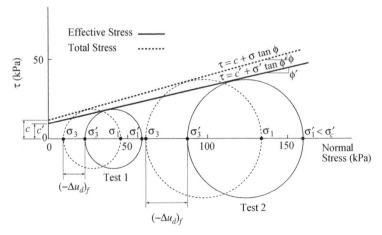

FIGURE 5.20   Total and effective-stress Mohr–Coulomb failure criteria from CU triaxial
test results: OC clay.

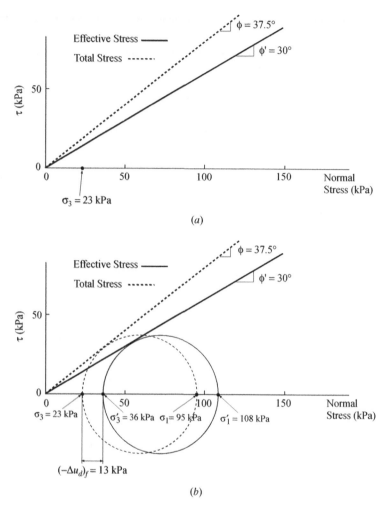

FIGURE 5.21   Consolidated undrained triaxial test results of a dense sand.

normal stress axis as shown in Figure 5.21b. Now we can draw a circle (by trial and error) that passes through $\sigma_3 = 23$ kPa and touches the total stress Mohr–Coulomb failure criterion. This is Mohr's circle, representing the total stresses at failure. The intersection of the circle with the horizontal axis is the major principal stress at failure. We can measure the location of this intersection to get $\sigma_{1f} \approx 95$ kPa.

(b) To determine the minor and major effective principal stresses at failure (σ'_{3f} and σ'_{1f}, respectively), we can draw an effective-stress Mohr's circle that touches the effective–stress Mohr–Coulomb failure criterion as shown in Figure 5.21b. This effective-stress Mohr's circle has the same diameter as the total stress Mohr's circle. The intersections of the effective–stress Mohr's circle with the horizontal axis can be measured from the figure to give us $\sigma'_{3f} \approx 36$ kPa and $\sigma'_{1f} \approx 108$ kPa.

(c) From Figure 5.21b we can measure the distance between σ_{1f} and σ'_{1f} (or between σ_{3f} and σ'_{3f}) to estimate $(\Delta u_d)_f$, which is approximately -13 kPa (negative = suction).

Analytical solution (a) From (5.6) we have

$$\sin\phi = \sin 37.5° = \frac{\sigma_{1f} - \sigma_{3f}}{\sigma_{1f} + \sigma_{3f}} = \frac{\sigma_{1f} - 23}{\sigma_{1f} + 23} \rightarrow \sigma_{1f} = 95 \text{ kPa}$$

(b) The effective-stress Mohr's circle has the same diameter as the total stress Mohr's circle:

$$\sigma'_{1f} - \sigma'_{3f} = \sigma_{1f} - \sigma_{3f} = 95 - 23 = 72 \text{ kPa}$$

From (5.5),

$$\sin\phi' = \frac{\sigma'_{1f} - \sigma'_{3f}}{\sigma'_{1f} + \sigma'_{3f}} \rightarrow \sigma'_{1f} + \sigma'_{3f} = \frac{\sigma'_{1f} - \sigma'_{3f}}{\sin\phi'} = \frac{72 \text{ kPa}}{\sin 30°} = 144 \text{ kPa}$$

or $\sigma'_{1f} + \sigma'_{3f} = 144$ kPa, but $\sigma'_{1f} - \sigma'_{3f} = 72$ kPa. Solving these two equations simultaneously, we get $\sigma'_{1f} = 108$ kPa and $\sigma'_{3f} = 36$ kPa.

(c) From the effective–stress equation and in reference to Figure 5.21b,

$$(\Delta u_d)_f = \sigma_{1f} - \sigma'_{1f} = 95 \text{ kPa} - 108 \text{ kPa} = -13 \text{ kPa}$$

Note that the graphical solution is done first to gain some insight into the problem. Doing so makes it easier to seek an analytical solution. Also note that the graphical solution is very simple and generally yields accurate answers if the drawing is done correctly and to scale, as illustrated above.

5.3.3 Unconsolidated–Undrained Triaxial Test

The unconsolidated–undrained (UU) triaxial test is usually performed on undisturbed saturated samples of fine-grained soils (clay and silt) to measure their undrained shear strength, c_u. The soil specimen is not allowed to consolidate in stage I under the confining pressure applied. It is also not allowed to drain during shearing in stage II. Identical soil specimens exhibit the same shear strength under different confining pressures, as indicated in Figure 5.22. It seems that applying more confining pressure to the soil specimen does not cause any increase in its shear strength! This can be explained as follows: When a fully saturated soil specimen is subjected to additional confining pressure (total stress), it generates an equal excess pore water pressure, which means that the additional confinement does not cause additional effective confining pressure. The effective-stress principle indicates that the shear strength of the soil specimen depends on the effective confining pressure.

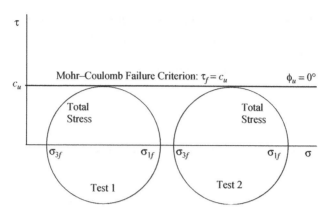

FIGURE 5.22 Typical results of an unconsolidated undrained triaxial test.

Therefore, since there was no increase in the effective confining pressure, there would be no increase in shear strength.

As shown in Figure 5.22, the Mohr–Coulomb failure criterion is horizontal ($\phi_u = 0$) and it intersects the vertical axis at $\tau = c_u$. Note that c_u is the undrained shear strength of a soil and is equal to the radius of the total stress Mohr's circle, [i.e., $c_u = \frac{\sigma_{1f} - \sigma_{3f}}{2}$]. The undrained shear strength is used appropriately to describe the strength of fine-grained soils subjected to rapid loading, during which drainage is not allowed; therefore, no dissipation of excess pore water pressures is possible. Examples of that include rapid construction of embankments on clay deposits or rapid loading of shallow foundations constructed on clay.

5.3.4 Unconfined Compression Test

The unconfined compression test is performed on unconfined cylindrical specimen of a cohesive soil to measure its unconfined compression strength, q_u. The undrained shear strength, c_u, is equal to one-half of the unconfined compression strength, q_u, as indicated in Figure 5.23. This test is a special case of the unconsolidated–undrained triaxial test when performed without applying any confining pressure. The undrained shear strength obtained by the two tests for the same cohesive soil is theoretically identical. The unconfined compression test is generally performed on undisturbed specimens of cohesive soils at their natural water contents. It is not possible to perform this test on cohesionless soils since they do not have shear strength at zero confinement.

5.4 FIELD TESTS

The following is a brief description of field tests that can be used to measure the in situ shear strength of soil.

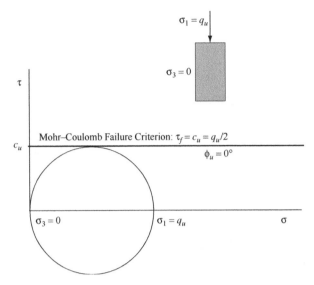

FIGURE 5.23   Typical results of an unconfined compression test.

5.4.1   Field Vane Shear Test

The vane shear test consists of a four-bladed vane that is pushed into the undisturbed cohesive soil at the bottom of a borehole and rotated to determine the torsional force required to cause a cylindrical soil specimen to be sheared by the vane. The torsional force is then correlated to the undrained shear strength, c_u. This test provides a direct and reliable measurement of the in situ undrained shear strength. A smaller handheld version of the vane shear test can be used to measure the undrained shear strength of samples recovered from a test boring. This is done by inserting the shear vane into the soil sample and twisting until failure. The undrained shear strength of the soil is captured by an indicator mounted on the apparatus.

5.4.2   Cone Penetration Test

The cone penetration test (CPT) consists of a cylindrical probe with a cone tip that is pushed continuously into the ground at a slow rate. The probe is instrumented with strain gauges to measure the tip and side resistance while the probe is advancing into the ground. The data are gathered continuously using a computer. The measured tip resistance is correlated with the undrained shear strength of the soil at various depths.

5.4.3   Standard Penetration Test

The standard penetration test (SPT) consists of driving a sampler (a thin hollow cylinder) into the bottom of a borehole using a standard hammer dropped from a standard distance. The number of blows required to drive the sampler a distance of 30 cm into the ground is defined as the SPT blow count (N). The blow count is correlated to the undrained shear strength for cohesive soils and the friction angle for granular soils.

5.5   DRAINED AND UNDRAINED LOADING CONDITIONS VIA FEM

When saturated coarse-grained soils (sand and gravel) are loaded slowly, volume changes occur, resulting in excess pore pressures that dissipate rapidly, due to high permeability. This is called *drained loading.* On the other hand, when fine-grained soils (silts and clays) are loaded, they generate excess pore pressures that remain entrapped inside the pores because these soils have very low permeabilities. This is called *undrained loading.*

Both drained and undrained conditions can be investigated in a laboratory triaxial test setup. Consider a soil specimen in a consolidated–drained (CD) triaxial test. During the first stage of the triaxial test (Figure 5.11) the specimen is subjected to a constant confining stress, σ_3, and allowed to consolidate by opening the drainage valves. In the second stage of the CD triaxial test (Figure 5.12) the specimen is subjected, by means of the loading ram, to a monotonically increasing deviator stress, $\sigma_1 - \sigma_3$, while the valves are kept open. The deviator stress is applied very slowly to ensure that no excess pore water pressure is generated during this stage—hence the term *drained.* Typical CD test results at failure can be presented using Mohr's circle as shown in Figures 5.13 and 5.14. The drained (or long-term) strength parameters of a soil, c' and ϕ', can be obtained from the Mohr–Coulomb failure criterion as indicated in the figures. Note that $c' = 0$ for sands and normally consolidated clays. These parameters must be used in drained (long-term) analysis of soils.

Now let's consider a soil specimen in a consolidated–undrained (CU) triaxial test (Figure 5.17). The first stage of this test is the same as the CD test—the specimen is subjected to a constant confining stress, σ_3, and allowed to consolidate by opening the drainage valves. In the second stage of the CU test, however, the specimen is subjected to a monotonically increasing deviator stress, $\sigma_1 - \sigma_3$, while the drainage valves are closed—hence the term *undrained.* The undrained condition means that there will be no volumetric change in the soil specimen (i.e., volume remains constant). It also means that excess pore water pressure will be developed inside the soil specimen throughout the test. Measurement of the pore water pressure allows for effective stress calculations. Typical CU test results (at failure) for a normally consolidated clay can be presented using Mohr's circles as shown in Figure 5.24. The undrained (or short-term) strength parameter of a soil, c_u, is the radius of the total stress Mohr's circle (note that $\phi_u = 0$). The parameter c_u is termed the *undrained shear strength* of the soil and must be used in undrained (short-term) analysis of fine-grained soils. The effective-stress Mohr–Coulomb failure criterion can be used to estimate the drained (or long-term) strength parameters c' and ϕ', as shown in Figure 5.24 ($c' = 0$ for NC clay).

In a finite element scheme, the drained (or long-term) behavior of a soil can be simulated using coupled analysis, where the pore water pressure is calculated for a

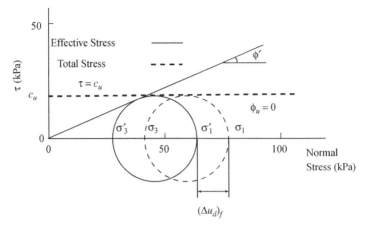

FIGURE 5.24 Drained and undrained strength parameters.

given load increment in each soil element and then subtracted from the total stresses to estimate the effective stresses in the element. These effective stresses control the deformation and shear strength of the soil element according to the effective-stress principle. Constitutive models such as the cap and Cam clay models can be used within the finite element framework to determine the deformation caused by these effective stresses.

Four main measures must be considered for a successful finite element analysis of soils considering their long-term (drained) behavior: (1) the initial conditions of the soil strata (initial geostatic stresses, initial pore water pressures, and initial void ratios), which will determine the initial stiffness and strength of the soil strata, must be estimated carefully and implemented in the analysis; (2) the boundary conditions must be defined carefully: pervious or impervious; (3) the long-term strength parameters of the soil must be used in an appropriate soil model; and (4) loads must be applied very slowly to avoid the generation of excess pore water pressure throughout the analysis.

For undrained (or short-term) analyses, the aforementioned measures apply with the exception of the last measure—the load can be applied very fast instead. This is one of the most attractive aspects of coupled analysis. Drained and undrained analyses (Examples 5.6 and 5.7, respectively) differ only in the way we apply the load: Very slow loading allows the excess pore water pressure generated to dissipate and the long-term-strength parameters to be mobilized, whereas fast loading does not allow enough time for the pore water pressure to dissipate, thus invoking the short-term strength of the soil. This means that there is no need to input the short-term-strength parameters because the constitutive model will react to fast loading in an "undrained" manner.

As an alternative method, undrained analysis can utilize the undrained shear strength parameter, c_u, directly in conjunction with a suitable constitutive model. If the cap model is used, the cap parameters β and d can be calculated using c_u

and $\phi_u (= 0)$. This procedure can be used only for undrained (short-term) analysis. Unlike the preceding procedure, this alternative procedure cannot be used for drained analysis and for cases falling between the drained and the undrained cases (partially drained).

Example 5.6 Hypothetical consolidated drained and consolidated undrained triaxial test results for a clayey soil are given in Figures 5.25 and 5.26, respectively. The isotropic consolidation curve for the same soil is given in Figure 5.27. Using the finite element method, simulate the consolidated–drained triaxial behavior of this soil when subjected to a 210-kPa confining pressure. Consider a cylindrical soil specimen with $D = 5$ cm and $H = 5$ cm as shown in Figure 5.28. The top and bottom surfaces of the specimen are open and permeable. The loading plate and the pedestal are smooth; therefore, their interfaces with the clay specimen can be assumed frictionless. The clay specimen is normally consolidated and has a constant permeability $k = 0.025$ m/s and initial void ratio $e_0 = 0.889$, corresponding to its 210-kPa confining pressure. Assume that the clay is elastoplastic, obeying the extended Cam clay model (Chapter 2).

SOLUTION(filename: Chapter5_Example6.cae): Let's determine the Cam clay model parameters from the tests results provided in Figures 5.25 to 5.27. From the e–log p' curve shown in Figure 5.27, we can determine the values of the compression index $C_c = 0.4$ and the swelling index $C_s = 0.06$. Note that the mean

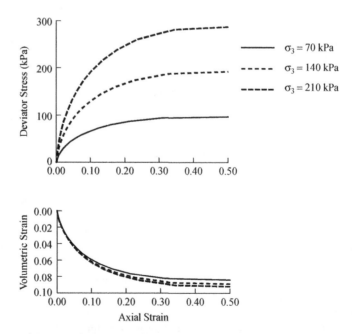

FIGURE 5.25 Hypothetical consolidated drained triaxial test results for a clayey soil.

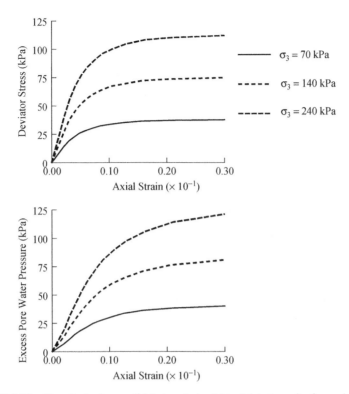

FIGURE 5.26   Hypothetical consolidated undrained triaxial test results for a clayey soil.

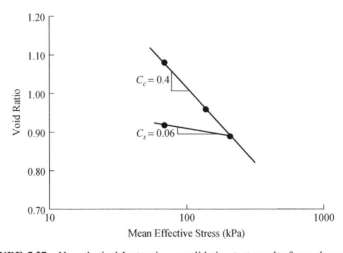

FIGURE 5.27   Hypothetical Isotropic consolidation test results for a clayey soil.

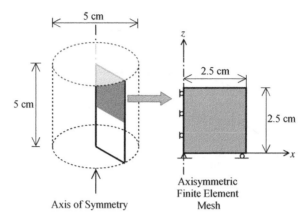

FIGURE 5.28   Axisymmetric finite element mesh (one element).

effective stress in this figure is defined as $p' = (\sigma'_1 + 2\sigma'_3)/3$ ($p' = \sigma'_3$ for isotropic consolidation). The parameter κ defines the elastic behavior of the soil in the Cam clay model, and it is related to the swelling index through the equation $\kappa = C_s/2.3 = 0.06/2.3 = 0.026$. The parameter λ is related to the compression index through $\lambda = C_c/2.3 = 0.4/2.3 = 0.174$.

The effective-stress Mohr's circles for the three consolidated drained triaxial tests and the three consolidated undrained tests are given in Figure 5.29. The effective-stress Mohr–Coulomb failure criterion is a straight line that is tangent to the circles as shown in the figure. The slope of this line is the effective friction angle of the soil $\phi' = 25.4°$. The Cam clay strength parameter M is related to the

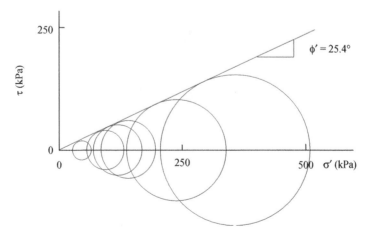

FIGURE 5.29   Effective-stress Mohr–Coulomb failure criterion for a clayey soil.

internal friction angle of the soil, ϕ', as follows:

$$M = \frac{6 \sin \phi'}{3 - \sin \phi'} \tag{5.7}$$

Therefore,

$$M = \frac{6 \sin 25.4°}{3 - \sin 25.4°} = 1$$

An alternative procedure for determining M is to plot the at-failure stresses of all six triaxial tests in the $p' = (\sigma'_{1f} + 2\sigma'_{3f})/3$ versus $q = \sigma'_{1f} - \sigma'_{3f}$ plane as shown in Figure 5.30. The data points are best fitted with a straight line called the *critical-state line*. The slope of the critical-state line shown in the figure is the Cam clay strength parameter M.

In the Cam clay model, the yield surface size is fully described by the parameter $p' = (\sigma'_1 + 2\sigma'_3)/3$. The evolution of the yield surface is governed by the volumetric plastic strain ε^p_{vol}, which is a function of p'. The relationship between ε^p_{vol} and p' can be deduced easily from an e–$\log p'$ line. The consolidation curve (the e–$\log p'$ line) is defined completely by its slope C_c ($= 2.3\lambda$), and the initial conditions p'_0 ($= \sigma'_0$) and e_0. Note that λ, σ'_0, and e_0 are part of the input parameters required in the finite element program used herein.

The preconsolidation pressure, σ'_c, is also a required parameter. It is specified by the size of the initial yield surface as shown in Table 5.1. Consider the initial yield surface that corresponds to $p'_0 = 210$ kPa (or $p'_0/2 = 105$ kPa), where $p'_0 = 210$ kPa is equal to the confining pressure σ'_3. In this example, σ'_c is assumed to be equal to the confining pressure, indicating that the clay is normally consolidated. If the clay were overconsolidated, its σ'_c can be set equal to its preconsolidation pressure,

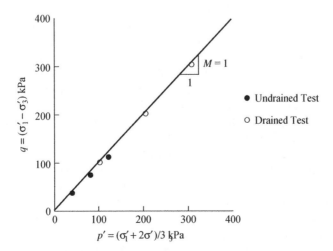

FIGURE 5.30 Critical-state line for a clayey soil.

TABLE 5.1   Cam Clay Model Parameters

General		Plasticity	
ρ (kg/m^3)	1923	λ	0.174
k (m/sec)	0.025	Stress ratio, M	1
γ_w (kN/m^3)	9.81	Initial yield surface	
e_0	0.889	size = $p'_0/2$ (kPa)	105
Elasticity		Wet yield surface	
κ	0.026	size	1
υ	0.28	Flow stress rate	1

which is greater than its confining pressure. A summary of the Cam clay model parameters is given in Table 5.1.

A two-dimensional axisymmetric mesh is used with one element only, as shown in Figure 5.28. The element chosen is a pore fluid/stress eight-node axisymmetric quadrilateral element with biquadratic displacement, bilinear pore pressure, and reduced integration. The boundary conditions of the finite element mesh shown in Figure 5.28 are as follows. On the bottom side, the vertical component of displacement is fixed ($u_z = 0$). The left-hand side of the mesh is a symmetry line ($u_x = 0$). On the top surface a uniform downward displacement of 0.5 cm is applied very slowly. During load application the top surface of the finite element mesh is made pervious. This means that the top and bottom of the soil specimen are allowed to drain due to symmetry about a plane passing through the midheight of the soil cylinder.

Similar to an actual consolidated–drained triaxial test, this finite element analysis is carried out in two steps: a consolidation step and a shearing step. The first step is a single increment of analysis with drainage allowed across the top surface. In this step the confining pressure of 210 kPa is applied at the top surface and the right side of the mesh. During step 1, the "geostatic" command is invoked to make sure that equilibrium is satisfied within the soil specimen. The geostatic option makes sure that the initial stress condition in the clay specimen falls within the initial yield surface of the Cam clay model. Step 2 is the shearing step with duration of 10^9 seconds. In this step the loading plate is forced to displace downward at a very small rate (5×10^{-10} cm/s). This low rate of displacement is used to ensure that the excess pore water pressure within the clay specimen is always zero. Automatic time stepping with a maximum pore water pressure change of 0.007 kPa per time increment is used. This procedure is useful for loading steps with very long duration. When the loading starts in the beginning of the step, the rate of pore water pressure change is high, therefore, small time increments are used. Later, when the rate of pore water pressure change decreases, larger time increments are used. As an alternative, one can use a shorter duration for this loading step with fixed-time stepping instead. However, we need to make sure that the excess pore water pressure during this step is kept equal to zero. This can be done easily by plotting the pore water pressure at the center of the soil specimen as a function of time. If not zero, the duration of the shearing step can be increased.

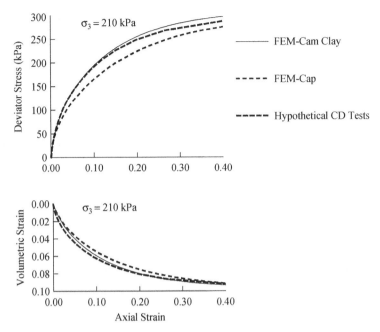

FIGURE 5.31 Comparison between hypothetical CD triaxial test results and finite element prediction using Cam clay and cap models.

The calculated consolidated–drained stress–strain behavior of the clay specimen is compared with the hypothetical test data as shown in Figure 5.31. The axial strain versus volumetric strain is also compared to the experimental data in the same figure. Excellent agreement is noted in the figure. This means that the Cam clay model embodied in the finite element method is capable of describing the consolidated–drained triaxial behavior of normally consolidated clays.

Example 5.7 Using the same hypothetical test data and assumptions given in Example 5.6, predict the consolidated–undrained triaxial behavior of a triaxial soil specimen subjected to a 210-kPa confining pressure.

SOLUTION(filename: Chapter5_Example7.cae): As described in Example 5.6, we have estimated the parameters of the Cam clay model from the hypothetical test data provided. Those parameters are used in this example as well. The solution of this problem is identical to the solution of Example 5.6 with two exceptions: (1) the top and bottom of the soil specimen are made impervious (undrained), and (2) the shearing load is applied faster.

We will use the same finite element mesh (Figure 5.28) that was used in Example 5.6. Similar to an actual consolidated–undrained triaxial test, this finite element analysis is carried out in two steps: a consolidation step and a shearing step. In the first step, the consolidation step, an all-around confining pressure of 210 kPa is applied while drainage is permitted across the top surface. During this step, the "geostatic" command is invoked to make sure that equilibrium is satisfied within the clay specimen. The geostatic option makes sure that the initial stress condition in the clay specimen falls within the initial yield surface of the Cam clay model.

The second step, the undrained shearing step, has a duration of 100 seconds. In the beginning of this step the pervious boundary at the top surface is removed, making the top surface impervious by default. In this step the loading plate is forced to displace downward a distance of 0.127 cm at a displacement rate of 1.27×10^{-3} cm/s. This rate of displacement, along with the impervious boundaries, will cause the excess pore water pressure to generate during shearing. Automatic time stepping with a maximum pore water pressure change of 0.7 kPa per time increment is used.

The consolidated–undrained stress–strain behavior calculated for the clay specimen is compared with the hypothetical test data as shown in Figure 5.32. The axial strain versus pore water pressure is also compared to the hypothetical test data in the same figure. Excellent agreement is noted in the figure.

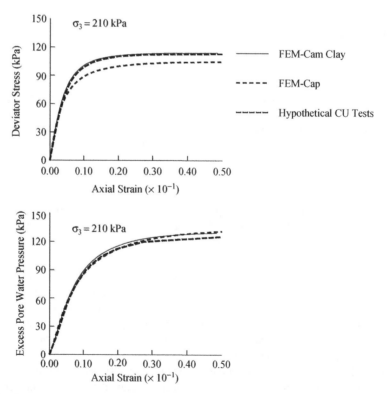

FIGURE 5.32 Comparison between hypothetical CU triaxial test results and finite element prediction using Cam clay and cap models.

Example 5.8   Assuming that the clayey soil described in Example 5.6 is elasto-plastic, obeying the cap model (Chapter 2), predict the (a) consolidated–drained and (b) consolidated–undrained triaxial behavior of the soil when subjected to a 210-kPa confining pressure. Consider a cylindrical specimen with $D = 5$ cm and $H = 5$ cm as shown in Figure 5.28. The loading plate and the pedestal are smooth; therefore, their interfaces with the clay specimen can be assumed frictionless. The clay specimen is normally consolidated and has a constant permeability $k = 0.025$ m/s and initial void ratio $e_0 = 0.889$, corresponding to its 210-kPa confining pressure.

SOLUTION: (a) *Consolidated–drained triaxial condition* (filename: Chapter5_Example8a.cae) We need to determine the cap model parameters from the hypothetical test results provided in Figures 5.25 to 5.27. We use the same finite element mesh (Figure 5.28) that was used in Example 5.6.

Six effective-stress Mohr's circles corresponding to failure stresses obtained from the hypothetical triaxial test results (Figures 5.25 and 5.26) are plotted in Figure 5.29. Subsequently, the effective-stress Mohr–Coulomb failure criterion is plotted as a straight line that is tangential to the six circles. The soil strength parameters $\phi' = 25.4°$ and $c' = 0$ kPa are obtained from the slope and intercept of the Mohr–Coulomb failure criterion.

For triaxial stress conditions, the Mohr–Coulomb parameters ($\phi' = 25.4°$ and $c' = 0$ kPa) can be converted to Drucker–Prager parameters (Chapter 2) as follows:

$$\tan\beta = \frac{6\sin\phi'}{3 - \sin\phi'} \qquad \text{for } \phi' = 25.4° \rightarrow \beta = 45°$$

$$d = \frac{18c\cos\phi'}{3 - \sin\phi'} \qquad \text{for } c' = 0 \rightarrow d = 0$$

An alternative procedure for determining β and d is to plot the at-failure stresses of all six triaxial tests in the $p' = (\sigma'_{1f} + 2\sigma'_{3f})/3$ versus $q = \sigma'_{1f} - \sigma'_{3f}$ plane as shown in Figure 5.33. The data points are best fitted with a straight line whose slope is equal to $\tan\beta = 1$; thus, $\beta = 45°$. The line intersects with the vertical axis at $d = 0$. The cap eccentricity parameter is chosen as $R = 1.2$. The initial cap position, which measures the initial consolidation of the specimen, is taken as $\varepsilon^{pl}_{vol(0)} = 0.0$, which corresponds to $p' = 210$ kPa. The transition surface parameter $\alpha = 0.05$ is used. These parameters are summarized in Table 5.2.

The cap hardening curve shown in Figure 5.34 was obtained from the isotropic consolidation test results shown in Figure 5.27. From Figure 5.27 we can calculate the plastic volumetric strain as

$$\varepsilon^p_v = \frac{\lambda - \kappa}{1 + e_0}\ln\frac{p'}{p'_0} = \frac{C_c - C_s}{2.3(1 + e_0)}\ln\frac{p'}{p'_0}$$

For this clayey soil we have $\lambda = 0.174$, $\kappa = 0.026$, $p'_0 = 210$ kPa, and $e_0 = 0.889$; therefore,

$$\varepsilon^p_v = \frac{0.174 - 0.026}{1 + 0.889}\ln\frac{p'}{210} = 0.07834\ln\frac{p'}{210}$$

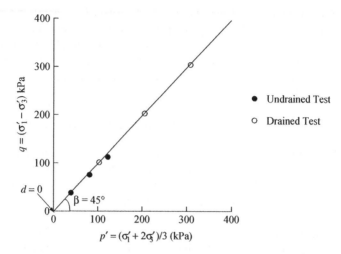

FIGURE 5.33 Determination of the cap model parameters d and β for a clayey soil.

TABLE 5.2 Cap Model Parameters

General		Plasticity	
ρ (kg/m^3)	1923	d	0
e_0	0.889	β (deg)	45
Elasticity		R	1.2
E (MPa)	182	Initial yield	0.0
υ	0.28	α	0.05
		K	1

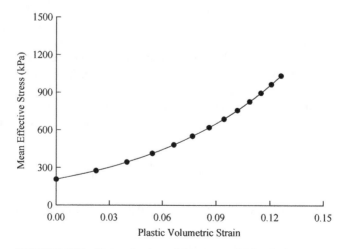

FIGURE 5.34 Determination of the cap model hardening curve.

This equation describes the evolution of the plastic volumetric strain (the hardening parameter) with the mean effective stress (Figure 5.34).

We will use the cap model parameters above to reproduce the stress–strain curves of the clayey soil subjected to a confining pressure of 210 kPa. This is done by using one axisymmetric finite element in the same manner described in Example 5.6. The calculated consolidated–drained stress–strain behavior of the clay specimen is compared with the hypothetical test data as shown in Figure 5.31. The axial strain versus volumetric strain is also compared to the hypothetical test data in the same figure. It can be noted from Figure 5.31 that the cap model can adequately simulate the behavior of the clayey soil on the "elemental" level.

(b) *Consolidated–undrained triaxial condition* (filename:Chapter5_Example8b. cae) As described above, we have estimated the parameters of the cap model from the hypothetical test data. Those parameters will also be used here to predict the consolidated–undrained triaxial behavior of the clayey soil. The solution of this problem is identical to the solution of the consolidated–drained triaxial test with two exceptions: (1) the top and bottom of the soil specimen are made impervious (undrained), and (2) the shearing load is applied faster.

The finite element mesh shown in Figure 5.28 is used here, too. The analysis is carried out in two steps: a consolidation step and a shearing step. In the first step, the consolidation step, a confining pressure of 210 kPa is applied while drainage is permitted across the top surface of the finite element mesh. During this step, the "geostatic" command is invoked to ensure equilibrium. The second step, the undrained shearing step, has a duration of 100 seconds, during which the pervious boundary at the top surface is canceled. The loading plate is forced to displace downward a distance of 0.127 at a displacement rate of 1.27×10^{-3} cm/s. This loading will cause the excess pore water pressure to generate during shearing. Automatic time stepping with a maximum pore water pressure change of 0.7 kPa per time increment is used.

The calculated consolidated–undrained stress–strain behavior of the clay specimen is compared with the triaxial test data as shown in Figure 5.32. The axial strain versus pore water pressure is also compared to the test data in the same figure. Excellent agreement is noted in the figure.

Example 5.9: *Back-Calculation of CD Triaxial Test Results Using Lade's Model*
Use Lade's model parameters for the dense silty sand, obtained in Example 2.3 (Table 2.5), along with the finite element method to back-calculate the stress–strain behavior of the soil under CD triaxial conditions.

SOLUTION(filename: Chapter5_example9.cae): Using the finite element method, we simulate the consolidated–drained triaxial behavior of this soil under three confining pressures. A two-dimensional axisymmetric mesh is used with one element only (similar to Example 5.8). The element chosen is a four-node axisymmetric quadrilateral element. As in Example 5.8, the left-hand side of the mesh is a symmetry line, and the bottom of the finite element mesh is on rollers. A uniform downward displacement is applied slowly on the top surface of the mesh (strain-controlled triaxial test).

The analysis is carried out in two steps: a consolidation step and a shearing step. The first step is a single increment of analysis in which the confining pressure is applied at the top surface and the right side of the mesh. During step 1, the "geostatic" command is invoked to make sure that equilibrium is satisfied within the soil specimen. The geostatic option makes sure that the initial stress condition in the soil specimen falls within the initial yield surface of Lade's model. Step 2 is a shearing step in which the loading plate is forced to displace downward at a small rate.

In this simulation we use Lade's model parameters for the dense silty sand, obtained in Example 2.3 (Table 2.5). Figure 5.35 presents the soil's triaxial behavior under three confining pressures. The calculated stress–strain behavior is compared with the experimental data. Reasonable agreement is noted in the figure. Of particular interest is the dilative behavior in the volumetric strain versus axial strain plane where the calculated results (Lade's model) captured the important dilation phenomenon. In Example 2.3 we presented a similar comparison between the experimental data and spreadsheet calculations based on Lade's model. It is noteworthy that the finite element results obtained in the present example are in perfect agreement with the results of the spreadsheet analysis. This implies that the implementation of Lade's model in the finite element program used herein is done correctly.

Example 5.10 *Plane Strain Test Simulation Using Lade's Model* Use Lade's model parameters for the dense silty sand, obtained in Example 2.3 (Table 2.5) along with the finite element method to predict the behavior of the soil specimen shown in Figure 5.36a. Assume that the soil specimen is infinitely long in the z-direction and that a plane strain condition applies. Also assume that the specimen is subjected to a constant confining pressure of 172.37 kPa.

SOLUTION(filename: Chapter5_example10.cae): The two-dimensional plane strain finite element mesh used in this analysis is shown in Figure 5.36b. The mesh has 154 elements. The element chosen is a four-node bilinear plane strain quadrilateral element. As shown in Figure 5.36b, the bottom of the finite element mesh is on rollers, simulating a frictionless interface between the soil and the bottom surface of the plane strain test apparatus. A uniform downward displacement is applied slowly on the top surface of the mesh (strain-controlled shear test).

The analysis is carried out in two steps: a consolidation step and a shearing step. The first step is a single increment of analysis in which the confining pressure is applied at the top surface and the sides of the mesh. During step 1, the "geostatic" command is invoked to make sure that equilibrium is satisfied within the soil specimen. The geostatic option makes sure that the initial stress condition in the soil specimen falls within the initial yield surface of Lade's model. Step 2 is a shearing step in which the loading plate is forced to displace downward at a small rate.

In this simulation we use Lade's model parameters for the dense silty sand, obtained in Example 2.3 (Table 2.5). Figure 5.37 presents the deformed shape of

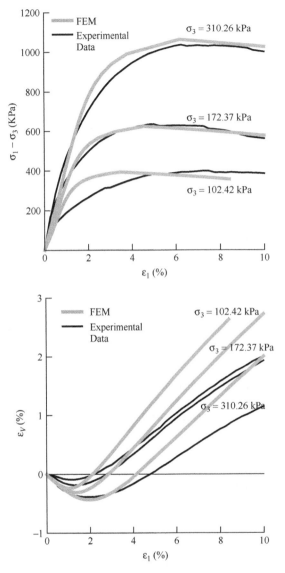

FIGURE 5.35 Stress–strain behavior of a dense silty sand: Lade's model versus experimental data.

the plane strain soil model at an advanced stage of loading (postpeak behavior). Figure 5.38 shows the distribution of the vertical strains at an advanced stage of loading. Note the presence of a zone of concentrated strains that resembles a failure plane. The severe distortion within this concentrated strain zone (also known as *shear band*) is evident in Figure 5.37. Note that the zones above and below the shear band have a minimal amount of distortion.

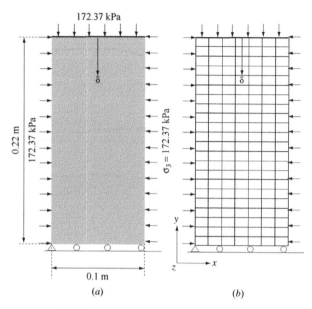

FIGURE 5.36   Plane strain test configuration.

PROBLEMS

5.1 A direct shear test is performed on a sand specimen at a constant normal stress of 40 kPa. The measured shear stress at failure is 28.5 kPa. Calculate the shear strength parameters of the sand. What is the shear stress at failure if the normal stress were 60 kPa?

5.2 The results of direct shear tests performed on a clay specimen indicated that $c' = 30$ kPa and $\phi' = 22°$. What is the shear strength of this soil at $\sigma' = 0$ and $\sigma' = 100\,\text{kPa}$?

5.3 Refer to Figure 5.8. Due to the self-weight of the wedge a normal stress of 1000 kPa and a shear stress of 620 kPa are applied to the inclined plane. What is the minimum friction angle between the wedge and the inclined plane required to maintain equilibrium?

5.4 Express the Mohr–Coulomb failure criterion in terms of the minor and major principal stresses. [*Hint*: In reference to Figure 5.13, on the failure plane at failure we have

$$\sigma' = \frac{\sigma'_1 + \sigma'_3}{2} - \frac{\sigma'_1 - \sigma'_3}{2}\sin\phi' \quad \text{and} \quad \tau = \frac{\sigma'_1 - \sigma'_3}{2}\cos\phi']$$

5.5 A CD triaxial compression test conducted on a sand specimen revealed that $\phi' = 22°$. What is the deviator stress at failure if the confining pressure is 50 kPa? Solve the problem graphically and analytically.

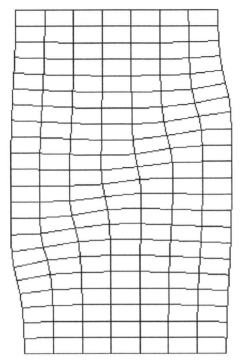

FIGURE 5.37 Deformed shape at an advanced stage (post failure).

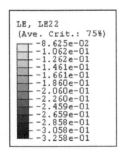

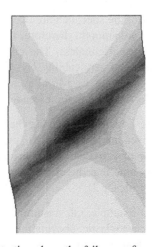

FIGURE 5.38 Strain concentration along the failure surface.

5.6 A CD triaxial compression test conducted on a clay specimen revealed that $c' = 5\,\text{kPa}$ and $\phi' = 22°$. What is the major principal stress at failure if the confining pressure is 35 kPa? Solve the problem graphically and analytically.

5.7 A consolidated–undrained triaxial test was performed on a sand specimen at a confining pressure $\sigma_3 = 40\,\text{kPa}$. The consolidated–undrained friction angle of the sand is $\phi = 28°$, and the effective friction angle is $\phi' = 33°$. Calculate **(a)** the major principal stress at failure, σ_{1f}; **(b)** the minor and the major effective principal stresses at failure, σ'_{3f} and σ'_{1f}; and **(c)** the excess pore water pressure at failure, $(\Delta u_d)_f$. Solve the problem graphically and analytically.

5.8 A CD triaxial compression test was conducted on a sand specimen using a confining pressure of 50 kPa. Failure occurred at a deviator stress of 120 kPa. Calculate the normal and shear stresses on the failure plane at failure. Solve the problem graphically and confirm your solution analytically.

5.9 A CU triaxial test was performed on a dense sand specimen at a confining pressure $\sigma_3 = 40\,\text{kPa}$. The consolidated–undrained friction angle of the sand is $\phi = 39°$, and the effective friction angle is $\phi' = 34°$. Calculate **(a)** the major principal stress at failure, σ_{1f}; **(b)** the minor and the major effective principal stresses at failure, σ'_{3f} and σ'_{1f}; and **(c)** the excess pore water pressure at failure, $(\Delta u_d)_f$. Solve the problem graphically and analytically.

5.10 Refer to Example 5.6. Hypothetical consolidated–drained and consolidated–undrained triaxial test results for a clayey soil are given in Figures 5.25 and 5.26, respectively. The isotropic consolidation curve for the same soil is given in Figure 5.27. Using the finite element method, simulate the consolidated–drained triaxial behavior of this soil when subjected to a 140-kPa confining pressure. Consider a cylindrical soil specimen with $D = 5$ cm and $H = 5$ cm as shown in Figure 5.28. The top and bottom surfaces of the specimen are open and permeable. The loading plate and the pedestal are smooth, therefore, their interfaces with the clay specimen can be assumed frictionless. The clay specimen is normally consolidated and has a constant permeability $k = 0.025\,\text{m/s}$ and initial void ratio $e_0 = 0.959$, corresponding to its 140-kPa confining pressure. Assume that the clay is elastoplastic obeying the extended Cam clay model (Chapter 2).

5.11 Refer to Example 5.7. Using the same hypothetical test data and assumptions given in Problem 5.10, predict the consolidated–undrained triaxial behavior of a triaxial soil specimen subjected to 140-kPa confining pressure. Assume that the clay is elastoplastic, obeying the extended Cam clay model (Chapter 2).

5.12 Refer to Example 5.8. Assuming that the clayey soil described in Problem 5.10 is elastoplastic, obeying the Cap model (Chapter 2), predict the consolidated–drained and the consolidated–undrained triaxial behavior of the soil when subjected to a 140-kPa confining pressure.

5.13 Obtain Lade's model parameters (Chapter 2) for the soil described in Problem 5.10. Use Lade's model parameters, along with the finite element method to predict the consolidated–drained triaxial behavior of the soil when subjected to a 140-kPa confining pressure.

5.14 The at-failure results of several CD and CU triaxial tests of a silty sand are presented in the $p' - q$ plane as shown in Figure 5.39. The results of an isotropic compression test performed on the same soil are shown in Figure 5.40. Estimate the Cam clay model parameters M, λ, and κ. Note that you can obtain the compression index and the swelling index from Figure 5.40 (not λ and κ).

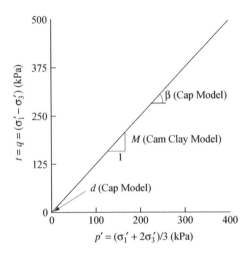

FIGURE 5.39 p' versus q' curve for silty sand.

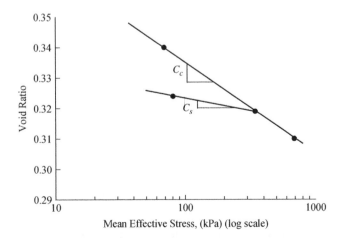

FIGURE 5.40 e–log p' curve for silty sand.

5.15 Using the finite element method and the Cam clay model parameters for the silty sand (Problem 5.14), predict the consolidated–drained triaxial behavior of this soil when subjected to a confining pressure of 70 kPa. Note that the initial void ratio that corresponds to this confining pressure is 0.34 (Figure 5.40).

5.16 Repeat Problem 5.15 for consolidated–undrained triaxial test conditions.

5.17 The at-failure results of several CD and CU triaxial tests of a silty sand are presented in the $p' - t$ plane as shown in Figure 5.39. The results of an isotropic compression test performed on the same soil are shown in

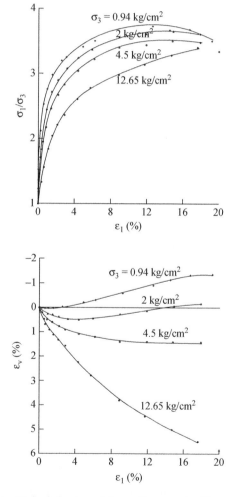

FIGURE 5.41   Stress–strain behavior of loose Sacramento River sand. (Adapted from Lade, 1977.)

Figure 5.40. (**a**) Calculate the soil's angle of friction, β, and its cohesion, d, in the $p' - t$ plane. (**b**) Using the results of the isotropic compression test performed on the same soil (Figure 5.40), calculate the hardening curve for the cap model assuming the initial conditions $p'_0 = 70\,\text{kPa}$ and $e_0 = 0.34$.

5.18   Using the finite element method and the cap model parameters for the silty sand (Problem 5.17), predict the consolidated–drained triaxial behavior of this soil when subjected to a confining pressure of 70 kPa. Note that the initial void ratio that corresponds to this confining pressure is 0.34 (Figure 5.40).

5.19   Repeat Problem 5.18 for consolidated–undrained triaxial test conditions.

5.20   The results of four consolidated–drained triaxial tests and one isotropic compression test on loose Sacramento River sand are shown in Figures 5.41 and 5.42, respectively. Estimate Lade's model parameters following the procedure discussed in Chapter 2.

5.21   Using the finite element method and Lade's model parameters obtained in Problem 5.20, predict the consolidated drained triaxial behavior of a loose Sacramento River sand specimen subjected to a confining pressure of 2 kg/cm^2.

5.22   From the results of the consolidated drained triaxial tests and the isotropic compression test on loose Sacramento River sand shown in Figures 5.41 and 5.42, respectively, estimate the Cam clay model parameters.

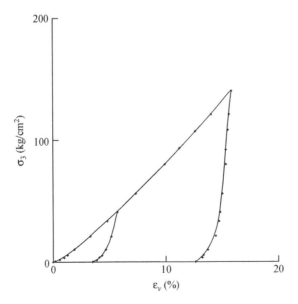

FIGURE 5.42   Isotropic compression behavior of loose Sacramento River sand. (Adapted from Lade, 1977.)

5.23 Using the finite element method and the Cam clay model parameters obtained in Problem 5.22, predict the consolidated–drained triaxial behavior of a loose Sacramento River sand specimen subjected to a confining pressure of 2 kg/cm^2.

5.24 Obtain the cap model parameters using the results of the consolidated–drained triaxial tests and the isotropic compression test on loose Sacramento River sand shown in Figures 5.41 and 5.42, respectively.

5.25 Using the finite element method and the cap model parameters obtained in Problem 5.24, predict the consolidated–drained triaxial behavior of a loose Sacramento River sand specimen subjected to a confining pressure of 2 kg/cm^2.

CHAPTER 6

SHALLOW FOUNDATIONS

6.1 INTRODUCTION

Shallow foundations are structural members that convert the concentrated super-structural loads into pressures applied to the supporting soil. Square, circular, strip, and mat foundations are common shapes of shallow foundations. Each of these shapes is suitable for a specific type of structure: A square foundation is used under a column, a circular foundation is used for cylindrical structures such as water tanks, a strip foundation is used under retaining walls, and a mat (raft) foundation is used under an entire building. A foundation is considered shallow if $D_f \leq B$ as proposed by Terzaghi (1943), where B is the foundation width and D_f is the foundation depth, as shown in Figure 6.1. Others proposed that foundations with greater depths (up to $4B$) can be considered shallow foundations. When designing a shallow foundation, two aspects must be considered: (1) the applied foundation pressure should not exceed the bearing capacity of the supporting soil; and (2) the foundation settlement should not be excessive due to the applied foundation pressure.

6.2 MODES OF FAILURE

There are three possible modes of soil failure, depending on soil type and foundation size and depth. The first mode, *general shear failure*, is usually encountered in dense sands and stiff clays underlying a shallow foundation. In reference to Figure 6.2, when the load Q is increased gradually, the corresponding foundation pressure, q, will increase. The foundation settlement will also increase, with increasing pressure until the ultimate bearing capacity, q_u, is reached. A sudden increase in settlement is noted immediately after reaching q_u, indicating severe loss of support. The general shear failure mode is accompanied by the occurrence of a failure surface (Figure 6.2) and the inability to maintain the applied pressure. There is a distinctive peak in the pressure versus settlement curve shown in the figure, which corresponds to the ultimate bearing capacity, q_u.

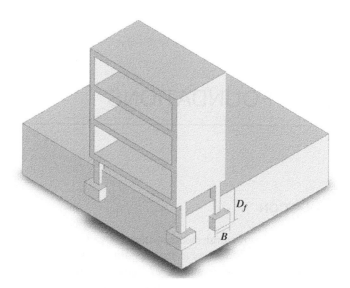

FIGURE 6.1   Shallow foundations.

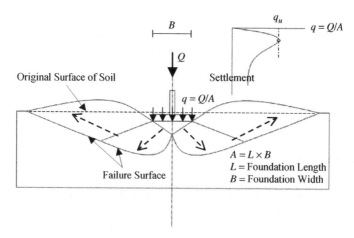

FIGURE 6.2   General shear failure.

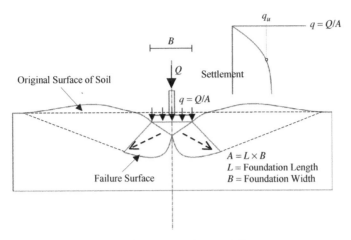

FIGURE 6.3 Local shear failure.

The second failure mode, *local shear failure*, is encountered in medium-dense sands and medium-stiff clays. It is characterized by the lack of a distinct peak in the pressure versus settlement curve, as shown in Figure 6.3. In the case of local shear failure, determination of the ultimate bearing capacity is usually governed by excessive foundation settlements, as indicated in the figure. The local shear failure mode is accompanied by a progressive failure surface that may extend to the ground surface after q_u is reached (Figure 6.3).

The third mode of failure, *punching shear failure*, usually occurs in loose sands and soft clays. This type of failure is accompanied by a triangular failure surface directly under the foundation. As in local shear failure, punching failure is also characterized by the lack of a distinctive ultimate bearing capacity. Thus, the ultimate bearing capacity in this case is taken as the pressure corresponding to excessive foundation settlements.

6.3 TERZAGHI'S BEARING CAPACITY EQUATION

Terzaghi presented his bearing capacity equation for shallow foundations in 1943. The equation was derived for a continuous (strip) foundation with general shear failure. The supporting material was assumed to be a thick layer of a homogeneous soil. A continuous foundation is a foundation with a finite width B and infinite length L; thus, $B/L \approx 0$.

Figure 6.4 shows the assumed failure surface underlying the foundation. There are three distinct failure zones of soil under the footing: a triangular zone, DEH, immediately under the footing; two radial zones, DHG and EIH; and two Rankine passive zones, DGC and EFI. The soil above the foundation level, having a thickness of D_f, is replaced by an overburden pressure of $q = \gamma D_f$ to simplify the equilibrium analysis. Note that the angle α (Figure 6.4) is assumed to be equal

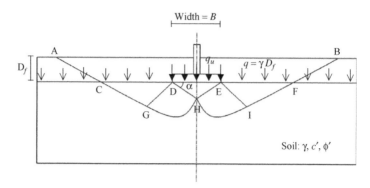

FIGURE 6.4 General shear failure of a strip foundation: Terzaghi's assumptions.

to the soil angle of internal friction ϕ', and that the soil shear resistance along CA and FB is neglected. In his limit equilibrium analysis, Terzaghi assumed that the bearing capacity of the foundation is the pressure of the foundation that will cause the triangular zone to be in a downward impending motion condition. For that to happen, the triangular zone will push the radial shear zones to the left and right away from the footing, and in turn, the radial shear zones will push the Rankine passive zones upward. The impending motion condition is assumed to take place in all zones simultaneously. Based on this assumption, Terzaghi derived the following equation for a strip foundation and general shear failure:

$$q_u = c'N_c + qN_q + \tfrac{1}{2}\gamma BN_\gamma \tag{6.1}$$

where c' is the cohesion intercept of soil, q the overburden pressure at foundation depth ($q = \gamma D_f$), γ the unit weight of soil, B the foundation width, and N_c, N_q, and N_γ are nondimensional bearing capacity factors that are functions of soil friction angle ϕ'. The bearing capacity factors N_c, N_q, and N_γ are given by

$$N_q = e^{\pi \tan \phi'} \tan^2 \left(45 + \frac{\phi'}{2} \right) \tag{6.2}$$

$$N_c = (N_q - 1)\ \cot \phi' \tag{6.3}$$

$$N_\gamma = (N_q - 1)\ \tan 1.4\phi' \tag{6.4}$$

Equation (6.1) can be modified to estimate the bearing capacity for a square foundation:

$$q_u = 1.3c'N_c + qN_q + 0.4\gamma BN_\gamma \tag{6.5}$$

and for a circular foundation:

$$q_u = 1.3c'N_c + qN_q + 0.3\gamma BN_\gamma \tag{6.6}$$

Bearing capacity equations for local shear and punching shear modes of failure for strip, square, and circular foundations are available in many geotechnical and foundation engineering books (e.g., Das, 2004).

Example 6.1 *Bearing Capacity of a Strip Foundation*   Using Terzaghi's equation, calculate the bearing capacity of a 0.6-m-wide strip foundation on a thick homogeneous layer of Ottawa sand with $c' = 0$ and $\phi' = 37°$. The foundation is situated at a depth $= 0.38$ m, as shown in Figure 6.5. The unit weight of Ottawa sand is 18.14 kN/m³.

SOLUTION: Use (6.1) (for a strip foundation):

$$c' = 0$$

$$\gamma = 18.14 \text{ kN/m}^3, D_f = 0.38 \text{ m} \rightarrow q = \gamma D_f = (18.14)(0.38) = 6.89 \text{ kPa}$$

$$\phi' = 37° \rightarrow$$

$$N_q = e^{\pi \tan 37°} \tan^2 \left(45° + \frac{37°}{2} \right) = 42.92$$

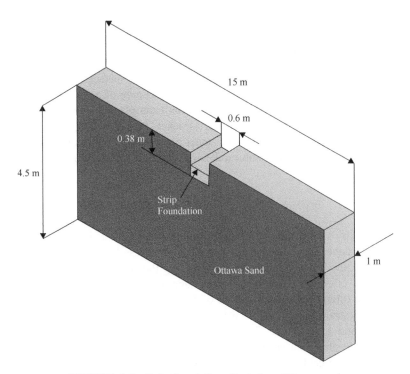

FIGURE 6.5   Strip foundation situated on Ottawa sand.

$$N_c = (42.92 - 1) \ \cot 37° = 55.63$$

$$N_\gamma = (42.92 - 1) \ \tan[1.4(37°)] = 53.27$$

Apply (6.1):

$$q_u = 0 + (6.89)(42.92) + \tfrac{1}{2}(18.14)(0.6)(53.27) = 586 \ \text{kPa}$$

Example 6.2 *Finite Element Application—Bearing Capacity Failure of a Strip Foundation* Solve Example 6.1 using the finite element method.

SOLUTION: (filename: Chapter6_Example2.cae): This example presents a limit equilibrium solution for a layer of Ottawa sand loaded by a rigid, perfectly rough strip footing. The solution is obtained using the finite element method, in which the behavior of Ottawa sand is simulated using the cap model (modified Drucker–Prager model with a cap, Chapter 2) with parameters matched to the classical Mohr–Coulomb yield model parameters c' and ϕ'. The problem geometry, boundary conditions, and materials are identical to those of Example 6.1, providing a direct means to compare the finite element analysis results with Terzaghi's equation.

As described in Chapter 2, the Drucker–Prager/cap model adds a cap yield surface to the modified Drucker–Prager model. The cap surface serves two main purposes: It bounds the yield surface in hydrostatic compression, thus providing an inelastic hardening mechanism to represent plastic compaction; and it helps control volume dilatancy when the material yields in shear by providing softening as a function of the inelastic volume increase created as the material yields on the Drucker–Prager shear failure and transition yield surfaces. The model uses associated flow in the cap region and a particular choice of nonassociated flow in the shear failure and transition regions. In this example we show how to match the parameters of a corresponding linear Drucker–Prager model, β and d, to the Mohr–Coulomb parameters, ϕ' and c', under plane strain conditions. As described in Chapter 5, the Mohr–Coulomb model is a classical failure model for soils and is written as

$$(\sigma'_1 - \sigma'_3) - (\sigma'_1 + \sigma'_3) \sin \phi' - 2c' \cos \phi' = 0$$

where σ'_1 and σ'_3 are the major and minor principal stresses, ϕ' is the friction angle, and c' is the cohesion. The intermediate principal stress has no effect on yield in this model. Experimental evidence suggests that the intermediate principal stress does have an effect on yield; nonetheless, laboratory data characterizing granular materials are often presented as values of ϕ' and c'.

The plane strain model analyzed is shown in Figure 6.6. The layer of sand is 4.5 m deep and 15 m wide. The foundation is a 0.3-m-thick rigid and perfectly rough concrete plate that spans a central portion 0.6 m wide. The model assumes symmetry about a center plane. The foundation is assumed to be in perfect contact with the soil (i.e., they share the nodal points in between). This means that relative

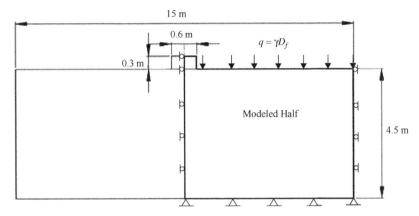

FIGURE 6.6 Idealization of the plane strain strip foundation problem.

displacement between the foundation and soil is not permitted. Reduced-integration bilinear plane strain quadrilateral elements are used for the sand and the concrete foundation. The base of the sand layer is fixed in both the horizontal and vertical directions. The right and left vertical boundaries are fixed in the horizontal direction but free in the vertical direction. The finite element mesh used in the analysis is shown in Figure 6.7. It is noted that the mesh is finer in the vicinity of the foundation since that zone is the zone of stress concentration. No mesh convergence studies have been performed. However, the depth and width of the sand layer are chosen such that the boundary effect on foundation behavior is minimized.

For a "fair" comparison with Terzaghi's equation, the 0.38-m-thick soil layer in the idealized section (Figure 6.6) is replaced with an overburden pressure of $q = \gamma D_f = (18.14)(0.38) = 6.89$ kPa. That way, the shear resistance of the 0.38-m-thick soil layer will not be considered in the finite element analysis—as you

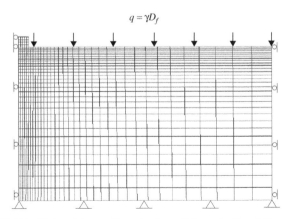

FIGURE 6.7 Finite element discretization of the strip foundation problem.

recall, this was one of the assumptions Terzaghi used for the derivation of his equation. Ordinarily, the foundation should be modeled with beam or shell elements that consider bending resistance. However, quadrilateral elements (which have no bending resistance) are used in the current analysis for simplicity. Thus, the foundation was discretized into four layers of elements in the vertical direction to capture its bending resistance.

The concrete foundation is assumed to be linear elastic with a Young's modulus of 1435 MPa and a Poisson ratio of 0.2. The elastic response of Ottawa sand is assumed to be linear and isotropic, with a Young's modulus of 182 MPa and a Poisson ratio of 0.3. Young's modulus is estimated from the initial slope of the stress–strain triaxial test results shown in Figure 6.8. Two Mohr circles corresponding to failure stresses obtained from the triaxial test results (Figure 6.8) can be plotted. Subsequently, the Mohr–Coulomb failure criterion can be plotted as a straight line that is tangential to the two circles. The soil strength parameters $\phi' = 37°$ and $c' = 0$ MPa can be obtained from the slope and intercept of the Mohr–Coulomb failure criterion.

For plane strain conditions, the Mohr–Coulomb parameters ($\phi' = 37°$ and $c' = 0$ MPa) can be converted to Drucker–Prager parameters as follows:

$$\tan \beta = \frac{3\sqrt{3} \tan \phi'}{\sqrt{9 + 12 \tan^2 \phi'}} \qquad \text{for } \phi' = 37° \rightarrow \beta = 44.56°$$

$$d = \frac{3\sqrt{3}c'}{\sqrt{9 + 12 \tan^2 \phi'}} \qquad \text{for } c' = 0 \rightarrow d = 0$$

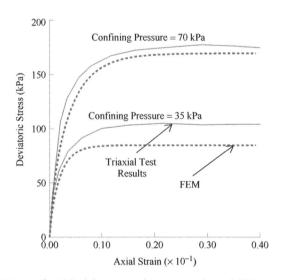

FIGURE 6.8 Triaxial compression test results and FEM results.

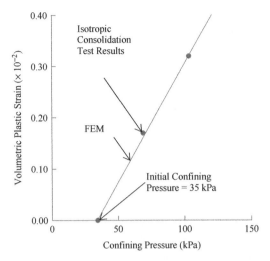

FIGURE 6.9 Isotropic consolidation test results and FEM results.

The cap eccentricity parameter is chosen as $R = 0.4$. The initial cap position (which measures the initial consolidation of the specimen) is taken as $\varepsilon^{pl}_{vol(0)} = 0.0$, and the cap hardening curve is as shown in Figure 6.9 as obtained from an isotropic consolidation test on Ottawa sand. The transition surface parameter $\alpha = 0.05$ is used. These parameters, summarized in Table 6.1, were used to reproduce the stress–strain curves of Ottawa sand under two confining pressures as shown in Figure 6.8. This was done by using one axisymmetric finite element in the same manner described in Chapter 5. The parameters were also used to reproduce the results of an isotropic consolidation test on Ottawa sand as shown in Figure 6.9. It can be noted from Figures 6.8 and 6.9 that the cap model can adequately simulate the behavior of Ottawa sand on the "elemental" level.

We are interested primarily in obtaining the limit foundation pressure and in estimating the vertical displacement under the foundation as a function of load. The analysis is performed using two different numerical formulations: (1) static analysis and (2) dynamic-explicit analysis with very slow loading. Static analysis is the obvious choice since the loads in the current problem are applied statically

TABLE 6.1 Cap Model Parameters

General		Plasticity	
ρ (kg/m^3)	1923	d	10^{-5}
e_0	1.5	β (deg)	44.56 (plane strain)
Elasticity			55.7 (3-D)
E (MPA)	182	R	0.4
ν	0.28	Initial yield	0.0
		α	0.05
		K	1

without inducing any dynamic effects in the model. However, for problems with expected severe distortions, such as the problem in hand, a static analysis may terminate when a few soil elements near the edge of the foundation are distorted excessively. For this class of problems, it is possible to analyze progressive failure and postfailure conditions if dynamic–explicit formulations are used with caution. When the dynamic–explicit mode is used, the loads must be applied very slowly to avoid "exciting" the finite element model. Explicit analyses use very small time increments to ensure stability, making them computationally expensive.

In the beginning of the analysis, gravity loads and surcharge loads are applied to the sand layer. These loads are very important because they will determine the initial stresses in all soil elements. As you know, the soil behavior is dependent on the confining stresses, and the cap model used herein considers this important fact (Chapter 2). In both analyses the foundation load is applied using a constant-downward-velocity boundary condition at the top surface of the foundation with $v = 3.7$ cm/s for a duration of 10 seconds. Other methods of load application are usually available in finite element programs. This may include applying increasing pressures, concentrated loads, or applying forced vertical displacements on the foundation.

The constant-velocity boundary condition applied to the top surface of the foundation causes the foundation to settle at a constant rate. Theoretically, the foundation pressure can be increased gradually up to the failure point (termed the *bearing capacity*) at which a failure surface, similar to the one in Figure 6.4, develops. In the present finite element analysis, such a failure surface is evident when the plastic shear strains are plotted for the at-failure condition as shown in Figure 6.10. From this figure one can immediately notice the presence of a triangular zone directly under the foundation, a radial zone, and a Rankine passive zone resembling the three zones assumed by Terzaghi (Figure 6.4).

The pressure–settlement curve for the static analysis and the dynamic–explicit analysis are shown in Figure 6.11. Both analyses agree well until some soil element in the vicinity of the corner of the foundation have severely distorted. The static analysis was automatically terminated at that stage, whereas the dynamic–explicit analysis continued until a much more advanced stage. Nonetheless, both analyses predicted approximately the same bearing capacity of 440 kPa. For reference, the bearing capacity of 586 kPa predicted by Terzaghi's equation (Example 6.1) is also shown in Figure 6.11. Note that the finite element prediction of bearing capacity is significantly smaller than Terzaghi's bearing capacity. This difference can be attributed to several causes, the most important of which is that Terzaghi's equation assumes that the soil is a rigid–perfectly plastic material that fails abruptly when the bearing capacity of the soil is reached. In contrast, the present finite element analysis assumes that the soil is an elastoplastic material with hardening. Such a material will deform under applied loads, as opposed to a rigid material that does not deform. Also, the soil can yield in a progressive manner due to the nature of the finite element formulation—elements can yield gradually and progressively: A yielding element causes an element next to it to yield, and so on until a shear surface similar to the one shown in Figure 6.10 is realized.

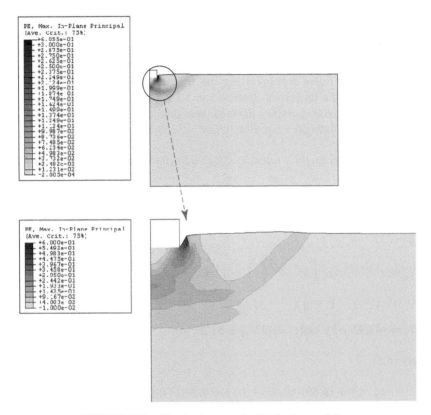

FIGURE 6.10   Plastic shear strain distribution at failure.

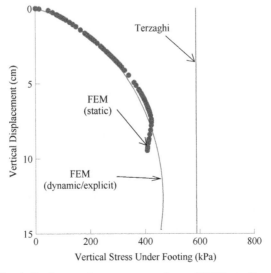

FIGURE 6.11   Load–displacement curve: comparison of FEM results with Terzaghi calculation.

Example 6.3 Bearing Capacity of a Square Foundation (a) Using Terzaghi's equation, calculate the bearing capacity of a 3 m × 3 m foundation on a 12-m-thick homogeneous layer of Ottawa sand ($c' = 0$ and $\phi' = 37°$) underlain by bedrock. The foundation is situated at a depth $D_f = 0.38$ m. The unit weight of soil is 18.14 kN/m^3. (b) Solve part (a) using the finite element method. Assume that Ottawa sand has the properties described in Example 6.2. Use the cap model to simulate sand behavior. Compare finite element prediction of bearing capacity with that predicted by Terzaghi's equation. (*Hint*: Use three-dimensional simulation.)

SOLUTION: (a) *Terzaghi's solution* Use (6.5) (for a square foundation):

$$c' = 0$$

$$\gamma = 18.14 \text{ kN/m}^3, D_f = 0.38 \text{ m} \rightarrow q = \gamma D_f = (18.14)(0.38) = 6.89 \text{ kPa}$$

$$\phi' = 37° \rightarrow$$

$$N_q = e^{\pi \tan 37°} \tan^2 \left(45° + \frac{37°}{2} \right) = 42.92$$

$$N_c = (42.92 - 1) \cot 37° = 55.63$$

$$N_\gamma = (42.92 - 1) \tan[(1.4)(37°)] = 53.27$$

Apply (6.5):

$$q_u = 0 + (6.89)(42.92) + (0.4)(18.14)(3.0)(53.27) = 1474 \text{ kPa}$$

(b) *Finite element solution* (filename: Chapter6_Example3.cae) This example presents a limit equilibrium solution for a layer of Ottawa sand loaded by a rigid, perfectly rough square footing. The cap model (modified Drucker–Prager model with a cap, Chapter 2), with parameters matched to the classical Mohr–Coulomb yield model parameters c' and ϕ', is used to simulate the behavior of Ottawa sand. The problem geometry, boundary conditions, and materials are identical to those of part (a) of Example 6.3, providing a direct means to compare the finite element analysis results with Terzaghi's solution.

The three-dimensional model analyzed is shown in Figure 6.12. The sand layer is 50 m deep and 100 m × 100 m in plan. The loaded area is 3 m × 3 m. The model considers only one-fourth of the sand layer and the loaded area, taking advantage of symmetry as indicated in the figure. The loaded area simulates a foundation with perfect contact with the soil. Reduced-integration, eight-node linear brick elements are used for the sand layer. The base of the sand layer is fixed in all directions. All vertical boundaries are fixed in the horizontal direction but free in the vertical direction. The finite element mesh used in the analysis is shown in Figure 6.13. It is noted that the mesh is finer in the vicinity of the foundation since that zone is the zone of stress concentration. No mesh convergence studies have been performed. However, the dimensions of the sand layer are chosen in a way that the boundary effect on foundation behavior is minimized.

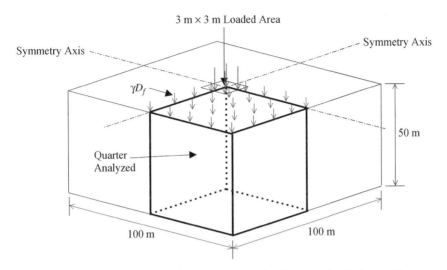

FIGURE 6.12   Idealization of the three-dimensional square foundation problem.

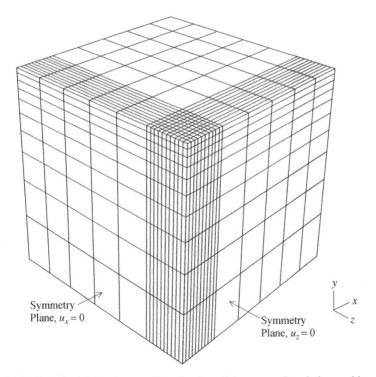

FIGURE 6.13   Finite element discretization of the square foundation problem.

To compare with Terzaghi's equation, the 0.38-m-thick soil layer (foundation depth, D_f) is replaced by an overburden pressure of $q = \gamma D_f = (18.14)(0.38) = 6.89$ kPa. That way the shear resistance of the 0.38-m-thick soil layer will not be considered in the finite element analysis. This is one of the assumptions that Terzaghi used for the derivation of his equation. The elastic response of Ottawa sand is assumed to be linear and isotropic, with a Young's modulus of 182 MPa and a Poisson ratio of 0.3. Young's modulus is estimated from the initial slope of the stress–strain triaxial test results shown in Figure 6.8. The soil strength parameters $\phi' = 37°$ and $c' = 0$ MPa can be obtained from the slope and intercept of the Mohr–Coulomb failure criterion as described in Example 6.2.

In this example, the parameters of a linear Drucker–Prager model, β and d, are matched to the Mohr–Coulomb parameters, ϕ' and c', under triaxial stress conditions—appropriate for the stress conditions of this three-dimensional problem. For triaxial stress conditions the Mohr–Coulomb parameters ($\phi' = 37°$ and $c' = 0$ MPa) can be converted to Drucker–Prager parameters as follows:

$$\tan \beta = \frac{6 \sin \phi'}{3 - \sin \phi'} \qquad \text{for } \phi' = 37° \rightarrow \beta = 55.7°$$

$$d = \frac{18c' \cos \phi'}{3 - \sin \phi'} \qquad \text{for } c' = 0 \rightarrow d = 0$$

The cap eccentricity parameter is chosen as $R = 0.4$. The initial cap position (which measures the initial consolidation of the specimen) is taken as $\varepsilon_{vol(0)}^{pl} = 0.0$, and the cap hardening curve is as shown in Figure 6.9, as obtained from an isotropic consolidation test on Ottawa sand. The transition surface parameter $\alpha = 0.05$ is used. These parameters, summarized in Table 6.1, were used to reproduce the stress–strain curves of Ottawa sand under two confining pressures, as shown in Figure 6.8.

In this example we establish the load–displacement relationship for the 3 m × 3 m footing. The bearing capacity of the footing can be obtained from the load–displacement curve. The analysis is performed using two different numerical formulations: (1) static analysis and (2) dynamic–explicit analysis with very slow loading. Static analysis is generally used for problems with a statically applied load, such as the present problem. It is also possible to analyze the problem at hand using the dynamic–explicit option. This has to be done carefully. The loads must be applied very slowly to avoid exciting the model. Note that explicit analyses use very small time increments automatically to ensure stability, making them computationally expensive.

In the beginning of the analysis, gravity loads and surcharge loads are applied to the sand layer. These loads are very important because they will determine the initial stresses in all soil elements. The soil behavior is stress dependent, and the cap model used herein considers this important fact (Chapter 2). In both analyses the foundation load is applied using a constant downward velocity boundary condition at the top surface of the foundation with $v = 6$ cm/s for a duration of 10 seconds.

The pressure–settlement curve for the static analysis and the dynamic–explicit analysis are shown in Figure 6.14. Both analyses agree well up to the failure load. Both analyses predicted approximately the same bearing capacity of 1350 kPa. For reference, the bearing capacity of 1474 kPa predicted by Terzaghi's equation [part (a) of Example 6.3] is also shown in Figure 6.14. It is noted that the finite element prediction of bearing capacity is slightly smaller than Terzaghi's bearing capacity. The reasons for this difference were described in Example 6.2. Figure 6.15 shows

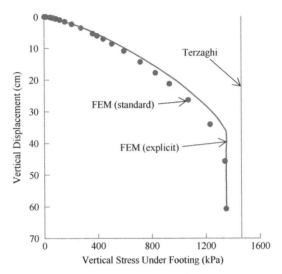

FIGURE 6.14 Load–displacement curve: comparison of FEM results with Terzaghi calculation.

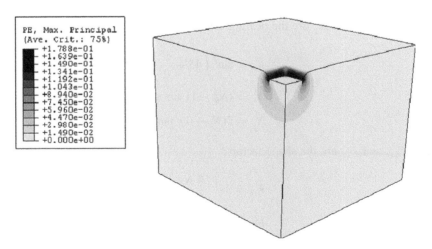

FIGURE 6.15 Plastic shear strain distribution at failure.

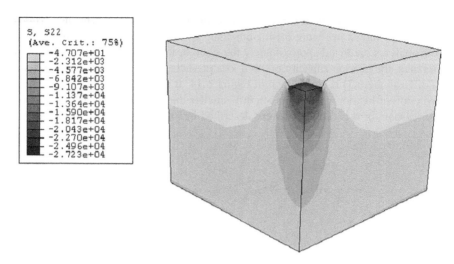

FIGURE 6.16 Vertical stress distribution at failure.

the plastic shear strain contours in the sand layer at failure load, and Figure 6.16 shows the contours of vertical stresses in the sand layer at failure load.

6.4 MEYERHOF'S GENERAL BEARING CAPACITY EQUATION

Meyerhof (1963) developed a generalized bearing capacity equation that includes correction factors for foundation depth, foundation shape, and for inclined loads:

$$q_u = c' N_c F_{cs} F_{cd} F_{ci} + q N_q F_{qs} F_{qd} F_{qi} + \tfrac{1}{2} \gamma B N_\gamma F_{\gamma s} F_{\gamma d} F_{\gamma i} \tag{6.7}$$

where N_c, N_q, and N_γ are the bearing capacity factors:

$$N_q = \tan^2 \left(45 + \frac{\phi'}{2} \right) e^{\pi \tan \phi'}$$

$$N_c = (N_q - 1) \, \cot \phi'$$

$$N_\gamma = 2(N_q + 1) \, \tan \phi'$$

F_{cs}, F_{qs}, and $F_{\gamma s}$ are the shape factors:

$$F_{cs} = 1 + \frac{B}{L} \frac{N_q}{N_c}$$

$$F_{qs} = 1 + \frac{B}{L} \tan \phi'$$

$$F_{\gamma s} = 1 - 0.4 \frac{B}{L}$$

where L and B are the length and width of the foundation, respectively. F_{cd}, F_{qd}, and $F_{\gamma d}$ are the depth factors:

For $D_f/B \leq 1$, use

$$F_{cd} = 1 + 0.4 \frac{D_f}{B}$$

$$F_{qd} = 1 + 2 \tan \phi'(1 - \sin \phi')^2 \frac{D_f}{B}$$

$$F_{\gamma d} = 1$$

For $D_f/B > 1$, use

$$F_{cd} = 1 + 0.4 \tan^{-1} \frac{D_f}{B}$$

$$F_{qd} = 1 + 2 \tan \phi'(1 - \sin \phi')^2 \tan^{-1} \frac{D_f}{B}$$

$$F_{\gamma d} = 1$$

F_{ci}, F_{qi}, and $F_{\gamma i}$ are the inclination factors:

$$F_{ci} = F_{qi} = \left(1 - \frac{\beta^\circ}{90^\circ}\right)^2$$

$$F_{\gamma i} = \left(1 - \frac{\beta}{\phi'}\right)^2$$

where β is the inclination of the load with respect to the vertical as shown in Figure 6.17.

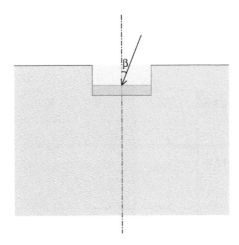

FIGURE 6.17   Inclined load.

Example 6.4 *Bearing Capacity of a Strip Footing with Inclined Load* (a) Using Meyerhof's equation, calculate the bearing capacity of a 2-m-wide strip foundation on a 14.5-m-thick homogeneous layer of Ottawa sand ($c' = 0$ and $\phi' = 37°$) underlain by bedrock. The foundation is subjected to an inclined load making a $15°$ angle with the vertical as shown in Figure 6.18. The foundation is situated at a depth $D_f = 0.5$ m. The unit weight of soil is 19 kN/m³. (b) Solve part (a) using the finite element method. Use the cap model to simulate sand behavior. Assume that Ottawa sand has the properties described in Example 6.2. Compare the finite element prediction of bearing capacity with that predicted by Meyerhof's equation.

SOLUTION: (a) *Meyerhof's solution*   Let's calculate the bearing capacity factors N_c, N_q, and N_γ:

$$N_q = \tan^2\left(45 + \frac{37°}{2}\right) e^{\pi \tan 37°} = 42.9$$

$$N_c = (42.9 - 1)\ \cot 37° = 55.6$$

$$N_\gamma = 2(42.9 + 1)\ \tan 37° = 66.2$$

For this strip foundation, calculate the shape factors F_{cs}, F_{qs}, and $F_{\gamma s}$:

$$F_{cs} = 1 + \left(\frac{2}{\infty}\right)\left(\frac{42.9}{55.6}\right) = 1$$

$$F_{qs} = 1 + \frac{2}{\infty}\tan 37° = 1$$

$$F_{\gamma s} = 1 - 0.4\left(\frac{2}{\infty}\right) = 1$$

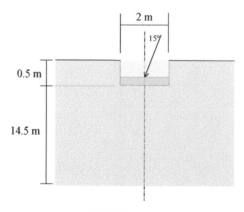

FIGURE 6.18

For $D_f/B \leq 1$, the depth factors F_{cd}, F_{qd}, and $F_{\gamma d}$ are

$$F_{cd} = 1 + 0.4 \left(\frac{0.5}{2} \right) = 1.1$$

$$F_{qd} = 1 + 2(\tan 37°)(1 - \sin 37°)^2 \left(\frac{0.5}{2} \right) = 1.0597$$

$$F_{\gamma d} = 1$$

The inclination factors F_{ci}, F_{qi}, and $F_{\gamma i}$ are

$$F_{ci} = F_{qi} = \left(1 - \frac{15°}{90°} \right)^2 = 0.69$$

$$F_{\gamma i} = \left(1 - \frac{15°}{37°} \right)^2 = 0.35$$

The bearing capacity can be calculated using (6.7):

$$q_u = (0)(55.6)(1)(1.1)(0.69) + (19)(0.5)(42.9)(1)(1.0597)(0.69)$$

$$+ \left(\tfrac{1}{2} \right)(19)(2)(66.2)(1)(1)(0.35) = 738 \text{ kPa}$$

(b) *Finite element solution* (filename: Chapter6_Example4.cae) This example presents a limit equilibrium solution for a layer of sand loaded by a rigid, perfectly rough strip footing with inclined loading. The cap model (modified Drucker–Prager model with a cap, Chapter 2), with parameters matched to the classical Mohr–Coulomb yield model parameters c' and ϕ', is used to simulate the behavior of the sand. The elastic parameters and the cap model parameters of the sand used in this analysis are identical to those of Ottawa sand used in Table 6.1. The concrete foundation is assumed to be linear elastic with a Young's modulus of 1435 MPa and a Poisson ratio of 0.2. The problem geometry, boundary conditions, and materials are identical to those of part (a) of Example 6.4, providing a direct means to compare the finite element analysis results with Meyerhof's solution.

The plane strain finite element model analyzed is shown in Figure 6.19. The sand layer is 15 m deep and 30 m wide. The strip foundation is 2 m wide and 0.5 m thick. The model considers the entire sand layer and foundation because of the inclined load that makes the system asymmetric. The strip foundation has perfect contact with the soil—they share the same nodes at the sand–foundation interface. Reduced-integration, bilinear, plane strain quadrilateral elements are used for the sand and the concrete foundation. The base of the sand layer is fixed in all directions. All vertical boundaries are fixed in the horizontal direction but free in the vertical direction. It is noted that the mesh is finer in the vicinity of the foundation since that zone is the zone of stress concentration. No mesh convergence studies have been performed. However, the dimensions of the sand layer are chosen in a way that the boundary effect on foundation behavior is minimized.

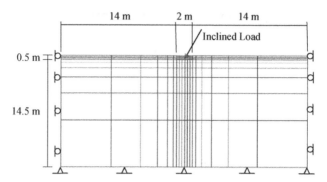

FIGURE 6.19 Finite element discretization.

In the beginning of the analysis, gravity loads are applied to the sand layer. These loads are very important because they will determine the initial stresses in all soil elements. The soil behavior is stress dependent, and the cap model used herein considers this important fact (Chapter 2). During this step of analysis, the "geostatic" command is invoked to make sure that equilibrium is satisfied within the sand layer. The geostatic option makes sure that the initial stress condition in any element within the sand layer falls within the initial yield surface of the cap model. A monotonically increasing point load having a horizontal and a vertical component is applied at the center of the foundation. The resultant of these two components is inclined at a 15° angle from the vertical.

In this example we establish the load–displacement relationship for the 2-m-wide strip footing. The bearing capacity of the footing can be obtained from the load–displacement curve as shown in Figure 6.20. In this figure, the initial and final straight-line portions of the curve are extended. The bearing capacity of the footing is located at the point where those two extensions meet. As shown in Figure 6.20, the finite element analyses predicted a bearing capacity of approximately 675 kPa. The bearing capacity of 738 kPa predicted by Meyerhof's equation [part (a) of Example 6.4] is also shown in the figure. It is noted that the finite element prediction of bearing capacity is slightly smaller than Meyerhof's bearing capacity. This difference can be attributed to several causes, the most important of which is that Meyerhof's equation assumes that the soil is a rigid–perfectly plastic material that fails abruptly when the bearing capacity of the soil is reached. In contrast, the present finite element analysis assumes that the soil is an elastoplastic material that deforms under applied loads, as opposed to a rigid material that does not deform. Furthermore, the soil can yield in a progressive manner, due to the nature of the finite element formulation, where elements can yield in a gradual and progressive manner: A yielding element causes an element next to it to yield, and so on until a slip surface is realized.

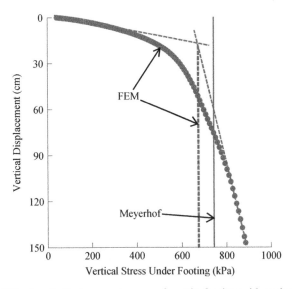

FIGURE 6.20   Load–displacement curve of a strip footing with an inclined load.

6.5   EFFECTS OF THE WATER TABLE LEVEL ON BEARING CAPACITY

Both Terzaghi's equation and Meyerhof's equation need to be adjusted when the water table level is close to the foundation. There are three cases to be considered: Case 1 is when the water table level is above the foundation level, case 2 is when the water table level is between the foundation level and a distance B (= width of foundation) below the foundation level, and case 3 is when the water level is lower than a distance B below the foundation level. The three cases are illustrated in Figure 6.21.

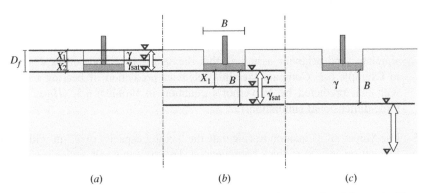

FIGURE 6.21   Effects of groundwater table level on bearing capacity: (a) case 1; (b) case 2; (c) case 3.

For case 1, two adjustments need to be made in (6.1) and (6.7). First, the surcharge $q = \gamma D_f$ in the second term of both equations needs to be replaced by the effective surcharge $q = X_1\gamma + X_2(\gamma_{sat} - \gamma_w)$, where X_1 and X_1 are as defined in Figure 6.21a. Second, the unit weight of soil, γ, needs to be replaced by $\gamma' = \gamma_{sat} - \gamma_w$ in the third term of both equations.

For case 2 we need to make only one adjustment in (6.1) and (6.7). The unit weight of soil, γ, in the third term of both equations needs to be replaced by $\overline{\gamma} = \gamma' + (X_1/B)(\gamma - \gamma')$, where $\gamma' = \gamma_{sat} - \gamma_w$. The distance X_1 is defined in Figure 6.21b.

For case 3 (Figure 6.21c) we do not need to make any adjustments. In this case the water table is too deep to have an effect on the bearing capacity.

PROBLEMS

6.1   Using Terzaghi's equation, calculate the bearing capacity of a 2.0-m-wide strip foundation on a thick homogeneous layer of sand with $c' = 0$ and $\phi' = 37°$. The foundation is situated at a depth of 2 m. The unit weight of the sand is 18.14 kN/m^3. The groundwater table is very deep.

6.2   Solve Problem 6.1 using the finite element method. Compare the finite element analysis results with the analytical solution obtained in Problem 6.1. The concrete foundation is assumed to be linear elastic with a Young's modulus of 1435 MPa and a Poisson ratio of 0.2. The elastic response of the sand is assumed to be linear and isotropic, with a Young's modulus of 182 MPa and a Poisson ratio of 0.3. The elastoplastic behavior of the sand can be simulated using the cap model. Assume the same cap model parameters as those used in Table 6.1.

6.3   Use Meyerhof's equation to calculate the bearing capacity of a 2 m × 4 m foundation on a 30-m-thick homogeneous layer of sand, with $c' = 0$ and $\phi' = 37°$, underlain by bedrock. The foundation is situated at a depth $D_f = 2$ m. The unit weight of the sand is 19 kN/m^3. The groundwater table is very deep.

6.4   Solve Problem 6.3 using the finite element method. Use the cap model to simulate sand behavior assuming that the sand has the properties described in Example 6.2. Compare the finite element prediction of bearing capacity with that predicted by Meyerhof's equation in Problem 6.3. (*Hint*: Use a three-dimensional simulation.)

6.5   Use Meyerhof's equation to calculate the bearing capacity of a 2-m-wide strip foundation on a 20-m-thick homogeneous layer of sandy silt, with $c' = 20$ kPa and $\phi' = 30°$, underlain by bedrock. The foundation is situated at a depth $D_f = 1$ m and subjected to an inclined loading making 20° angle with the vertical. The unit weight of the sand is 18 kN/m^3. The groundwater table is very deep.

6.6 Solve Problem 6.5 using the finite element method. Compare the finite element analysis results with the analytical solution obtained in Problem 6.5. The concrete foundation is assumed to be linear elastic with a Young's modulus of 2000 MPa and a Poisson ratio of 0.2. The elastic response of the sandy silt is assumed to be linear and isotropic, with a Young's modulus of 200 MPa and a Poisson ratio of 0.3. The elastoplastic behavior of the sandy silt can be simulated using the cap model. Calculate d and β for plane strain condition using $c' = 20$ kPa and $\phi' = 30°$. Assume that $R = 0.3$, $e_0 = 1$, $K = 1$, and $\alpha = 0$. Use the hardening parameters given in Figure 6.9. (*Hint*: Use a plane strain condition.)

6.7 For the 3.0-m-wide strip foundation shown in Figure 6.22, calculate the bearing capacity considering the three cases indicated. In the winter the water table is 1.5 m below the ground surface. The water table drops to 4 m below the ground surface in the spring, and to 6.5 m in the summer. The foundation is situated at a depth of 3 m and is supported by a thick homogeneous layer of sandy silt with $c' = 15$ kPa and $\phi' = 33°$. The saturated unit weight of the sandy silt is 19 kN/m^3, and its dry unit weight is 17.5 kN/m^3.

6.8 Calculate the bearing capacity of a 2-m-wide strip foundation situated on top of a 4-m-high embankment shown in Figure 6.23. The embankment soil is a compacted sandy silt with $c' = 40$ kPa and $\phi' = 20°$. A thick layer of a very stiff clay lies below the embankment. The concrete foundation is assumed to be linear elastic with a Young's modulus of 2000 MPa and a Poisson ratio of 0.2. The elastic response of the sandy silt is assumed to be linear and

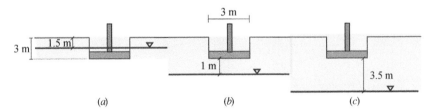

FIGURE 6.22 Groundwater level in (*a*) winter, (*b*) spring, and (*c*) summer.

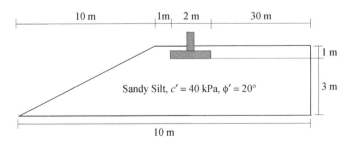

FIGURE 6.23

isotropic, with a Young's modulus of 200 MPa and a Poisson ratio of 0.3. The elastoplastic behavior of the sandy silt can be simulated using the cap model. Calculate d and β for plane strain condition using $c' = 40$ kPa and $\phi' = 20°$. Assume that $R = 0.3$, $e_0 = 1$, $K = 1$, and $\alpha = 0$. Use the hardening parameters given in Figure 6.9. The groundwater table is very deep. (*Hint*: Use a plane strain condition.)

CHAPTER 7

LATERAL EARTH PRESSURE AND RETAINING WALLS

7.1 INTRODUCTION

Retaining walls are structural members used to support vertical or nearly vertical soil backfills. This is usually needed when there is a change of grade. A retaining wall, for example, can be used to retain the backfill required to widen a roadway as shown in Figure 7.1a or to retain a backfill that is used to support a structure as shown in Figure 7.1b. There are several types of retaining walls, including the traditional gravity, semigravity, cantilever, and counterfort retaining walls that are made of plain and reinforced concrete (Figure 7.2). Retaining walls can also be constructed from other materials, such as gabions, reinforced earth, steel, and timber. Nevertheless, all retaining walls have to be designed to resist the external forces applied, and that includes lateral earth pressure, surcharge load, hydrostatic pressure, and earthquake loads.

There are three possible types of lateral earth pressure that can be exerted on a retaining wall: at-rest pressure, active pressure, and passive pressure. *At-rest earth pressure* occurs when the retaining wall is not allowed to move away or toward the retained soil. *Active earth pressure* occurs when the retaining wall is permitted to move away from the retained soil. The active earth pressure produces a destabilizing earth force behind retaining walls. *Passive earth pressure*, on the other hand, develops when the retaining wall is forced to move toward the retained soil.

Granular soils are usually used as backfill materials behind retaining walls because of their high permeability (freely draining materials). Using granular fills along with weep holes (small openings in the retaining wall) prevents the accumulation of hydrostatic pressures during a rainfall, for example. Other types of

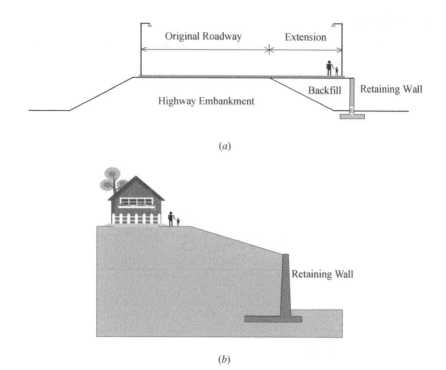

(a)

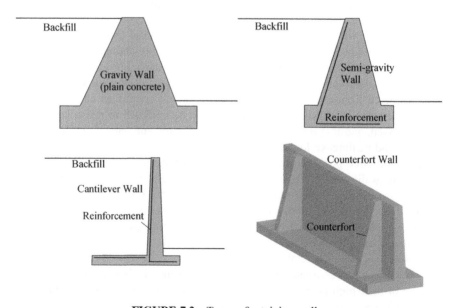

(b)

FIGURE 7.1   Retaining walls.

FIGURE 7.2   Types of retaining walls.

drains are available, including perforated pipes that are commonly installed at the bottom of the granular fill directly behind the heel of the retaining wall to facilitate drainage. Soils such as sand, gravel, silty sand, and sand with gravel are good backfill materials. Cohesive soils are not desirable as backfill materials because of their low permeability.

In Chapter 3 you have learned how to calculate the variation of the effective vertical stress in a soil strata. Now we can calculate the effective lateral earth pressure by multiplying the effective vertical stress by an appropriate lateral earth pressure coefficient: at-rest, active, or passive. The at-rest lateral earth pressure coefficient, K_0, can be estimated using the empirical equation $K_0 = 1 - \sin \phi'$ (Jaky, 1944), where ϕ' is the internal friction angle of the backfill soil. The active lateral earth pressure coefficient, K_a, can be calculated using the equation $K_a = (1 - \sin \phi')/(1 + \sin \phi')$, whereas the passive lateral earth pressure coefficient, K_p, can be calculated using the equation $K_p = (1 + \sin \phi')/(1 - \sin \phi')$. It can be seen from these equations that both the at-rest earth pressure coefficient and the active earth pressure coefficient are smaller than 1, and the passive earth

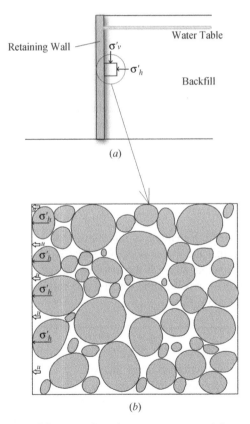

FIGURE 7.3 Lateral earth pressure on a retaining wall.

pressure coefficient is greater than 1. This means that in soils the lateral stress is not equal to the vertical stress for any possible condition: at-rest, active, and passive. This is unlike the stresses inside a fluid, which are equal in all directions, as you have learned in fluid mechanics.

If water is present in the backfill behind a retaining wall, the water pressure should be added to the effective lateral earth pressure exerted on the wall. Figure 7.3 shows details of a soil element in contact with a retaining wall. Two types of pressure are exerted on the wall by this fully saturated soil element located at some distance below the water table. The first pressure is the one exerted by the grains and denoted as σ'_h in the figure. This is by definition the intergranular stress, known also as the effective stress. The other type of pressure is the pore water pressure caused by the water filling the interconnected voids (pores) and denoted by u in the same figure. The retaining wall "feels" both σ'_h and u, therefore, these two pressures should be considered when calculating the lateral earth pressure exerted on a retaining wall.

7.2 AT-REST EARTH PRESSURE

Consider a soil element in contact with a retaining wall that is restricted from lateral movement (Figure 7.4a). The at-rest lateral earth pressure is calculated by multiplying the effective vertical stress by the coefficient of lateral earth pressure at rest, K_0. This coefficient can be determined experimentally for a given soil by restraining the soil from lateral movement and subjecting it to an effective vertical stress σ'_v. This can be done using a rigid container filled with soil. The resulting effective horizontal stress, σ'_h, exerted on the sides of the rigid container can be measured by means of a load or pressure cell. Consequently, the coefficient of lateral earth pressure at-rest can be calculated as

$$K_0 = \frac{\sigma'_h}{\sigma'_v} \tag{7.1}$$

The stress state in the soil element is represented by a Mohr's circle as shown in Figure 7.4b. Note that the major principal stress is the effective vertical stress, whereas the minor principal stress is the effective lateral stress because K_0 is less that 1. In the absence of experimental results, one can use the empirical equation by Jaky (1944) to estimate K_0:

$$K_0 = 1 - \sin \phi' \tag{7.2}$$

where ϕ' is the internal friction angle of the soil.

Let's consider the retaining wall supporting a dry fill of a height H as shown in Figure 7.5a. The wall is restricted from any lateral motion: it is not allowed to move away or toward the backfill. Therefore, the lateral earth pressure exerted by the backfill against the retaining wall can be regarded as at-rest lateral earth pressure

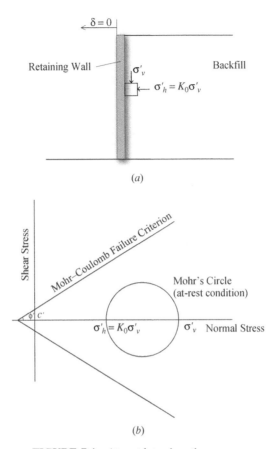

FIGURE 7.4   At-rest lateral earth pressure.

that can be calculated using (7.1). Note that the effective stresses in this case are the same as the total stresses because of the absence of water (the water table is deep). The pressure felt by the wall is the effective lateral earth pressure only.

From (7.1) we can write $\sigma'_h = K_0\sigma'_v$. But $\sigma'_v = \gamma z$, where γ is the unit weight of the backfill soil and z is the depth measured from the top surface of the back-fill. Combining the two equations, we get $\sigma'_h = K_0\gamma z$. This equation indicates that the effective lateral earth pressure increases linearly with depth as shown in Figure 7.5b. At the top surface, $\sigma'_h = 0$ because $z = 0$. At the bottom of the retain-ing wall, $\sigma'_h = K_0\gamma H$ because $z = H$. The at-rest lateral earth pressure distribution is therefore triangular, as shown in Figure 7.5b. The at-rest lateral force (P_0) exerted on the wall is the area of the lateral earth pressure triangle: $P_0 = \frac{1}{2}K_0\gamma H^2$. The point of application of P_0 is located at a distance of $H/3$ from the bottom of the retaining wall because of the triangular pressure distribution.

When water is present within the backfill, as may happen during a rainfall with the drains clogged, the effect of pore water pressure needs to be accounted

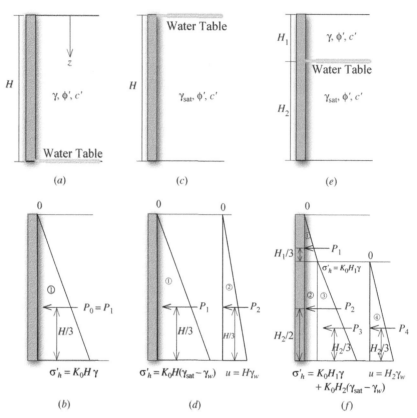

FIGURE 7.5   At-rest lateral earth pressure for a backfill without surcharge: (a, b) low water table; (c, d) high water table; (e, f) intermediate water table.

for. Figure 7.5c shows a retaining wall supporting a backfill with a water table that is coincident with the top surface of the fill. In this case the lateral pressure exerted on the wall consists of two parts: the effective lateral earth pressure (at rest) and the pore water pressure. This is illustrated in Figure 7.5d, in which the effective lateral earth pressure and the pore water pressure are plotted separately. The effective lateral earth pressure is $\sigma'_h = K_0(\gamma_{sat} - \gamma_w)z$. This equation indicates that the effective lateral earth pressure increases linearly with depth. At the top surface, $\sigma'_h = 0$ because $z = 0$. At the bottom of the retaining wall, $\sigma'_h = K_0(\gamma_{sat} - \gamma_w)H$ because $z = H$. The pore water pressure (u) varies linearly from $u = 0$ at $z = 0$ to $u = \gamma_w H$ at $z = H$. The at-rest lateral force (P_0) exerted on the wall is the sum of the area of the effective lateral earth pressure triangle and the area of the pore water pressure triangle: $P_0 = \frac{1}{2}K_0(\gamma_{sat} - \gamma_w)H^2 + \frac{1}{2}\gamma_w H^2$. The point of application of P_0 is located at a distance of $H/3$ from the bottom of the retaining wall because of the triangular distribution of both the effective lateral earth pressure and the pore water pressure.

Figure 7.5*e* shows a retaining wall supporting a fill with a water table that is at a distance H_1 below the top surface of the fill. In this case we can assume that the soil below the water table is fully saturated. The distribution of the effective lateral earth pressure and of the pore water pressure are shown in Figure 7.5*f*.

At $z = 0$:

$$\sigma'_h = 0 \quad \text{and} \quad u = 0$$

At $z = H_1$:

$$\sigma'_h = K_0 \gamma H_1 \quad \text{and} \quad u = 0$$

At $z = H_1 + H_2 = H$:

$$\sigma'_h = K_0 \gamma H_1 + K_0 (\gamma_{\text{sat}} - \gamma_w) H_2 \quad \text{and} \quad u = \gamma_w H_2$$

To calculate the lateral force, we can simplify the pressure distribution using triangular and rectangular shapes as shown in Figure 7.5*f*. The area of each shape can be calculated separately to give us the lateral force that must be applied at the centroid of the respective shape. The sum of these lateral forces is the total lateral force exerted on the wall. The point of application of the total lateral force can be calculated by taking the sum of the moments of all forces about the heel of the retaining wall and equating that to the moment of the total lateral force (resultant) about the heel. This will be illustrated in Example 7.1.

A uniform pressure (surcharge) is sometimes applied at the top surface of the backfill. For example, it is customary to apply a uniform pressure of 15 to 20 kPa to simulate the effect of traffic loads on a retaining wall. A uniform pressure q applied at the top surface causes an increase in the at-rest lateral earth pressure equal to $K_0 q$. Figure 7.6 shows how to account for the surcharge effects for a retaining wall with a dry backfill, a submerged backfill, and a partially submerged backfill.

Example 7.1 As shown in Figure 7.7*a*, a 4.5-m-high retaining wall has a sandy backfill with $c' = 0$, $\phi' = 37°$, $\gamma = 17$ kN/m^3, and $\gamma_{\text{sat}} = 19$ kN/m^3. The water table is 1.5 m below the surface of the backfill. Calculate the at-rest lateral force exerted on a 1-m-long section of the retaining wall. Assume that the sand below the water table is fully saturated.

SOLUTION: Figure 7.7*b* shows the distribution of the effective lateral earth pressure and the distribution of the pore water pressure against the retaining wall. The coefficient of lateral earth pressure at rest is calculated using Jaky's equation, $K_0 = 1 - \sin \phi' = 1 - \sin 37° = 0.4$. Thus:

At $z = 0$:

$$\sigma'_h = 0 \quad \text{and} \quad u = 0$$

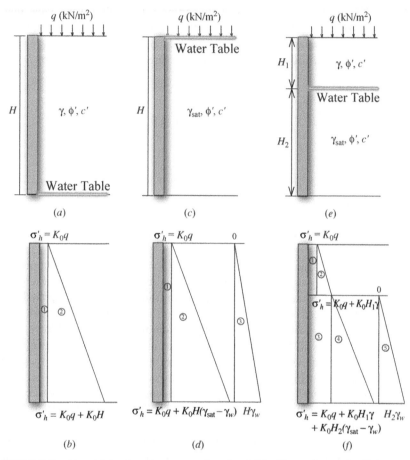

FIGURE 7.6 At-rest lateral earth pressure for a backfill with surcharge: (a, b) low water table; (c, d) high water table; (e, f) intermediate water table.

At $z = 1.5$ m:

$$\sigma'_h = K_0 \gamma H_1 = (0.4)(17)(1.5) = 10.2 \text{ kN/m}^2 \quad \text{and} \quad u = 0$$

At $z = 4.5$ m:

$$\sigma'_h = K_0 \gamma H_1 + K_0(\gamma_{\text{sat}} - \gamma_w)H_2 = (0.4)(17)(1.5) + 0.4(19 - 9.81)(3)$$

$$= 21.2 \text{ kN/m}^2 \quad \text{and} \quad u = \gamma_w H_2 = (9.81)(3) = 29.4 \text{ kN/m}^2$$

The at-rest lateral earth pressure diagram is divided into two triangles and one rectangle, as indicated in Figure 7.7b. The pore water pressure distribution is triangular by nature (does not need further simplification). The lateral forces P_1, P_2, P_3,

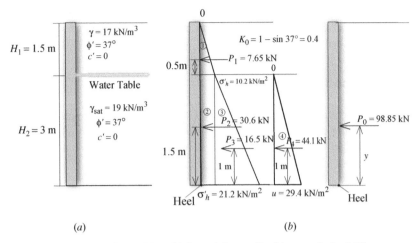

FIGURE 7.7 A 4.5-m-high retaining wall with a sandy backfill.

and P_4 are equal to their respective areas denoted as areas, 1, 2, 3, and 4 in the figure:

$P_1 = (1/2)(10.2 \text{ kN/m}^2)(1.5\text{m})(1\text{m}) = 7.65 \text{ kN}$ (for a 1-m-long section)

$P_2 = (10.2 \text{ kN/m}^2)(3\text{m})(1\text{m}) = 30.6 \text{ kN}$

$P_3 = (1/2)(11 \text{ kN/m}^2)(3\text{m})(1\text{m}) = 16.5 \text{ kN}$

$P_4 = (1/2)(29.4 \text{ kN/m}^2)(3\text{m})(1\text{m}) = 44.1 \text{ kN}$

The at-rest lateral force is

$$P_0 = P_1 + P_2 + P_3 + P_4 = 7.65 + 30.6 + 16.5 + 44.1 = 98.85 \text{ kN}$$

The location (y) of the point of application of the resultant P_0, measured from the heel of the retaining wall, can be calculated by taking the moment of P_0 about the heel and equating that to the sum of the moments of P_1, P_2, P_3, and P_4 about the same point:

$$(98.85)y = (7.65)(3 + 0.5) + (30.6)(1.5) + (16.5)(1) + (44.1)(1) \rightarrow y = 1.35 \text{ m}$$

7.3 ACTIVE EARTH PRESSURE

The active earth pressure is a destabilizing pressure that occurs when the retaining wall is allowed to move away from the retained soil. The active earth pressure condition will develop only if the wall moves a sufficient distance away from the backfill. The lateral outward movement required to mobilize the full active pressure condition is approximately $0.001H$ for loose sand and $0.04H$ for soft clay, where H is the height of the wall. The active earth pressure distribution can be calculated using Rankine theory or Coulomb theory.

7.3.1 Rankine Theory

The Rankine active earth pressure theory (Rankine, 1857) assumes that the soil behind a retaining wall is in a condition of incipient failure. Every element within the sliding wedge, shown in Figure 7.8a, is on the verge of failure as depicted by

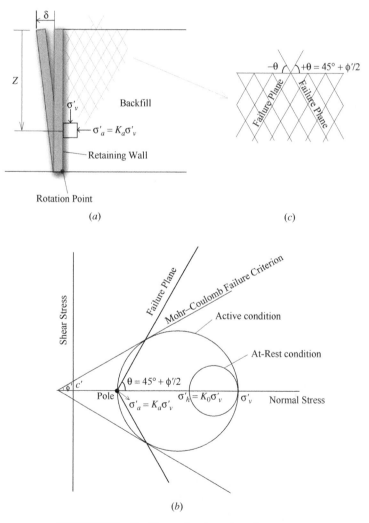

FIGURE 7.8 Rankine active lateral earth pressure.

the Mohr's circle, which is tangent to the Mohr–Coulomb failure criterion as shown in Figure 7.8b. This Mohr's circle represents the stress conditions in a soil element located at a depth z below the top surface of the backfill. For reference, Mohr's circle for the at-rest condition is also shown in the figure. The at-rest stress condition occurs when the wall is stationary. When the wall starts moving away from the backfill, the lateral stress, σ'_h, decreases while the vertical stress, σ'_v, remains essentially constant. The limiting condition occurs when the Mohr–Coulomb failure criterion becomes tangent to Mohr's circle, at which condition the lateral stress reaches its minimum possible value, termed the *active lateral stress* σ'_a.

The Mohr's circle shown in Figure 7.8b predicts two sets of failure planes with $\theta = \pm(45° + \phi'/2)$. These two families of failure planes are also shown in Figure 7.8a and c. To simplify the analysis, Rankine theory assumes that there is no adhesion or friction between the wall and the backfill soil, the retaining wall is vertical, and the backfill soil has a horizontal surface. From the limiting failure condition represented by Mohr's circle shown in Figure 7.9, we can write

$$\sin\phi' = \frac{OB}{OA} = \frac{(\sigma'_v - \sigma'_a)/2}{c'\cot\phi' + (\sigma'_v + \sigma'_a)/2} \tag{7.3}$$

After some manipulation of (7.3), we can show that

$$\sigma'_a = \sigma'_v \tan^2\left(45° - \frac{\phi'}{2}\right) - 2c'\tan\left(45° - \frac{\phi'}{2}\right) \tag{7.4}$$

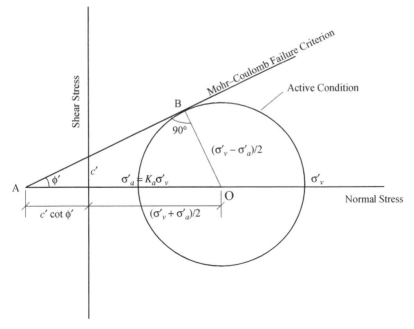

FIGURE 7.9 Mohr's circle for active lateral earth pressure conditions.

or

$$\sigma_a' = K_a\sigma_v' - 2c'\sqrt{K_a} \tag{7.5}$$

where K_a is the active earth pressure coefficient, given by

$$K_a = \tan^2\left(45° - \frac{\phi'}{2}\right) \tag{7.6}$$

For the case of a moist backfill soil with a deep water table (Figure 7.10a), the effective vertical stress can be calculated as $\sigma_v' = \gamma z$, where γ is the moist unit weight of the backfill soil. Substituting this equation into (7.5) yields

$$\sigma_a' = K_a\gamma z - 2c'\sqrt{K_a} \tag{7.7}$$

Figure 7.10b shows the active earth pressure distribution exerted on the retaining wall. The pressure is divided into two parts: a positive part, $K_a\gamma z$, and a negative part, $-2c'\sqrt{K_a}$ [consistent with (7.7)]. The total earth pressure is the sum of the two. Note that there is a tension zone in the backfill soil near the top of the retaining wall, as indicated by the active earth pressure distribution shown in Figure 7.10b. This causes a tension crack that is usually observed at the top of a cohesive backfill immediately behind the retaining wall. The depth of the tension crack, Z_c, can be calculated by setting σ_a' equal to zero in (7.7):

$$Z_c = \frac{2c'}{\gamma\sqrt{K_a}} \tag{7.8}$$

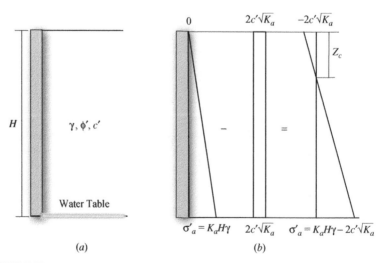

FIGURE 7.10 Rankine active lateral earth pressure for a cohesive backfill without surcharge.

Figure 7.11a shows a retaining wall supporting a fill with a water table located at a distance H_1 below the top surface of the fill. A uniform pressure q is applied at the top surface of the backfill. The wall is allowed to move away from the soil, thus, mobilizing the active earth pressure. Let's assume that the soil below the water table is fully saturated. The distribution of the effective lateral earth pressure and the pore water pressure are shown in Figure 7.11b.

At $z = 0$:

$$\sigma_a' = K_a q - 2c'\sqrt{K_a} \quad \text{and} \quad u = 0$$

At $z = H_1$:

$$\sigma_a' = K_a q + K_a H_1 \gamma - 2c'\sqrt{K_a} \quad \text{and} \quad u = 0$$

At $z = H_1 + H_2 = H$:

$$\sigma_h' = K_a q + K_a \gamma H_1 + K_a(\gamma_{\text{sat}} - \gamma_w)H_2 - 2c'\sqrt{K_a} \quad \text{and} \quad u = \gamma_w H_2$$

To calculate the lateral force, we can simplify the pressure distribution using triangular and rectangular shapes as was done in Example 7.1. The area of each shape can be calculated separately to give us the lateral force that must be applied at the centroid of the respective shape. The sum of these lateral forces is the total lateral force exerted on the wall. The point of application of the total lateral force can be

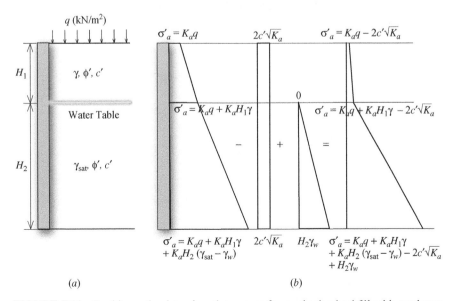

(a) (b)

FIGURE 7.11   Rankine active lateral earth pressure for a cohesive backfill with surcharge.

calculated by taking the sum of the moments of all forces about the heel of the retaining wall and equating that to the moment of the total lateral force (resultant) about the heel.

7.3.2 Coulomb Theory

The Coulomb active earth pressure theory (Coulomb, 1776) assumes that a soil wedge with a failure plane making a critical angle θ_{cr} with the horizontal is in a condition of incipient failure, as shown in Figure 7.12. The intact wedge is about to slide down the failure plane, thus generating active earth pressure against the retaining wall. From statics, the wedge can be treated as a particle that is subjected to three coplanar forces, W, P, and R, where W is the self-weight of the wedge, which can be determined easily by multiplying the area of the wedge with the unit weight of the backfill; P is the resultant of the wall reaction against the soil; and R is the soil reaction against the sliding wedge. The force P makes an angle δ with the normal to the back face of the wall (the face in contact with the backfill). The angle δ is the wall–soil interface friction angle, which can be determined from laboratory tests or can be assumed as a fraction of ϕ'; usually, you can assume that $\delta \approx \frac{2}{3}\phi'$. The force R makes an angle ϕ' with the normal to the failure plane. The internal friction angle ϕ' is assumed to be mobilized at the interface between the sliding wedge and the underlying soil.

It is to be noted that in Coulomb theory the backfill soil is assumed to be failing only along a failure plane, whereas in Rankine theory every element within the backfill soil is on the verge of failure, as discussed earlier. Also, Coulomb theory accounts for three factors that were not accounted for in the original Rankine theory: the wall–soil interface friction signified by the friction angle δ in Figure 7.12, the

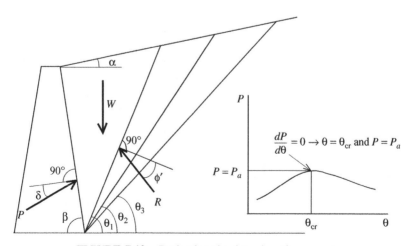

FIGURE 7.12 Coulomb active lateral earth pressure.

sloping backfill soil with an angle α, and the battered back face of the retaining wall with an angle β.

As shown in Figure 7.12, Coulomb theory considers a trial wedge with a failure plane making an angle θ with the horizontal. At failure, two equilibrium equations can be written (along the x and y axes) assuming an "impending motion" condition: that is, the wedge is about to slide down the failure plane. Those two equations involve three unknown: P, R, and θ; we need a third equation involving some (or all) of the three variables to solve for the three unknowns. Note that P, W, and R will vary if a different angle θ is used. Coulomb theory searches for the angle $\theta = \theta_{cr}$ that yields the maximum earth pressure $P = P_a$. Thus,

$$\frac{dP}{d\theta} = 0 \qquad (7.9)$$

The solution of (7.9) will yield the angle θ_{cr}, and subsequently, the corresponding maximum earth pressure P_a can be calculated.

Let's consider the simple case shown in Figure 7.13, in which the retaining wall has a vertical back face ($\beta = 90°$) with a frictionless surface ($\delta = 0$) and a horizontal backfill soil ($\alpha = 0$). Note that these are the assumptions used in Rankine theory, and the following Coulomb solution can be compared directly with the Rankine solution described in Section 7.3.1. The height of the wall is H and the friction angle of the granular backfill soil is ϕ' ($c' = 0$). Consider a wedge with an angle θ as shown in Figure 7.13. The closed force triangle shown in the same figure indicates equilibrium of the wedge that is on the verge of failure. The weight of the wedge is function of the angle θ as follows:

$$W = \frac{\gamma H^2}{2 \tan \theta} \qquad (7.10)$$

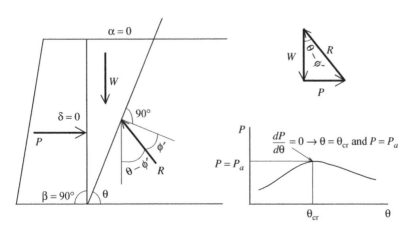

FIGURE 7.13 Coulomb active lateral earth pressure for a frictionless vertical wall and a horizontal backfill.

For horizontal and vertical equilibrium, respectively, the following equations must be satisfied:

$$P = R \sin(\theta - \phi') \tag{7.11}$$

$$W = R \cos(\theta - \phi') \tag{7.12}$$

Substitute (7.10) into (7.12):

$$\frac{\gamma H^2}{2 \tan \theta} = R \cos(\theta - \phi') \tag{7.13}$$

Dividing (7.11) by (7.13), we get

$$P = \frac{\gamma H^2}{2 \tan \theta} \tan(\theta - \phi') \tag{7.14}$$

Combining (7.9) and (7.14) yields

$$\frac{dP}{d\theta} = 0 \rightarrow \theta = \theta_{cr} = 45° + \frac{\phi'}{2} \tag{7.15}$$

Substituting (7.15) into (7.14) yields

$$P_a = \frac{\gamma H^2}{2} \tan^2\left(45° - \frac{\phi'}{2}\right) \tag{7.16}$$

But for a granular backfill, the active force P_a is given by

$$P_a = K_a \frac{\gamma H^2}{2} \tag{7.17}$$

Substituting (7.17) into (7.16), we get

$$K_a = \tan^2\left(45° - \frac{\phi'}{2}\right) \tag{7.18}$$

Equation (7.18) is identical to (7.6), which was obtained from Rankine theory.

The Coulomb active earth pressure coefficient for the general case shown in Figure 7.12 is a more complicated expression that depends on the angle of the back face of the wall, the soil–wall friction, and the angle of backfill slope:

$$K_a = \frac{\sin^2(\beta + \phi')}{\sin^2\beta \sin(\beta - \delta)\left[1 + \sqrt{\dfrac{\sin(\phi' + \delta)\sin(\phi' - \alpha)}{\sin(\beta - \delta)\sin(\alpha + \beta)}}\right]^2} \tag{7.19}$$

7.4 PASSIVE EARTH PRESSURE

Passive earth pressure occurs when a retaining wall is forced to move toward the retained soil. Passive earth pressure condition will develop if the wall moves a sufficient distance toward the backfill. The lateral inward movement required to mobilize the full passive pressure condition is approximately $0.01H$ for loose sand and $0.05H$ for soft clay, where H is the height of the wall. The passive earth pressure distribution can be calculated using Rankine theory or Coulomb theory.

7.4.1 Rankine Theory

Rankine theory assumes that there is no adhesion or friction between the wall and the backfill soil, the retaining wall is vertical, and the backfill soil has a horizontal surface. The Rankine passive earth pressure theory assumes that every soil element within the sliding wedge (Figure 7.14a) is on the verge of failure, as depicted by Mohr's circle shown in Figure 7.14b. This Mohr's circle represents the stress conditions in a soil element located at a depth z below the top surface of the backfill. For reference, Mohr's circle for the at-rest condition is also shown in the figure. The at-rest stress condition occurs when the wall is stationary. When the wall starts moving toward the backfill, the lateral stress, σ'_h, increases while the vertical stress, σ'_v, remains essentially constant. The limiting condition occurs when the Mohr–Coulomb failure criterion becomes tangent to Mohr's circle; at this condition the lateral stress reaches its maximum possible value, termed the *passive lateral stress* σ'_p. Mohr's circle predicts two sets of failure planes with $\theta = \pm(45° - \phi'/2)$, as indicated in Figure 7.14a and b.

From the limiting failure condition represented by the Mohr's circle shown in Figure 7.14b, we can show that

$$\sigma'_p = \sigma'_v \tan^2\left(45° + \frac{\phi'}{2}\right) + 2c' \tan\left(45° + \frac{\phi'}{2}\right) \tag{7.20}$$

or

$$\sigma'_p = K_p \sigma'_v + 2c'\sqrt{K_p} \tag{7.21}$$

where K_p is the passive earth pressure coefficient, given by

$$K_p = \tan^2\left(45° + \frac{\phi'}{2}\right) \tag{7.22}$$

For the case of a moist backfill soil with a deep water table (Figure 7.15a), the effective vertical stress can be calculated as $\sigma'_v = \gamma z$, where γ is the moist unit weight of the backfill soil. Substituting this equation into (7.21) yields

$$\sigma'_p = K_p \gamma z + 2c'\sqrt{K_p} \tag{7.23}$$

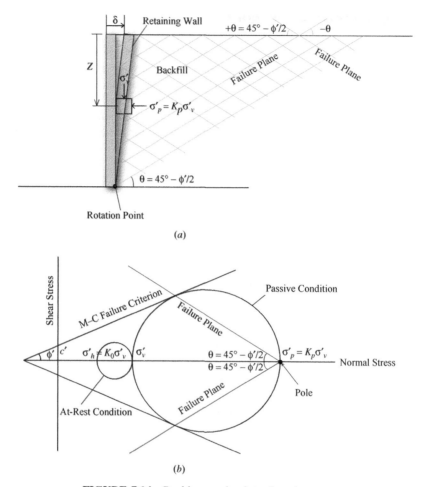

FIGURE 7.14   Rankine passive lateral earth pressure.

Figure 7.15*b* shows the passive earth pressure distribution exerted on the retaining wall. The pressure is divided into two parts: $K_p \gamma z$ and $2c'\sqrt{K_p}$ [consistent with (7.23)]. The total earth pressure is the sum of the two.

Figure 7.16*a* shows a retaining wall supporting a fill with a water table located a distance H_1 below the top surface of the fill. A uniform pressure q is applied at the top surface of the backfill. The wall is forced toward the soil, thus mobilizing passive earth pressure. Let's assume that the soil below the water table is fully saturated. The distribution of the effective lateral earth pressure and the pore water pressure are shown in Figure 7.16*b*.

At $z = 0$:

$$\sigma'_p = K_p q + 2c'\sqrt{K_p} \quad \text{and} \quad u = 0$$

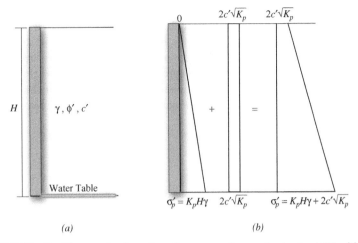

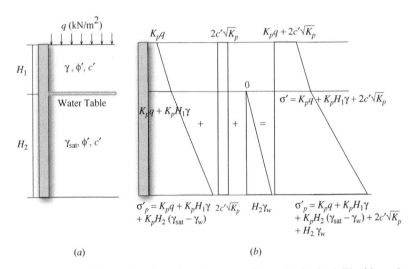

FIGURE 7.15   Rankine passive lateral earth pressure for a cohesive backfill without surcharge.

FIGURE 7.16   Rankine passive lateral earth pressure for a cohesive backfill with surcharge.

At $z = H_1$:

$$\sigma'_p = K_p q + K_p \gamma H_1 + 2c'\sqrt{K_p} \quad \text{and} \quad u = 0$$

At $z = H_1 + H_2 = H$:

$$\sigma'_p = K_p q + K_p \gamma H_1 + K_p(\gamma_{\text{sat}} - \gamma_w)H_2 + 2c'\sqrt{K_p} \quad \text{and} \quad u = \gamma_w H_2$$

To calculate the lateral force we can simplify the pressure distribution using triangular and rectangular shapes as was done in Example 7.1. The area of each shape can be calculated separately to give us the lateral force that must be applied at the centroid of the respective shape. The sum of these lateral forces is the total lateral force exerted on the wall. The point of application of the total lateral force can be calculated by taking the sum of the moments of all forces about the heel of the retaining wall and equating that to the moment of the total lateral force (resultant) about the heel.

7.4.2   Coulomb Theory

The Coulomb passive earth pressure theory assumes that a soil wedge with a failure plane making a critical angle θ_{cr} with the horizontal is in the condition of incipient failure as shown in Figure 7.17. The critical wedge is about to slide up the failure plane, thus generating a passive earth pressure against the retaining wall. The wedge can be treated as a particle that is subjected to three coplanar forces, W, P, and R, where W is the self-weight of the wedge, which can be determined easily by multiplying the area of the wedge with the unit weight of the backfill; P is the resultant of the wall reaction against the soil; and R is the soil reaction against the sliding wedge. The force P makes an angle δ with the normal to the back face of the wall (the face in contact with the backfill). The angle δ is the wall–soil interface friction angle, which can be determined from laboratory tests or can be assumed as a fraction of ϕ'; usually, you can assume that $\delta \approx \frac{2}{3}\phi'$. The force R makes an angle ϕ' with the normal to the failure plane. The internal friction angle ϕ' is assumed to be mobilized at the interface between the sliding wedge and the underlying soil.

It is to be noted that Coulomb theory accounts for three factors that were not accounted for in the original Rankine theory: the wall–soil interface friction signified by the friction angle δ in Figure 7.17, the sloping backfill soil with an angle

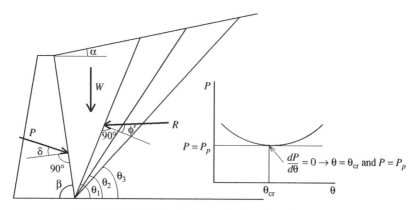

FIGURE 7.17   Coulomb passive lateral earth pressure.

α, and the battered back face of the retaining wall with an angle β. As shown in Figure 7.17, Coulomb theory considers a trial wedge with a failure plane making an angle θ with the horizontal. At failure, two equilibrium equations can be written (along the x and y axes) assuming an "impending motion" condition. Those two equations involve three unknowns, P, R, and θ; thus, there is a need for a third equation involving some (or all) of the three variables. Note that P, W, and R are functions of the angle θ. Coulomb theory searches for the angle $\theta = \theta_{cr}$ that yields the minimum earth pressure $P = P_p$. Thus,

$$\frac{dP}{d\theta} = 0 \qquad (7.24)$$

The solution of (7.24) will yield the angle θ_{cr}, and subsequently, the corresponding minimum earth pressure P_p can be calculated.

The Coulomb passive earth pressure coefficient for the general case shown in Figure 7.17 is dependent on the angle of the back face of the wall, the soil–wall friction, and the angle of backfill slope:

$$K_p = \frac{\sin^2(\beta - \phi')}{\sin^2 \beta \sin(\beta + \delta) \left[1 - \sqrt{\dfrac{\sin(\phi' + \delta) \sin(\phi' + \alpha)}{\sin(\beta + \delta) \sin(\alpha + \beta)}} \right]^2} \qquad (7.25)$$

7.5 RETAINING WALL DESIGN

The retaining wall designer needs to identify ahead of time the fundamental properties of the backfill soil and the soil under the base of the retaining wall (foundation soil). These soil properties include the unit weight and the shear strength parameters c' and ϕ'. Usually, the height H of the retaining wall is specified and the designer can proportion the wall (assume approximate dimensions) based on H. Figure 7.18 shows the approximate dimensions for a cantilever wall. The retaining wall needs to satisfy three external stability criteria: sliding (Figure 7.19a), overturning (Figure 7.19b), and bearing capacity/excessive settlement (Figure 7.19c and d). If any of these is not satisfied with a safety margin, the retaining wall needs to be reproportioned. This is an iterative procedure that requires some practice. In addition to the external stability, the internal stability has to be satisfied as well. In the case of a reinforced concrete cantilever retaining wall, for example, the dimensions of the wall and the amount of steel reinforcement should be adequate to resist structural failure of the base and the stem of the retaining wall.

For cantilever retaining walls, usually Rankine theory is used to calculate the active earth pressure. The Rankine active force can be calculated using (7.4). The active force is applied to a vertical plane passing through the heel of the retaining wall as shown in Figure 7.20a. The force is applied at $H'/3$ and assumed to be parallel to the backfill slope α. Note that H' is slightly greater than H and can be calculated from wall and backfill geometries. Also, the weight W_s of the soil

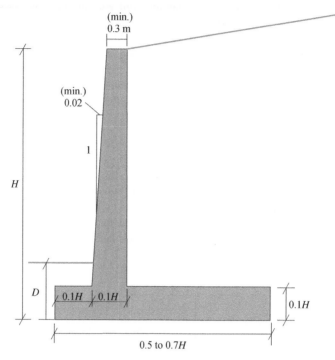

FIGURE 7.18   Proportioning a cantilever retaining wall based on its height H. (Adapted from Das, 2004.)

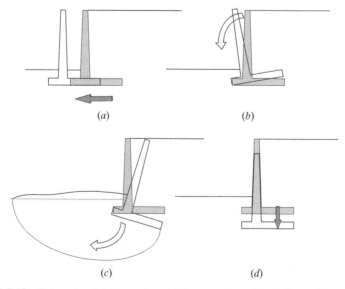

FIGURE 7.19   External stability of retaining walls: (a) sliding; (b) overturning; (c) bearing capacity; (d) excessive settlement.

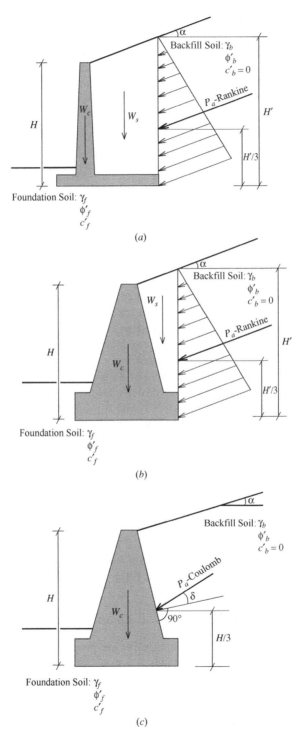

FIGURE 7.20 Application of Rankine and Coulomb lateral earth pressures on a retaining wall.

wedged between the vertical plane passing through the heel and the back face of the retaining wall should be regarded as a stabilizing force when we calculate the safety factors for both sliding and overturning. As mentioned earlier, the Rankine theory assumes that the backfill soil has a horizontal surface, and (7.6) can be used to calculate the active earth pressure coefficient for that condition. But for a granular backfill with an inclined surface, the following equation can be used instead:

$$K_a = \cos \alpha \frac{\cos \alpha - \sqrt{\cos^2 \alpha - \cos^2 \phi'}}{\cos \alpha + \sqrt{\cos^2 \alpha - \cos^2 \phi'}} \tag{7.26}$$

where α is the inclination angle of the backfill and ϕ' is the friction angle of the backfill soil.

For gravity retaining walls, either Rankine or Coulomb active earth pressure theory can be used. If Rankine theory is used, the Rankine active force is applied to the retained soil at a vertical plane passing through the heel of the retaining wall as shown in Figure 7.20b. The force is applied at $H'/3$ and assumed to be parallel to the backfill slope α. The height H' can be calculated from wall and backfill geometries. The weight W_s of the soil wedged between the vertical plane passing through the heel and the back face of the retaining wall should be regarded as a stabilizing force against sliding and overturning.

If Coulomb active force is used instead of Rankine's active force, the Coulomb force is applied directly to the back face of the gravity retaining wall. The line of action of the active force makes an angle δ with the normal to the back face of the retaining wall, and the point of application of the Coulomb active force is located at $H/3$, as shown in Figure 7.20c.

7.5.1   Factors of Safety

The design of a retaining wall is an iterative procedure. An initial wall geometry is assigned to the wall (proportioning the wall) and the resulting forces, such as the weight of the wall and the active and passive forces, are calculated. These forces are then checked using appropriate factors of safety and the geometry of the wall is then revised until satisfactory factors of safety are obtained.

7.5.2   Proportioning Walls

For external stability concerns, retaining walls are proportioned so that the width of the base, B, is equal to approximately 0.5 to 0.7 times the height of the wall, H. Other wall dimensions can also be proportioned, as illustrated in Figure 7.18. For example, a 10-m-high cantilever wall would have a base approximately 5 to 7 m wide, a base thickness of approximately 1 m, and a stem thickness varying from 1 m at the bottom to 0.3 m at the top.

7.5.3 Safety Factor for Sliding

A retaining wall has a tendency to move (slide) away from the backfill soil because of the active earth pressure exerted by the soil, as shown in Figure 7.21. This active force is called a *driving force*. The retaining wall resists sliding by the friction and adhesion mobilized between the base of the wall and the foundation soil. These forces are called *resisting forces*. Note that the frictional force is proportional to the total vertical force, which includes the weight of the retaining wall (W_c) and the weight of the soil (W_s) that is wedged between the wall and a vertical plane passing through the heel (if Rankine theory is used, Figure 7.20 *a* and *b*). The passive earth pressure in the fill in front of the wall is a resisting horizontal force acting opposite to the driving force.

In general, a *safety factor* is defined as the ratio of the sum of available resisting (stabilizing) forces to the sum of driving (destabilizing) forces. With respect to sliding, for which a factor of safety of 1.5 or better is required, the factor of safety is given by

$$\text{FS}_{\text{sliding}} = \frac{\text{resisting (stabilizing) forces}}{\text{driving (destabilizing) forces}} = \frac{\left(\sum V\right) \tan(k_1 \phi'_f) + B k_2 c'_f + P_p}{P_a \cos \alpha}$$

(7.27)

where $\sum V$ is the total vertical force, $\tan(k_1 \phi'_f)$ the coefficient of friction between the base of the wall and the foundation soil, k_1 and k_2 reduction factors ranging

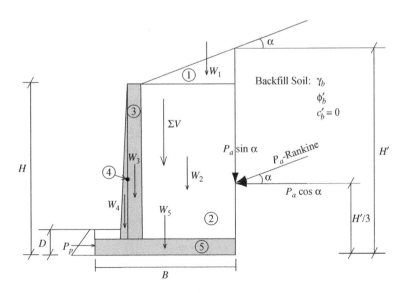

FIGURE 7.21 Calculations of safety factors for sliding and overturning.

from $\frac{1}{2}$ to $\frac{2}{3}$, ϕ'_f the friction angle of the foundation soil, c'_f the cohesion intercept (adhesion) of the foundation soil, P_p the horizontal passive earth pressure force, and $P_a(\cos \alpha)$ the horizontal component of the Rankine active force. For gravity walls, the Coulomb active force can be used instead. The horizontal component of the Coulomb active force can be calculated (Figure 7.20c) to replace the denominator in (7.27).

7.5.4 Safety Factor for Overturning

A retaining wall tends to rotate about the toe, due to the active earth pressure exerted by the backfill soil. The moment resulting from the active force is opposed by the moments resulting from the vertical forces produced by the wall, the soil, and the vertical component of the active force (Figure 7.21). With respect to overturning, for which a factor of safety of 2 to 3 is required, the factor of safety is defined as the sum of resisting moments divided by the overturning moment caused by the horizontal component of the active force as given by

$$\text{FS}_{\text{overturning}} = \frac{\sum M_R}{\sum M_O} \tag{7.28}$$

where $\sum M_R$ is the sum of the resisting moments about the toe of the wall and $\sum M_O$ is the sum of the overturning moments about the toe of the wall. Example 7.2 shows how to calculate $\sum M_R$ and $\sum M_O$ for a cantilever retaining wall.

7.5.5 Safety Factor for Bearing Capacity

The bearing capacity of the foundation soil must be adequate to support the base of the retaining wall. Because of the lateral earth pressure exerted on the stem of a retaining wall, the resulting load on the base of the wall (foundation) is both eccentric and inclined. The ultimate bearing capacity of the foundation soil, q_u, can be calculated using Meyerhof's equation (Chapter 6) because it accounts for eccentric and inclined loads:

$$q_u = c'_f N_c F_{cs} F_{cd} F_{ci} + q N_q F_{qs} F_{qd} F_{qi} + \tfrac{1}{2} \gamma_f B' N_\gamma F_{\gamma s} F_{\gamma d} F_{\gamma i} \tag{7.29}$$

where

$$N_q = \tan^2 \left(45° + \frac{\phi'_f}{2} \right) e^{\pi \tan \phi'_f}$$

$$N_c = (N_q - 1) \cot \phi'_f$$

$$N_\gamma = 2(N_q + 1) \tan \phi'_f$$

Also, $q = \gamma_f D$, where D is the height of the fill in front of the wall, γ_f is the unit weight of the foundation soil (if it is the same soil as that in front of the wall), ϕ'_f is the internal friction angle of the foundation soil, and $B' = B - 2e$, where e is the eccentricity, defined below.

For a retaining wall the base can be considered as a strip foundation in which $(B'/L) \approx 0$; therefore, the shape factors are

$$F_{cs} = 1 + \frac{B'}{L}\frac{N_q}{N_c} = 1 + 0 = 1$$

$$F_{qs} = 1 + \frac{B'}{L}\tan\phi'_f = 1 + 0 = 1$$

$$F_{\gamma s} = 1 - 0.4\frac{B'}{L} = 1 - 0 = 1$$

F_{cd}, F_{qd}, and $F_{\gamma d}$ are the depth factors:

$$F_{cd} = 1 + 0.4\frac{D}{B}$$

$$F_{qd} = 1 + 2\tan\phi'_f(1 - \sin\phi'_f)^2\frac{D}{B}$$

$$F_{\gamma d} = 1$$

where D is the height of the soil in front of the wall.

Finally, F_{ci}, F_{qi}, and $F_{\gamma i}$ are the inclination factors:

$$F_{ci} = F_{qi} = \left(1 - \frac{\eta^{\circ}}{90^{\circ}}\right)^2$$

$$F_{\gamma i} = \left(1 - \frac{\eta^{\circ}}{\phi'^{\circ}_f}\right)^2$$

where η is the inclination of the resultant R (Figure 7.22):

$$\eta = \tan^{-1}\left(\frac{P_a\cos\alpha}{\sum V}\right) \tag{7.30}$$

The reaction of the foundation soil against the base of the wall is nonuniform, as shown in Figure 7.22. This nonuniform pressure is the greatest (q_{max}) below the toe of the base and the least (q_{min}) below the heel:

$$q_{max} = \frac{\sum V}{B}\left(1 + \frac{6e}{B}\right) \tag{7.31}$$

$$q_{min} = \frac{\sum V}{B}\left(1 - \frac{6e}{B}\right) \tag{7.32}$$

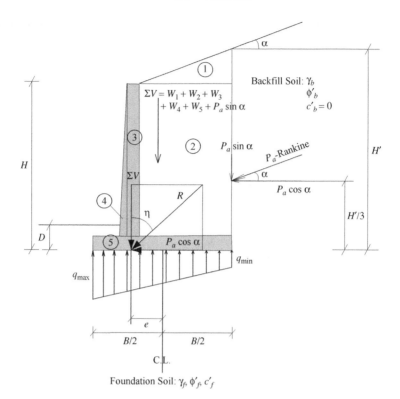

FIGURE 7.22   Calculations of safety factor for bearing capacity.

where e is the eccentricity:

$$e = \frac{B}{2} - \frac{\sum M_R - \sum M_O}{\sum V} \tag{7.33}$$

The factor of safety with respect to bearing capacity is given by (7.34). A bearing capacity factor of safety of 3 is required.

$$FS_{\text{bearing capacity}} = \frac{q_u}{q_{max}} \tag{7.34}$$

Note from (7.32) that if the eccentricity e is less than (or equal to) $B/6$, q_{min} remains positive (or equal to zero) indicating that the base is in complete contact with the soil. However, if the eccentricity e is greater than $B/6$, q_{min} becomes negative, indicating that the pressure under the heel is negative and the heel of the base is separated from the foundation soil. In such a case the retaining wall needs to be reproportioned.

When all safety factors of a retaining wall are satisfactory, the settlement of the base of the wall along with the global stability of the entire soil mass on which the wall is supported must be evaluated. The estimated settlement must be tolerable. Also, the global stability, using slope stability calculations, must be adequate. The theoretical background of slope stability is beyond the scope of this book.

Example 7.2 Calculate the factors of safety with respect to sliding, overturning, and bearing capacity for the 5-m-high cantilever wall shown in Figure 7.23. Use the Rankine method of analysis. The foundation soil is a granular soil with $\gamma_f = 17$ kN/m³ and $\phi'_f = 37°$. The backfill soil is also a granular soil with $\gamma_b = 17$ kN/m³ and $\phi'_b = 30°$. The concrete has a unit weight $\gamma_c = 24$ kN/m³. The groundwater table is well below the foundation.

SOLUTION: Let's start with proportioning the retaining wall. Based on the wall height $H = 5$ m, the base width $B = 0.5 \times H = 2.5$ m is selected. Other wall dimensions are selected as shown in Figure 7.23. The Rankine force, P_a, acts along a vertical plane passing through the heel of the wall. The active force is applied at a distance $H'/3$ from the base, and it has a vertical and a horizontal component as shown in Figure 7.23. Note that $H' = 5 + 1.5 \tan 10° = 5 + 0.26 = 5.26$ m.

$$K_a = \cos 10° \frac{\cos 10° - \sqrt{\cos^2 10° - \cos^2 30°}}{\cos 10° + \sqrt{\cos^2 10° - \cos^2 30°}} = 0.35$$

$$K_p = \tan^2\left(45° + \frac{\phi'_f}{2}\right) = \tan^2\left(45° + \frac{37°}{2}\right) = 4.0$$

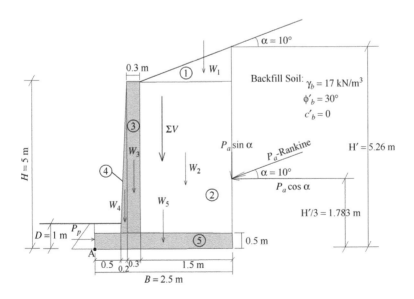

Foundation Soil: $\gamma_f = 17$ kN/m³, $\phi'_f = 37°$, $c'_f = 0$

FIGURE 7.23 A 5-m-high cantilever wall with a granular backfill.

$$P_a = \tfrac{1}{2}K_a\gamma_b H'^2 = \left(\tfrac{1}{2}\right)(0.35)(17)(5.26)^2 = 82.3 \text{ kN/m}$$

$$P_p = \tfrac{1}{2}K_p\gamma_f D^2 = \left(\tfrac{1}{2}\right)(4.0)(17)(1)^2 = 34 \text{ kN/m}$$

Let's divide the soil and the wall into triangular and rectangular areas (areas 1 to 5) as shown in Figure 7.23. Next, we can calculate the weight of each area per unit length, assuming that the wall is 1 m long. This is done by multiplying each area by its unit weight. The calculations are given in the second column of Table 7.1. The third column gives the moment arm of each weight vector. The moment arm is the perpendicular distance between the line of action of the weight vector (or force) and the toe of the wall (point A in Figure 7.23). The fourth column gives the resisting moments of the five weight vectors. It also includes the moment generated by the vertical component of the Rankine active force. The weights of areas 1 to 5 are added to the vertical component of the Rankine active force at the bottom of the second column to give us $\sum V$. Also, the resisting moments generated by areas 1 to 5 are added to the moment generated by the vertical component of the Rankine active force at the bottom of the fourth column to give us $\sum M_R$.

Sliding

$$\text{FS}_{\text{sliding}} = \frac{\left(\sum V\right)\tan(k_1\phi'_f) + Bk_2c'_f + P_p}{P_a\cos\alpha} = \frac{205.6\tan\left(\tfrac{2}{3}\times 37^\circ\right) + 0 + 34}{82.3\cos 10^\circ}$$

$$= 1.58 > 1.5 \qquad \text{okay!}$$

Overturning   From Figure 7.23 the overturning moment can be calculated as

$$\sum M_O = P_a\cos\alpha\frac{H'}{3} = 82.3\cos 10^\circ\left(\frac{5.26}{3}\right) = 142.1 \text{ kN/m}$$

$$\text{FS}_{\text{overturning}} = \frac{\sum M_R}{\sum M_O} = \frac{315}{142.1} = 2.2 > 2 \qquad \text{okay!}$$

TABLE 7.1

Area in Figure 7.23	$W = \gamma \cdot$ area (kN/m)	Moment Arm (m)	Resisting Moment, M_R(kN $\cdot$ m/m)
1	$\left(\tfrac{1}{2}\right)(1.5)(0.26)(17) = 3.32$	$1 + \left(\tfrac{2}{3}\right)(1.5) = 2$	6.64
2	$(1.5)(4.5)(17) = 114.75$	$1 + \left(\tfrac{1}{2}\right)(1.5) = 1.75$	200.8
3	$(0.3)(4.5)(24) = 32.4$	$0.5 + 0.2 + \left(\tfrac{1}{2}\right)(0.3) = 0.85$	27.5
4	$\left(\tfrac{1}{2}\right)(0.2)(4.5)(24) = 10.8$	$0.5 + \left(\tfrac{2}{3}\right)(0.2) = 0.633$	6.8
5	$(2.5)(0.5)(24) = 30$	$\left(\tfrac{1}{2}\right)(2.5) = 1.25$	37.5
Active force	$P_a\sin 10^\circ = 14.3$	2.5	35.8
	$\sum V = 205.6$		$\sum M_R = 315$

Bearing Capacity   First let's calculate the bearing capacity factors:

$$N_q = \tan^2\left(45° + \frac{\phi'_f}{2}\right) e^{\pi \tan \phi'_f} = \tan^2\left(45° + \frac{37°}{2}\right) e^{\pi \tan 37°} = 42.9$$

$$N_c = (N_q - 1)\cot\phi'_f = (42.9 - 1)\cot 37° = 55.6$$

$$N_\gamma = 2(N_q + 1)\tan\phi'_f = (2)(42.9 + 1)\tan 37° = 66.2$$

Note that all the shape factors are equal to 1 for a strip foundation. Now we can calculate the depth factors:

$$F_{cd} = 1 + 0.4\frac{D}{B} = 1 + 0.4\left(\frac{1}{2.5}\right) = 1.16$$

$$F_{qd} = 1 + 2\tan\phi'_f(1 - \sin\phi'_f)^2\frac{D}{B} = 1 + 2\tan 37°(1 - \sin 37°)^2\left(\frac{1}{2.5}\right) = 1.096$$

$$F_{\gamma d} = 1$$

The load inclination can be calculated as

$$\eta = \tan^{-1}\frac{P_a\cos\alpha}{\sum V} = \tan^{-1}\frac{81.05}{205.6} = 21.5°$$

Therefore, the inclination factors are

$$F_{ci} = F_{qi} = \left(1 - \frac{\eta°}{90°}\right)^2 = \left(1 - \frac{21.5°}{90°}\right)^2 = 0.579$$

$$F_{\gamma i} = \left(1 - \frac{\eta°}{\phi'_f°}\right)^2 = \left(1 - \frac{21.5°}{37°}\right)^2 = 0.175$$

Now calculate the eccentricity:

$$e = \frac{B}{2} - \frac{\sum M_R - \sum M_O}{\sum V} = \frac{2.5}{2} - \frac{315 - 142.1}{205.6} = 0.409 < \frac{B}{6} = \frac{2.5}{6}$$

$$= 0.417 \quad \text{okay!}$$

The maximum pressure under the toe:

$$q_{max} = \frac{\sum V}{B}\left(1 + \frac{6e}{B}\right) = \frac{205.6}{2.5}\left(1 + \frac{6 \times 0.409}{2.5}\right) = 163 \text{ kPa}$$

The minimum pressure under the heel:

$$q_{min} = \frac{\sum V}{B}\left(1 - \frac{6e}{B}\right) = \frac{205.6}{2.5}\left(1 - \frac{6 \times 0.409}{2.5}\right) = 1.5 \text{ kPa} > 0 \quad \text{okay!}$$

The effective foundation width:

$$B' = B - 2e = 2.5 - (2)(0.409) = 1.682 \text{ m}$$

The overburden pressure:

$$q = \gamma_f D = (17)(1) = 17 \text{ kPa}$$

Finally, we can calculate the ultimate bearing capacity:

$$q_u = c'_f N_c F_{cs} F_{cd} F_{ci} + q N_q F_{qs} F_{qd} F_{qi} + \tfrac{1}{2} \gamma_f B' N_\gamma F_{\gamma s} F_{\gamma d} F_{\gamma i}$$
$$q_u = 0 + (17)(42.9)(1)(1.096)(0.579) + \left(\tfrac{1}{2}\right)(17)(1.682)(66.2)(1)(1)(0.175)$$
$$= 629 \text{ kPa}$$

Therefore, the safety factor for bearing capacity is

$$\text{FS}_{\text{bearing capacity}} = \frac{q_u}{q_{\text{max}}} = \frac{629}{163} = 3.86 > 3 \qquad \text{okay!}$$

Note that all the three safety factors are satisfactory. This means that the wall design is also satisfactory.

Example 7.3 Lateral Earth Pressure Calculations Using FEM Figure 7.24 shows 3-m-high backfill sand supported by a concrete retaining wall. A 7 kPa surcharge pressure is applied at the surface of the backfill. (a) Calculate the at-rest earth pressure distribution on the 3-m-high retaining wall assuming that the wall is stationary. (b) Assuming that the wall moves away from the backfill by rotating about the heel (point A), calculate the active earth pressure and compare with Rankine's active earth pressure theory. (c) If the wall is forced toward the backfill by rotating about the heel (point A), calculate the passive earth pressure against the wall and compare with Rankine's passive earth pressure theory. Assume that the backfill soil is sand with $\phi' = 37°$ and $c' = 0$ MPa. The water table is well below the retaining wall base.

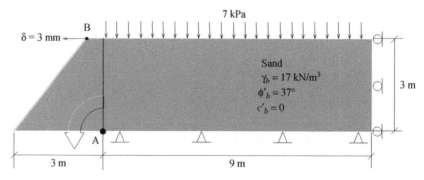

FIGURE 7.24 A 3-m-high backfill sand supported by a concrete retaining wall.

SOLUTION (filename: Chapter7_Example3.cae): This example presents a limit equilibrium solution for a layer of sand supported by a concrete retaining wall. The cap model (modified Drucker–Prager model with a cap, Chapter 2), with parameters matched to the classical Mohr–Coulomb yield model parameters c' and ϕ', is used to simulate the behavior of the backfill sand.

The two-dimensional plane strain model analyzed is shown in Figure 7.24. The sand layer is 3 m high. A 3-m-high retaining wall provides lateral support to the soil. Initially, the wall was restricted from lateral movements by fixing point B in Figure 7.24. The self-weight of the backfill soil was then applied using the "gravity" option. This condition simulates the at-rest condition in which the wall does not move laterally. After that the wall was forced to rotate away from the backfill to induce active earth pressure conditions. This was done by forcing point B (Figure 7.24) to displace away from the backfill using the loading history shown in Figure 7.25a. Also, the history of the gravity load is shown in Figure 7.25b. A separate analysis was carried out to simulate the passive earth pressure condition. In which case the wall was pushed inward by forcing point B to displace toward the backfill using the loading history shown in Figure 7.25c. Note that the wall is allowed to rotate about the z-axis passing through the heel at point A.

Figure 7.26 shows the finite element mesh along with the assumed boundary conditions. Reduced-integration, four-node linear plane strain quadrilateral elements are used for the sand layer and for the retaining wall. The base of the sand layer is fixed in all directions. The right-hand-side boundary of the sand layer is fixed in the horizontal direction but free in the vertical direction. Penalty-type interface elements are used between the wall and the backfill. The interface friction is assumed to be zero; thus, the interface elements provide frictionless contact between the wall and the backfill soil, and at the same time, they prevent the wall from penetrating the soil. It is noted that the mesh is finer in the vicinity of the retaining wall since that zone is the zone of stress concentration. No mesh convergence studies have been performed. However, the dimensions of the sand layer are chosen in a way that boundary effects are minimized.

The concrete retaining wall is assumed to be linear elastic with a Young's modulus of 21.3 GPa and a Poisson ratio of 0.2. The elastic response of the sand is assumed to be linear and isotropic, with a Young's modulus of 182 MPa and a Poisson ratio of 0.3. For plane strain conditions, the Mohr–Coulomb parameters ($\phi' = 37°$ and $c' = 0$ MPa) can be converted to Drucker–Prager parameters as follows:

$$\tan \beta = \frac{3\sqrt{3} \tan \phi'}{\sqrt{9 + 12 \tan^2 \phi'}} \qquad \text{for } \phi' = 37° \rightarrow \beta = 44.56°$$

$$d = \frac{3\sqrt{3} c'}{\sqrt{9 + 12 \tan^2 \phi'}} \qquad \text{for } c' = 0 \rightarrow d = 0$$

The cap eccentricity parameter is chosen as $R = 0.2$. The initial cap position (which measures the initial consolidation of the specimen) is taken as $\varepsilon^{pl}_{vol(0)} = 0.0$, and

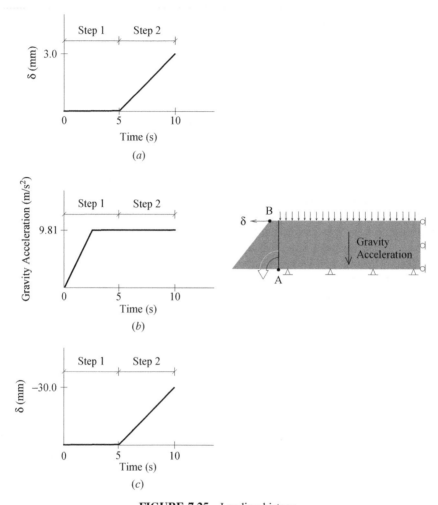

FIGURE 7.25 Loading history.

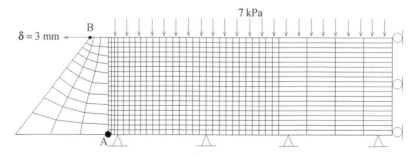

FIGURE 7.26 Finite element mesh.

the cap hardening curve is assumed to be the same as the one shown in Figure 6.9. The transition surface parameter $\alpha = 0.1$ is used. The cap model parameters are summarized in Table 7.2.

In the first analysis we calculate the lateral earth pressure exerted on the retaining wall for two conditions: at-rest and active. The second analysis is identical to the first analysis except for the direction of wall motion, which was simply reversed to induce passive earth pressure conditions. Both analyses were performed using the dynamic–explicit option with a "smooth-step" loading criterion. Dynamic–explicit analysis has to be carried out carefully. The loads must be applied very slowly to avoid exciting the model. For that reason, explicit analyses use very small time increments automatically to ensure stability, making them computationally expensive. But such analysis is desirable when failure and postfailure conditions are sought. Note that static analysis can be used for this problem (not done here).

The active earth pressure analysis consists of two steps. In step 1, point B is restricted from any lateral movements, and the gravity and surcharge loads are applied to the sand layer. The gravity load is particularly important because it will determine the initial stresses in all soil elements. The soil behavior is stress dependent, and the cap model used herein considers this important fact (Chapter 2). The surcharge load is applied using the instantaneous load option. The gravity load is applied using a smooth-step function that lasted 2.5 seconds, as indicated in Figure 7.25b. This was followed by another 2.5-second "resting" time period, as shown in the figure, to make sure that the model was not excited by the applied loads. The finite element results show that the model was indeed unexcited during and after gravity load application, as indicated in Figure 7.27 (oscillations are not detected). This figure shows the horizontal stress history of several soil elements adjacent to the wall. The calculated horizontal stresses in these elements at the end of the resting time period (5 seconds) represent the at-rest lateral earth pressure distribution against the wall since the wall was restricted from lateral motion. These stresses are shown in Figure 7.28, where they are compared with the theoretical at-rest lateral earth pressure distribution. Good agreement is noted between the theoretical and the finite element calculations.

In step 2 the wall is forced to rotate away from the backfill within 5 seconds, as indicated by the loading history of point B shown in Figure 7.25a. Consequently, the horizontal stresses in the elements adjacent to the wall decreased gradually, as

TABLE 7.2 Cap Model Parameters

General		Plasticity	
ρ (kg/m^3)	1923	d (kPa)	10^{-4}
e_0	1.5	β (deg)	44.56
Elasticity		R	0.2
E (MPa)	182	Initial yield	0.0
ν	0.3	α	0.1
		K	1

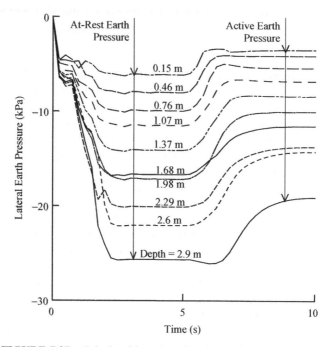

FIGURE 7.27 Calculated lateral earth pressure (at-rest and active).

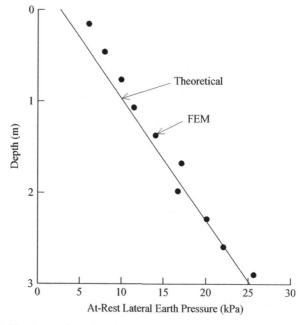

FIGURE 7.28 At-rest lateral earth pressure: comparison between theory and FEM.

shown in Figure 7.27, approaching their "limiting" values: the active earth pressure values. The active earth pressure distribution is compared with the Rankine active earth pressure distribution as shown in Figure 7.29. Again, the theoretical active earth pressure agrees well with the finite element results. It is to be noted that the limiting active earth pressure condition occurred after the formation of a failing soil wedge adjacent to the retaining wall. The plastic strain contours, shown in Figure 7.30, clearly illustrate the presence of a failure plane similar to the one assumed in the Coulomb active earth pressure theory. Figure 7.31 shows the displacement vectors of the retaining wall and the backfill soil at the end of step 2.

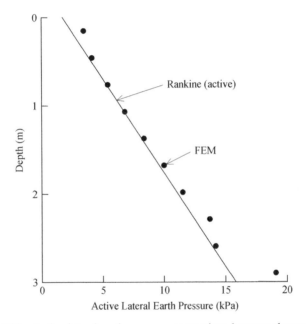

FIGURE 7.29 Active lateral earth pressure: comparison between theory and FEM.

FIGURE 7.30 Distribution of plastic strains in the backfill at active failure.

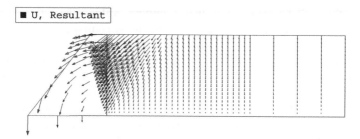

FIGURE 7.31 Displacement vectors in the backfill at active failure.

Again, the figure clearly shows a soil wedge that has displaced away from the backfill soil that remained essentially unaffected.

The analysis of passive earth pressure also consisted of two steps. This analysis is identical to the active earth pressure analysis, described above, except for the direction of the forced displacement of point B that was reversed to induce passive earth pressure. In step 2 of this analysis the wall is forced to rotate toward the backfill within 5 seconds, as indicated by the loading history of point B shown in Figure 7.25c. As a result, the horizontal stresses in the elements adjacent to the wall increased gradually, as shown in Figure 7.32, approaching their "limiting" values: the passive earth pressure values. The passive earth pressure distribution is compared with the Rankine passive earth pressure distribution as shown in Figure 7.33. The figure indicates that the theoretical passive earth pressure is greater than the

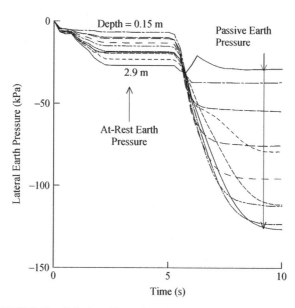

FIGURE 7.32 Calculated lateral earth pressure (at-rest and passive).

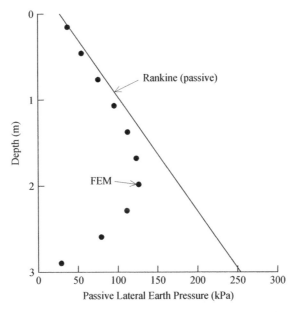

FIGURE 7.33   Passive lateral earth pressure: comparison between theory and FEM.

pressure calculated, especially near the bottom of the retaining wall. If the finite element results are considered "reasonable" in this analysis, the Rankine passive earth pressure is unsafe because it overestimates the passive resistance of the backfill soil.

7.6   GEOSYNTHETIC-REINFORCED SOIL RETAINING WALLS

Just like conventional retaining walls, geosynthetic-reinforced soil (GRS) walls are used when a sudden change of elevation is desired, as shown in Figure 7.34. A GRS wall consists of three components: soil (the main ingredient), geosynthetic reinforcement, and facing, as shown in the figure. Geosynthetics are made of polymers such as polyester, polyethylene, and polypropylene (all petroleum by-products). Two types of geosynthetics are widely used for soil reinforcement: (1) *geotextiles*, fabriclike sheets made of woven or nonwoven polymeric filaments, and (2) *geogrids*, netlike sheets made of polymers (using a tensile drawing manufacturing process).

The part of the backfill that contains the reinforcement is called the *reinforced soil*, and the part of the backfill behind that is called the *retained soil*. The entire reinforced zone that contains backfill soil, reinforcement, and the facing units act as a monolith whose main role is to resist the destabilizing active earth pressure exerted by the retained soil as shown in Figure 7.34. The self-weight of the reinforced zone is the predominant load in a GRS wall. Geosynthetics are high-strength polymer sheets that are typically used for soil reinforcement. They are resistant to

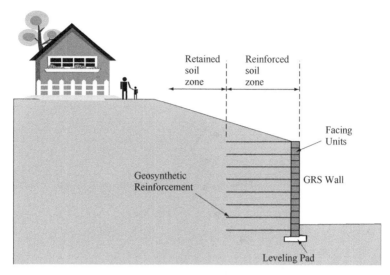

FIGURE 7.34   Geosynthetic-reinforced soil retaining walls.

corrosion but vulnerable to ultraviolet rays—they need to be protected from sunlight. This is done by covering the exposed geosynthetic reinforcements at the front of the GRS wall using concrete facing units as in segmental walls, for example.

In this section we describe design concepts of GRS walls, concepts similar to those of reinforced earth retaining walls that use steel strips as reinforcement instead of geosynthetics. The design of GRS walls involves satisfying external stability and internal stability. External stability refers to the stability of the reinforced soil mass as a whole in relation to the soil adjacent to it. Internal stability, on the other hand, refers to stability within the reinforced soil mass.

7.6.1   Internal Stability of GRS Walls

The internal stability of GRS walls requires that the wall be sufficiently stable against failure within the reinforced soil mass (i.e., the reinforcement is not overstressed and its length is adequately embedded). Internal failure modes include tensile rupture failure of reinforcement and pullout failure of reinforcement.

Destabilizing horizontal forces resulting from an assumed lateral earth pressure behind the reinforced soil are resisted by stabilizing horizontal forces provided by the reinforcement as indicated in Figure 7.35. Limiting equilibrium analysis is used to equate the horizontal forces with safety factors to assure adequate safety margins. Two independent safety factors are determined for each layer of reinforcement. The factor of safety for reinforcement rupture, FS_R, is the ratio of reinforcement strength to the lateral earth pressure thrust for the layer. The factor of safety for pullout, FS_P, is the ratio of pullout resistance to the lateral earth pressure thrust for the layer. The lateral earth pressure thrust for a reinforcement layer can be

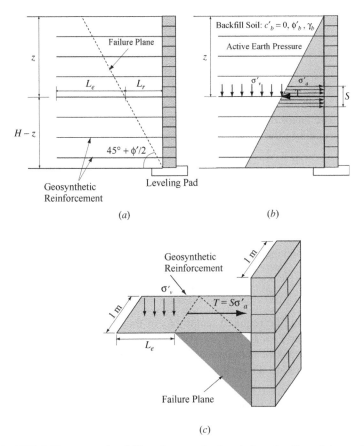

FIGURE 7.35 Internal stability of geosynthetic-reinforced soil retaining walls.

calculated as the product of the lateral earth pressure at the level of the layer and the "contributory" area $S \times 1$ m (Figure 7.35b). Note that S is equal to the spacing between two reinforcement layers.

A planar failure surface through the reinforced soil is assumed as shown in Figure 7.35a (Rankine active failure condition). For a wall with horizontal crest and subject to a uniform vertical surcharge, the failure surface slopes upward at an angle of $45° + \phi'_b/2$ from the horizontal (ϕ'_b is the angle of internal friction of the backfill). The reinforcements extend beyond the assumed failure surface and are considered to be tension-resistant tiebacks for the assumed failure wedge. As a result, this analysis is commonly referred to as *tieback wedge analysis*.

Let's discuss the equilibrium of a reinforcement layer located at a distance z measured from the top surface of the backfill (Figure 7.35b). Assuming a freely draining backfill soil, the active earth pressure at a depth z can be calculated as

$$\sigma'_a = K_a \sigma'_v = K_a \gamma_b z \tag{7.35}$$

in which

$$K_a = \tan^2\left(45° - \frac{\phi'_b}{2}\right) \tag{7.36}$$

where γ_b and ϕ'_b are the unit weight and the friction angle of the backfill soil, respectively. Consider a geosynthetic reinforcement layer with a rupture strength T_R (kN/m). The factor of safety for reinforcement rupture, FS_R, is the ratio of reinforcement strength to the lateral earth pressure thrust for the layer (Figure 7.35b):

$$FS_R = \frac{T_R}{\sigma'_a S} = \frac{T_R}{K_a \gamma_b z S} \tag{7.37}$$

where S is the spacing between two geosynthetic layers (Figure 7.35b). The factor of safety for reinforcement rupture, FS_R, is usually assumed to be 1.3 to 1.5 to account for various uncertainties. From (7.37) we can calculate the required spacing between two reinforcement layers:

$$S = \frac{T_R}{K_a \gamma_b z \cdot FS_R} \tag{7.38}$$

Now, let's determine the required length (L) of the reinforcement layer located at a depth z. Note that $L = L_r + L_e$, as indicated in Figure 7.35a, where L_r is the length within the Rankine's failure wedge and L_e is the extended length beyond Rankine's failure wedge. From Figure 7.35a one can write

$$\tan\left(45° + \frac{\phi'_b}{2}\right) = \frac{H - z}{L_r} \tag{7.39}$$

Therefore,

$$L_r = \frac{H - z}{\tan(45° + \phi'_b/2)} \tag{7.40}$$

In reference to Figure 7.35c, the length L_e can be calculated using the definition of the safety factor for pullout, FS_P, which is the ratio of pullout resistance to the lateral earth pressure thrust for the layer. Note from Figure 7.35c that the pullout resistance of the part of the reinforcement outside the Rankine's failure wedge is equal to $2A\gamma_b z \tan(\phi'_{int}) = 2(L_e \times 1)\gamma_b z \tan(\phi'_{int})$, where ϕ'_{int} is the "interface" friction angle between the backfill soil and the reinforcement. We can assume that $\phi'_{int} = \frac{2}{3}\phi'_b$ if no laboratory measurement of ϕ'_{int} is available. It follows that

$$FS_P = \frac{2L_e \gamma_b z \tan\phi'_{int}}{K_a \gamma_b z S} = \frac{2L_e \tan\phi'_{int}}{K_a S} \tag{7.41}$$

Therefore,

$$L_e = \frac{S K_a \cdot FS_P}{2 \tan\phi'_{int}} \tag{7.42}$$

Finally,

$$L = \frac{H - z}{\tan(45° + \phi_b'/2)} + \frac{SK_a \cdot FS_P}{2 \tan \phi_{int}'} \tag{7.43}$$

A safety factor for pullout of 1.3 to 1.5 can be assumed in (7.43).

The design procedure for a GRS wall of a height H starts with the selection of a commercially available geosynthetic reinforcement with a rupture strength T_R. Then the required spacing at various depths can be calculated using (7.38). The spacing calculated can be adjusted to fit the wall's particular geometry. For example, if we are designing a segmental wall with 0.3-m-high facing blocks, the adjusted spacing could be 0.3 m or one of its multiples as long as it is smaller than, or equal to, the spacing calculated.

The next step is to calculate the required length of each layer of reinforcement using (7.43). Note that the length L of any layer should not be less than 50% of the wall height (H). Also, note that it is desirable to use many layers of a lower-strength geosynthetic reinforcement (i.e., small spacing) than using fewer layers of a higher-strength reinforcement (i.e., large spacing). The reason of this recommendation is intuitive: If we have more reinforcement layers within the soil mass, there will be more interaction between the soil and the reinforcement, which means that the reinforcement is actually working. For the same reason, an experienced concrete designer would use a large number of small-diameter steel bars to reinforce a concrete beam as opposed to using fewer steel bars with a larger diameter, as long as the two provide the same total cross-sectional area of reinforcement.

7.6.2 External Stability of GRS Walls

The external stability of a GRS wall is generally evaluated by considering the reinforced soil mass as a rigid gravity retaining wall with earth pressure acting behind the wall (Figure 7.36). Using methods similar to those for conventional stability analysis of rigid earth retaining structures, the wall is checked for stability against three potential failure modes: sliding failure, overturning failure, and foundation bearing failure. Slope failure is also a potential failure mode that needs to be addressed.

For the simple GRS wall shown in Figure 7.36, the safety factor for sliding is defined as

$$FS_{sliding} = \frac{\text{resisting (stabilizing) forces}}{\text{driving (destabilizing) forces}} = \frac{W \tan(k_1 \phi_f') + Lk_2 c_f'}{P_a} \tag{7.44}$$

where k_1 and k_2 are reduction factors ranging from $\frac{1}{2}$ to $\frac{2}{3}$. The safety factor for overturning is

$$FS_{overturning} = \frac{\sum M_R}{\sum M_O} = \frac{W(L/2)}{P_a(H/3)} \tag{7.45}$$

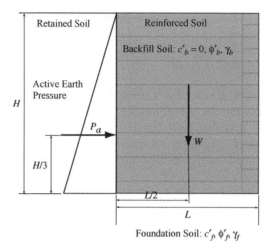

FIGURE 7.36   Lateral earth pressure against the reinforced soil zone.

The safety factor for bearing capacity is

$$\text{FS}_{\text{bearing capacity}} = \frac{q_u}{q} = \frac{q_u}{\gamma_b H} \tag{7.46}$$

in which

$$q_u = c'_f N_c + \tfrac{1}{2}\gamma_f L' N_\gamma \tag{7.47}$$

where

$$L' = L - 2e \tag{7.48}$$

and

$$e = \frac{L}{2} - \frac{\sum M_R - \sum M_O}{\sum V} = \frac{L}{2} - \frac{W(L/2) - P_a(H/3)}{W} \tag{7.49}$$

It is recommended that the safety factors for sliding, overturning, and bearing capacity be equal to or greater than 3.

Example 7.4   Design a 3-m-high geotextile-reinforced retaining wall (only for internal stability). The reinforcement has a rupture strength of 14 kN/m. The backfill is a sandy soil with $\gamma_b = 18.86$ kN/m³, $c'_b = 0$, and $\phi'_b = 37°$. Use $\text{FS}_R = \text{FS}_P = 1.5$.

SOLUTION: The design procedure described above is used herein. Table 7.3 is established (using a spreadsheet) to calculate at various depths the required spacing (S) using (7.38) and the required geosynthetic length (L) using (7.43).

In the first column the depth, z, varies from 0 (top) to 3.0 m (bottom) in 0.3-m increments. In the second column the required spacing is calculated using (7.38).

TABLE 7.3 Design of a Geotextile Reinforced Retaining Wall

z (m)	S (m) [Eq. (7.38)]	S Selected (m)	L (m) [Eq. (7.43)]	L Selected (m)	Force (kN/m)
0.3	6.643722	0.3	1.469713	2.13	0.421451
0.6	3.321861	0.3	1.319928	2.13	0.842901
0.9	2.214574	0.3	1.170143	2.13	1.264352
1.2	1.66093	0.3	1.020358	2.13	1.685802
1.5	1.328744	0.3	0.870573	2.13	2.107253
1.8	1.107287	0.3	0.720788	2.13	2.528703
2.1	0.949103	0.3	0.571003	2.13	2.950154
2.4	0.830465	0.3	0.421218	2.13	3.371604
2.7	0.738191	0.3	0.271433	2.13	3.793055
3	0.664372	0.3	0.121648	2.13	4.214505

In the third column the calculated spacing is changed to fit our specific application, in which we are using 0.3-m-thick soil lifts to facilitate compaction. We selected a uniform spacing of 0.3 m. Note that the selected spacing is less than the spacing required at the bottom of the wall. The fourth column includes the required geosynthetic layer length calculated using (7.43). We choose a uniform length of 2.13 m, which is greater than the required length at the top of the wall. The final design of the geotextile wall is shown in Figure 7.37.

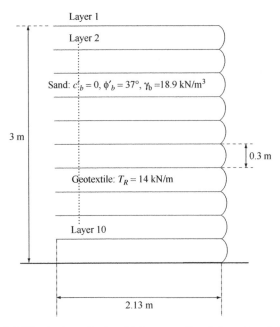

FIGURE 7.37 Design of a 3-m-high geotextile-reinforced retaining wall.

Example 7.5 *Sequential Construction of a Geotextile-Reinforced Soil Retaining Wall using FEM* Using the sequential construction procedure (layer by layer), analyze the 3-m-high geotextile-reinforced soil retaining wall shown in Figure 7.37 (same as Example 7.4). The geotextile reinforcement is 3 mm thick and has a rupture strength of 14 kN/m and an elastic modulus of 20,000 kPa. The backfill is a sandy soil with $\gamma_b = 18.86$ kN/m^3, $c'_b = 0$, and $\phi'_b = 37°$. The wall is founded on a rigid foundation soil. The wall is constructed in 10 equal layers each 0.3 m thick. Each layer is constructed in a 10-second period, during which the self-weight of the soil layer is applied. The total construction time is 100 seconds. Note that this problem is time independent, and time is used only to facilitate application of the sequential construction procedure. Assume that the backfill soil is elastoplastic obeying the cap model (Chapter 2). The cap model parameters for the backfill soil are given in Table 7.4 and Figure 6.9.

SOLUTION (filename: Chapter7_Example5.cae): A finite element mesh, shown in Figure 7.38, is constructed to simulate the sequential construction procedure of the geotextile wall. The mesh consists of three major components: (1) one part that includes 10 soil layers using solid elements, (2) 10 reinforcement layers using 3-mm-thick truss elements, and (3) 10 skin facing layers using 3-mm-thick beam elements. In the first calculation step, the entire wall is removed from the finite element mesh except for the bottom soil layer, the first reinforcement layer, and the first skin layer. Then the layers are added, one by one, in subsequent calculation steps. The self-weight of each layer is applied using the gravity option. When a new layer is added, it is situated on the deformed layer that was added previously. The new layer is assumed to be "strain-free" at the time of construction.

The soil is assumed to be elastoplastic, obeying the cap model (modified Drucker–Prager model with a cap, Chapter 2). Using such a model is essential for this class of analysis because we are concerned with the ability of the wall to withstand the stresses, due mainly to self-weight, during sequential construction. A model like this one can detect failure within the wall during construction. The elastic response of the sand is assumed to be linear and isotropic, with a Young modulus of 480 kPa and a Poisson ratio of 0.3.

The cap model parameters were matched to the classical Mohr–Coulomb yield model parameters c' and ϕ'. For plane strain conditions, the Mohr–Coulomb

TABLE 7.4 Cap Model Parameters

General		Plasticity	
ρ (kg/m^3)	1923	d (kPa)	0
e_0	1.5	β (deg)	44.56
Elasticity		R	0.55
E (kPa)	480	Initial yield	0.0
ν	0.3	α	0.01
		K	1

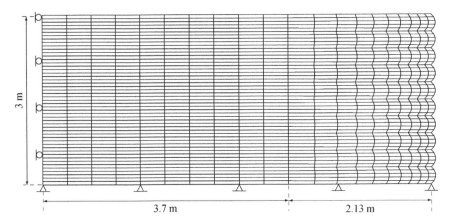

FIGURE 7.38 Finite element discretization of a geotextile-reinforced retaining wall.

parameters ($\phi' = 37°$ and $c' = 0$ MPa) can be converted to Drucker–Prager parameters as follows:

$$\tan \beta = \frac{3\sqrt{3} \tan \phi'}{\sqrt{9 + 12 \tan^2 \phi'}} \qquad \text{for } \phi' = 37° \rightarrow \beta = 44.56°$$

$$d = \frac{3\sqrt{3}c'}{\sqrt{9 + 12 \tan^2 \phi'}} \qquad \text{for } c' = 0 \rightarrow d = 0$$

The cap eccentricity parameter is chosen as $R = 0.55$. The initial cap position (which measures the initial consolidation of the specimen) is taken as $\varepsilon^{pl}_{vol(0)} = 0.0$, and the cap hardening curve is assumed to be the same as the one shown in Figure 6.9. The transition surface parameter $\alpha = 0.01$ is used. The cap model parameters are summarized in Table 7.4.

Figure 7.39 shows the contours of lateral displacements of the wall at the end of construction. Note that the maximum lateral displacement of approximately 80 mm is located below the middle of the wall. The figure indicates that the reinforced soil zone has experienced substantial lateral deformation during construction. The retained soil, however, remained essentially stationary.

Figure 7.40 shows the calculated lateral earth pressure distribution against the facing skin. This pressure is taken at a 0.3-m distance behind the facing to avoid the peculiar stress condition in the soil within the curved skin. Figure 7.40 also shows the distribution of the lateral earth pressure as calculated using Rankine's theory. The two pressure distributions are in good agreement except for the area near the bottom of the wall. At this location the lateral earth pressure from the finite element analysis is greater than that calculated using Rankine's analysis. This is mainly because of the "fixed" boundary conditions at the bottom of the wall that was assumed in the finite element analysis. Figure 7.41 presents the distribution of the maximum reinforcement force as calculated using FEM. The

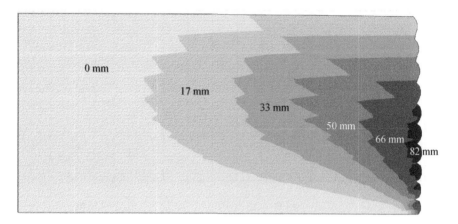

FIGURE 7.39 Displacement contours at end of construction of a geotextile-reinforced soil retaining wall.

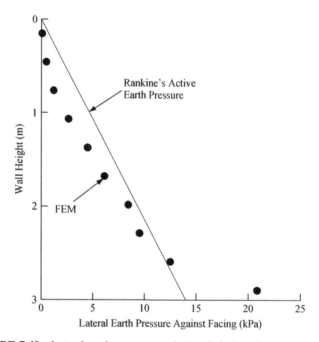

FIGURE 7.40 Lateral earth pressure against wall facing: theory versus FEM.

FEM reinforcement forces seem to vary with depth in a nonlinear manner as indicated in the figure. The calculated forces, using Rankine's theory (Table 7.3), are also included in the figure. Reasonable agreement between the finite element and theoretical results is noted in the figure.

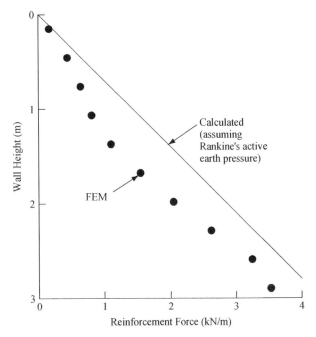

FIGURE 7.41   Reinforcement maximum axial force: theory versus FEM.

PROBLEMS

7.1   A 5-m-high retaining wall, shown in Figure 7.42, has a sandy backfill with $c' = 0$, $\phi' = 35°$, $\gamma = 18$ kN/m³, and $\gamma_{sat} = 19.5$ kN/m³. Calculate the magnitude and location of the *at-rest lateral force exerted* on a 1-m-long section of the retaining wall for three conditions: **(a)** the water table is 2.5 m below the backfill surface, **(b)** the water table is coincident with the backfill surface, and **(c)** the water table is at the bottom of the retaining wall. Assume that the sand below the water table is fully saturated. The wall is assumed to be frictionless.

7.2   Using the Rankine method of analysis, calculate the magnitude and location of the *active lateral force* exerted on the 5-m-high retaining wall for the three conditions described in Problem 7.1. The wall is assumed to be frictionless.

7.3   Using the Rankine method of analysis, calculate the magnitude and location of the *passive lateral force* exerted on the 5-m-high retaining wall for the three conditions described in Problem 7.1. The wall is assumed to be frictionless.

7.4   For the 5-m-high retaining wall shown in Figure 7.42, the friction angle between the wall and the backfill is $\delta = 2\phi'/3$, where ϕ' is the effective friction angle of the backfill soil. Calculate the magnitude and location of

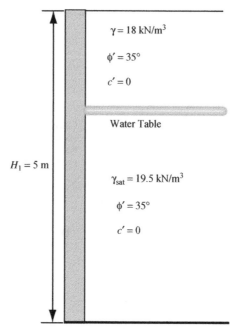

$H_1 = 5$ m

$\gamma = 18$ kN/m^3

$\phi' = 35°$

$c' = 0$

Water Table

$\gamma_{sat} = 19.5$ kN/m^3

$\phi' = 35°$

$c' = 0$

FIGURE 7.42

the *active lateral force* exerted on a 1-m-long section of the retaining wall for three conditions: **(a)** the water table is 2.5 m below the backfill surface, **(b)** the water table is coincident with the backfill surface, and **(c)** the water table is at the bottom of the retaining wall. Assume that the sand below the water table is fully saturated. Use the Coulomb method of analysis.

7.5 Using the Coulomb method of analysis, calculate the magnitude and location of the *passive lateral force* exerted on the 5-m-high retaining wall for the three conditions described in Problem 7.4.

7.6 Calculate the factors of safety with respect to sliding, overturning, and bearing capacity for the 7-m-high retaining wall shown in Figure 7.43. The foundation soil is a granular soil with $\gamma_f = 17.5$ kN/m^3 and $\phi'_f = 32°$. The backfill soil is also a granular soil with $\gamma_b = 17$ kN/m^3 and $\phi'_b = 37°$. The concrete has a unit weight $\gamma_c = 24$ kN/m^3. The friction angle between the wall and the backfill is $\delta_1 = 2\phi'_b/3$, and the friction angle between the wall and the foundation soil is $\delta_2 = 2\phi'_f/3$. The groundwater table is well below the foundation.

7.7 Figure 7.43 shows 7-m-high backfill sand supported by a concrete retaining wall. Using the finite element method, calculate the active earth pressure exerted on the retaining wall. The backfill sand obeys the elastoplastic cap model. The cap model parameters β and d can be calculated using strength

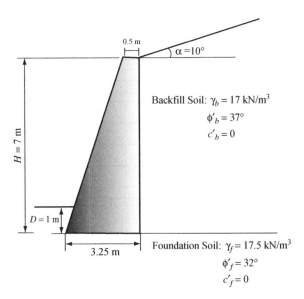

0.5 m

$\alpha = 10°$

Backfill Soil: $\gamma_b = 17$ kN/m³

$\phi'_b = 37°$

$c'_b = 0$

$H = 7$ m

$D = 1$ m

3.25 m

Foundation Soil: $\gamma_f = 17.5$ kN/m³

$\phi'_f = 32°$

$c'_f = 0$

FIGURE 7.43

parameters ϕ' and c' for each of the backfill soil and the foundation soil. The cap eccentricity parameter is $R = 0.2$, the initial cap position (which measures the initial consolidation of the specimen) is $\varepsilon^{pl}_{vol(0)} = 0.001$, the cap hardening curve is a straight line passing through two points ($p' = 1$ kPa, $\varepsilon^{pl}_{vol} = 0.0$) and ($p' = 500$ kPa, $\varepsilon^{pl}_{vol} = 0.01$), and the transition surface parameter is $\alpha = 0.1$. The foundation soil has $R = 0.2$, $\varepsilon^{pl}_{vol(0)} = 0.0$, the cap hardening curve is a straight line passing through two points [($p' = 1$ kPa, $\varepsilon^{pl}_{vol} = 0.0$) and ($p' = 500$ kPa, $\varepsilon^{pl}_{vol} = 0.025$)], and $\alpha = 0.1$.

7.8 Design a 3-m-high geosynthetic-reinforcement soil (GRS) retaining wall with segmental facing (Figure 7.44). The rupture strength of the reinforcement is 14 kN/m. The concrete block is 20 cm high, 20 cm wide (heel to toe), and 60 cm long. The backfill is a sandy soil with $\gamma_b = 18$ kN/m³, $c'_b = 0$, and $\phi'_b = 37°$. Use $FS_R = FS_P = 1.5$.

7.9 Using the sequential construction procedure (layer by layer), analyze the 3-m-high GRS retaining wall with segmental facing shown in Figure 7.44. The geotextile reinforcement is 3 mm thick, has a rupture strength of 14 kN/m, and has an elastic modulus of 20,000 kPa. The backfill is a sandy soil with $\gamma_b = 18$ kN/m³, $c'_b = 0$, and $\phi'_b = 37°$. The wall is founded on a rigid foundation soil. The wall is constructed in 15 equal layers 0.2 m thick each. Each layer can be constructed in a 10-second period, during which the self-weight of the soil layer is applied. Note that this problem is time independent, and time is used only to facilitate the application of the sequential construction procedure. Assume that the backfill soil is elastoplastic, obeying the

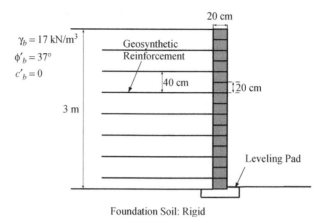

Foundation Soil: Rigid

FIGURE 7.44

cap model, with the following parameters: the cap eccentricity parameter $R = 0.2$, the initial cap position (which measures the initial consolidation of the specimen) is $\varepsilon_{vol(0)}^{pl} = 0.001$, the cap hardening curve is a straight line passing through two points $[(p' = 1 \text{ kPa}, \varepsilon_{vol}^{pl} = 0.0)$ and $(p' = 500 \text{ kPa}, \varepsilon_{vol}^{pl} = 0.01)]$, and the transition surface parameter is $\alpha = 0.1$.

7.10 Using the sequential construction procedure (layer by layer), analyze the 3-m-high GRS retaining wall with segmental facing shown in Figure 7.45. The 3-m-high GRS segmental wall (described in Problem 7.9) is constructed on a thick clay layer. The groundwater table is coincident with the top surface of the clay layer. The wall is constructed in 15 equal layers each 0.2 m thick.

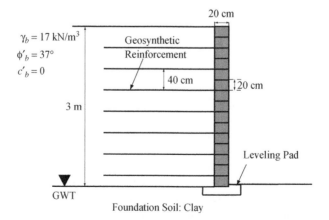

GWT

Foundation Soil: Clay

FIGURE 7.45

Each layer can be constructed in a one-day period, during which the self-weight of the soil layer is applied. Note that this is a consolidation problem (time dependent) because of the presence of the clay layer.

Assume that the backfill soil is elastoplastic, obeying the cap model, with the following parameters: the cap eccentricity parameter $R = 0.2$, the initial cap position (which measures the initial consolidation of the specimen) is $\varepsilon^{pl}_{vol(0)} = 0.001$, the cap hardening curve is a straight line passing through two points [($p' = 1$ kPa, $\varepsilon^{pl}_{vol} = 0.0$) and ($p' = 500$ kPa, $\varepsilon^{pl}_{vol} = 0.01$)], and the transition surface parameter is $\alpha = 0.1$. Also, assume that the clay layer obeys the Cam clay model with $M = 1.5$, $\lambda = 0.12$, $\kappa = 0.02$, $e_0 = 1.42$, $k = 10^{-6}$ m/s, and OCR $= 1.2$.

CHAPTER 8

PILES AND PILE GROUPS

8.1 INTRODUCTION

Piles are long, slender structural members that transmit superstructure loads to greater depths within the underlying soil. Piles are generally used when soil conditions are not suited for the use of shallow foundations. Piles resist applied loads through side friction (skin friction) and end bearing as indicated in Figure 8.1. *Friction piles* resist a significant portion of their loads by the interface friction developed between their surface and the surrounding soils. On the other hand, *end-bearing piles* rely on the bearing capacity of the soil underlying their bases. Usually, end-bearing piles are used to transfer most of their loads to a stronger stratum that exists at a reasonable depth.

Piles can be either driven or cast in place (bored piles). Pile driving is achieved by (1) impact dynamic forces from hydraulic and diesel hammers, (2) vibration, or (3) jacking. Concrete and steel piles are most common. Timber piles are less common. Driven piles with solid sections (e.g., concrete piles with square cross section) tend to displace a large amount of soil due to the driving process. These are *full-displacement piles*. Hollow piles such as open-ended pipe piles tend to displace a minimal amount of soil during the driving process. These are called *partial-displacement piles*. Cast-in-place (or bored) piles do not cause any soil displacement since no pile driving is involved; therefore, they are *nondisplacement piles*.

8.2 DRAINED AND UNDRAINED LOADING CONDITIONS

When saturated coarse-grained soils (sand and gravel) are loaded slowly, volume changes occur, resulting in excess pore pressures that dissipate rapidly due to high

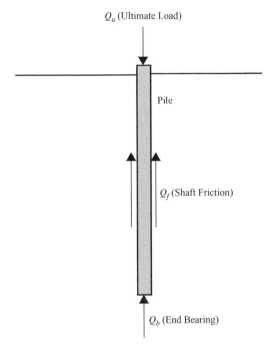

FIGURE 8.1   Pile's side friction (skin friction) and end bearing.

permeability. This is called *drained loading*. On the other hand, when fine-grained soils (silts and clays) are loaded, they generate excess pore pressures that remain entrapped inside the pores because these soils have very low permeabilities. This is called *undrained loading*.

Both drained and undrained conditions can be investigated in a laboratory tri-axial test setup (Chapter 5). Consider a soil specimen in a consolidated–drained (CD) triaxial test, as shown in Figure 8.2. During the first stage of the triaxial test, the specimen is subjected to a constant confining stress, σ_3, and allowed to con-solidate by opening the drainage valve. In the second stage of the CD triaxial test, the specimen is subjected, via the loading ram, to a monotonically increasing devi-atoric stress, $\sigma_1 - \sigma_3$, while the valve is kept open. The deviatoric stress is applied very slowly to ensure that no excess pore water pressure is generated during this stage—hence the term *drained*. Typical CD test results at failure can be presented using Mohr's circle as shown in Figure 8.3. The drained (or long-term) strength parameters of a soil, c' and ϕ', can be obtained from the Mohr–Coulomb failure criterion as indicated in the figure. Note that $c' = 0$ for sands and normally consoli-dated clays. These parameters must be used in drained (long-term) analysis of piles.

Now let's consider a soil specimen in a consolidated–undrained (CU) triaxial test (Figure 8.2). The first stage of this test is the same as the CD test–the specimen is subjected to a constant confining stress, σ_3, and allowed to consolidate by opening the drainage valve. In the second stage of the CU test, however, the specimen

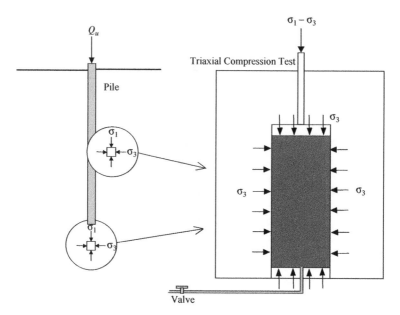

FIGURE 8.2   Drained and undrained loading conditions.

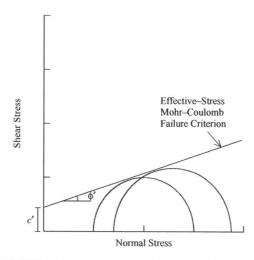

FIGURE 8.3   Long-term (drained) shear strength parameters.

is subjected to a monotonically increasing deviatoric stress, $\sigma_1 - \sigma_3$, while the valve is closed—hence the term *undrained*. The undrained condition means that there will be no volumetric change in the soil specimen (i.e., volume remains constant). It also means that pore water pressure will be developed inside the soil specimen throughout the test. Measurement of the pore water pressure allows for

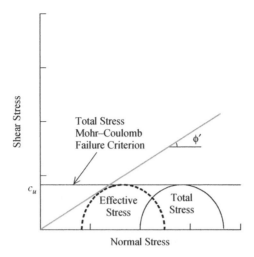

FIGURE 8.4   Short-term (undrained) shear strength parameters.

effective stress calculations. Typical CU test results (at failure) for a normally consolidated clay can be presented using Mohr's circles as shown in Figure 8.4. The undrained (or short-term) strength parameter of a soil, c_u, can be obtained from the total stress Mohr–Coulomb failure criterion as indicated in the figure (note that $\phi_u = 0$). The parameter c_u is termed the *undrained shear strength* of the soil and must be used in undrained (short-term) analysis of piles. Also, the effective-stress Mohr–Coulomb failure criterion can be used to estimate the drained (or long-term) strength parameters, c' and ϕ', as shown in the same figure ($c' = 0$ for NC clay).

In a finite element scheme, the drained (or long-term) behavior of a soil can be simulated using coupled analysis where the pore water pressure is calculated in each soil element, for a given load increment, and then subtracted from the total stresses to estimate the effective stresses in the element. These effective stresses control the deformation and shear strength of the soil element according to the effective stress principle. Constitutive models such as the cap and Cam clay models can be used within the finite element framework to determine the deformation caused by these effective stresses.

Four main measures must be considered for a successful finite element analysis of soils considering their long-term (drained) behavior: (1) the initial conditions of the soil strata (initial geostatic stresses, initial pore water pressures, and initial void ratios) must be estimated carefully and implemented in the analysis. The initial conditions will determine the initial stiffness and strength of the soil strata; (2) the boundary conditions must be defined carefully as being pervious or impervious; (3) the long-term strength parameters of the soil must be used in an appropriate soil model; and (4) loads must be applied very slowly to avoid the generation of excess pore water pressure throughout the analysis.

For undrained (or short-term) analyses, the aforementioned measures apply with the exception of the last measure—the load can be applied very fast instead. This is one of the most attractive aspects of coupled analysis. Drained and undrained analyses (Examples 8.4 and 8.2, respectively) differ only in the way we apply the load: Very slow loading allows the generated excess pore water pressure to dissipate and the long-term strength parameters to be mobilized, whereas fast loading does not allow enough time for the pore water pressure to dissipate, thus invoking the short-term strength of the soil. This means that there is no need to input the short-term strength parameters because the constitutive model will react to fast loading in an "undrained" manner. This is illustrated by the following example.

Consider a 16-m-thick homogeneous clay layer with the long-term strength parameters $c' = 0$ and $\phi' = 30°$ (Figure 8.5). Unlike long-term strength parameters, the undrained shear strength parameter c_u increases with depth due to the increase in soil confinement. Here we show how to calculate c_u at various depths using the long-term strength parameters c' and ϕ' along with the cap model within the finite element framework.

The 16-m-thick clay layer is divided into four equal sublayers as shown in Figure 8.5a. The mean effective stress, $p' = (\sigma_1' + 2\sigma_3')/3$, in the middle of each soil sublayer is calculated. To calculate the undrained shear strength in the middle of each sublayer, an axisymmetric soil element (one element) was used to simulate a CU triaxial compression test with a confining pressure that corresponds to the mean effective stress at the center of each of the four soil sublayers. The cap model is used to describe the behavior of the clay (Cam clay can be used for the same purpose). The long-term strength parameters $c' = 0$ and $\phi' = 30°$ are used to

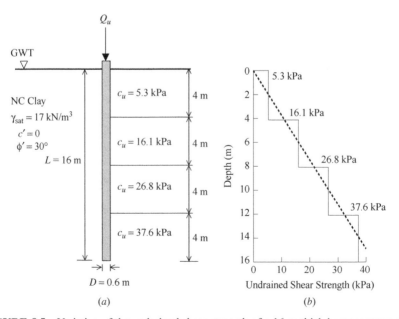

FIGURE 8.5   Variation of the undrained shear strength of a 16-m-thick homogeneous clay layer.

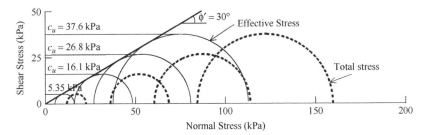

FIGURE 8.6   Determination of undrained shear strength parameters.

calculate the strength parameters for the cap model: $d = 0$ and $\beta = 50.2°$. The CU triaxial test simulation consists of two stages. The first stage involves application of the confining pressure, and the second stage involves application of an increasing vertical stress until failure. Drainage is allowed during the first stage and suppressed during the second stage. Mohr's circles representing failure conditions of these CU tests, using both total and effective stresses, are shown in Figure 8.6. The undrained shear strength at the center of each soil sublayer corresponds to the radius of the respective Mohr's circle as shown in the figure.

As an alternative method, undrained analysis can utilize the undrained shear strength parameter, c_u, directly in conjunction with a suitable constitutive model. If the cap model is used, the cap parameters β and d can be calculated using c_u and $\phi_u(= 0)$. This procedure can be used only for undrained (short-term) analysis (see Example 8.8). Unlike the preceding procedure, this alternative procedure cannot be used for drained analysis and for cases falling between the drained and undrained cases (partially drained).

8.3   ESTIMATING THE LOAD CAPACITY OF PILES

Pile load carrying capacity depends on various factors, including (1) pile characteristics such as pile length, cross section, and shape; (2) soil configuration and short- and long-term soil properties; and (3) pile installation method. Two widely used methods for pile design will be described: the α and β methods. The former method is used to calculate the short-term load capacity of piles in cohesive soils, and the latter method is used to calculate the short- and long-term load capacity of piles in both cohesive and cohesionless soils.

8.3.1 α-Method

The α-method is used to calculate the load capacity of piles in cohesive soils. This method is based on the undrained shear strength of cohesive soils; thus, it is well suited for short-term pile load capacity calculations. The ultimate load capacity of a pile is the sum of its friction capacity, Q_f, and end-bearing capacity, Q_b (Figure 8.1).

Friction Capacity: α-Method   The interface shear stress, f_s, between the pile surface and the surrounding soil determines the value of skin friction, Q_f (or shaft friction.) In this method the interface shear stress is assumed to be proportional to the undrained shear strength, c_u, of the cohesive soil as follows:

$$f_s = \alpha c_u \tag{8.1}$$

where α is a factor that can be obtained from one of several semiempirical equations available in the literature (e.g., API, 1984; Semple and Ridgen, 1984; Fleming et al., 1985). For brevity, only one of these equations is given herein. Other equations may be found in Budhu (1999). The equation by API (1984) suggests values for α as a function of c_u as follows:

$$\alpha = \begin{cases} 1 - \dfrac{c_u - 25}{90} & \text{for } 25 \text{ kPa} < c_u < 70 \text{ kPa} \\ 1.0 & \text{for } c_u \leqslant 25 \text{ kPa} \\ 0.5 & \text{for } c_u \geqslant 70 \text{ kPa} \end{cases} \tag{8.2}$$

Now, let's calculate the friction force between the pile surface and soil:

$$Q_f = f_s(\text{contact area}) = \alpha c_u \times \text{perimeter} \times \text{length} \tag{8.3}$$

For a pile with variable diameter that is embedded in a layered system containing n layers, one can generalize (8.3) as follows:

$$Q_f = \sum_{i=1}^{i=n} \left[\alpha_i (c_u)_i \times \text{perimeter}_i \times \text{length}_i \right] \tag{8.4}$$

End-Bearing Capacity: α-Method   The bearing capacity of the base of a pile is called *end-bearing capacity*. For cohesive soils it can be shown, using Terzaghi's bearing capacity equation, that the bearing capacity at the base of the pile is

$$f_b = (c_u)_b N_c \tag{8.5}$$

where $(c_u)_b$ is the undrained shear strength of the cohesive soil under the base of the pile, and N_c is the bearing capacity coefficient that can be assumed equal to 9.0 (Skempton, 1959). The corresponding load capacity is

$$Q_b = f_b A_b = (c_u)_b N_c A_b \tag{8.6}$$

where A_b is the cross-sectional area of the tip of the pile.

Ultimate Load Capacity The ultimate load capacity of a pile is the sum of its friction capacity and end-bearing capacity:

$$Q_{ult} = Q_f + Q_b \tag{8.7}$$

Example 8.1 Consider a concrete-filled pipe pile with diameter $D = 0.6$ m and embedded length $L = 16$ m in a thick homogeneous clay layer as shown in Figure 8.5*a*. The undrained shear strength of the clay layer varies with depth as shown in Figure 8.5*b*. Calculate the ultimate load capacity of the pile using the α-method.

SOLUTION: *Friction capacity* The friction capacity of this pile can be calculated using (8.4). To simplify the calculation of the friction capacity, let's divide the soil into four 4-m-thick sublayers, each having a constant undrained shear strength as shown in Figure 8.5*b*. From (8.2), we can calculate $\alpha = 0.98$ for the third sublayer (from the top) using $c_u = 26.8$ kPa and $\alpha = 0.86$ for the fourth sublayer (also from top) using $c_u = 37.6$ kPa. The other two sublayers have $\alpha = 1$ since they have undrained shear strength below 25 kPa. From (8.4),

$$Q_f = [(1)(5.3) + (1)(16.1) + (0.98)(26.8) + (0.86)(37.6)][\pi(0.6)](4) = 603.2 \text{ kN}.$$

End-bearing capacity The end-bearing capacity of this pile can be calculated using (8.6), in which A_b is the cross-sectional area of the closed-end base: $A_b = \pi D^2/4$. Thus,

$$Q_b = (c_u)_b N_c A_b = (c_u)_b N_c \pi D^2/4 = (37.6)(9)\frac{\pi(0.6^2)}{4} = 95.7 \text{ kN}.$$

Ultimate load capacity Equation (8.7): $Q_{ult} = Q_f + Q_b = 603.2 + 95.7 = 699$ kN.

Example 8.2 Consider a concrete-filled pipe pile with diameter $D = 0.6$ m and embedded length $L = 16$ m in a thick homogeneous clay layer as shown in Figure 8.5 (same as Example 8.1). The undrained shear strength of the clay layer varies with depth as shown in Figure 8.5*b*. Calculate the short-term load capacity of the pile using the finite element method assuming undrained loading conditions. Compare the results of the finite element analysis with the analytical solution obtained in Example 8.1 (using the α-method).

SOLUTION: *Finite element solution* (filename: Chapter8_Example2.cae) In this example a limit equilibrium solution is sought for a thick layer of clay loaded in undrained conditions by a 16-m-long concrete-filled pipe pile with $D = 0.6$ m. The problem geometry, boundary conditions, and materials are identical to those of Example 8.1, providing a direct means to compare the finite element analysis results with the analytical solution obtained in Example 8.1 using the α-method.

The pile in this example is cylindrical in shape and loaded in the axial direction only; therefore, the finite element mesh of the pile and the surrounding soil can take advantage of this axisymmetric condition. This simplification cannot be used for piles that are loaded with horizontal loads. Such piles should be treated as three-dimensional objects. It also should be noted that the finite element mesh of a soil–pile system must include interface elements that are capable of simulating the frictional interaction between the pile surface and the soil.

It is very difficult, but possible, to simulate the pile–soil interaction during pile driving. Such simulation is not attempted here. Instead, the pile is assumed to be embedded in perfect contact with the soil before applying pile loads. Excess pore water pressures caused by pile driving are assumed to have been dissipated completely before the application of pile loads.

The two-dimensional axisymmetric model and the finite element mesh analyzed are shown in Figure 8.7. The clay layer is 22.7 m deep and 15 m wide. The model considers only one-half of the pile taking advantage of symmetry as indicated in the figure. The pile is initially in perfect contact with the soil. The interaction between the pile and the soil is simulated using a penalty-type interface between the pile and the soil with a friction factor of 0.385. This type of interface is capable of describing the frictional interaction between the pile surface and the soil in contact.

Four-node axisymmetric quadrilateral, bilinear displacement, bilinear pore water pressure elements are used for the clay layer. The elements used for the pile are four-node bilinear axisymmetric quadrilateral reduced-integration elements (without pore water pressure). The base of the clay layer is fixed in the horizontal and vertical directions. The vertical boundary on the left side is a symmetry line, and the vertical boundary on the right side is fixed in the horizontal direction but free in the vertical direction. It is noted that the mesh is finer in the vicinity of the pile since that zone is the zone of stress concentration. No mesh convergence studies have been performed. However, the dimensions of the clay layer are chosen in a way that the boundary effect on pile behavior is minimized.

The elastic response of the clay is assumed to be linear and isotropic, with a Young's modulus of 68.9×10^3 kPa and a Poisson ratio of 0.3. The cap model, with parameters $d = 0$ and $\beta = 50.2°$ that were matched to the Mohr–Coulomb failure criterion parameters $c' = 0$ and $\phi' = 30°$, is used to simulate the undrained behavior of the clay. This procedure was explained in details in Section 8.2. The undrained shear strength of the clay layer, as estimated by this procedure, varies with depth as shown in Figure 8.5b. Note that $\tan \beta = 6 \sin \phi'/(3 - \sin \phi')$ for $\phi' = 30° \rightarrow \beta = 50.2°$. Also, $d = c'\sqrt{3}$ for $c' = 0 \rightarrow d = 0$.

The cap eccentricity parameter is chosen as $R = 0.4$. The initial cap position (which measures the initial consolidation of the specimen) is taken as $\varepsilon^{pl}_{vol(0)} = 0.0$, and the cap hardening curve is assumed to be a straight line passing through two points [($p' = 0.57$ kPa, $\varepsilon^{pl}_{vol} = 0.0$) and ($p' = 103$ kPa, $\varepsilon^{pl}_{vol} = 0.0032$)]. The transition surface parameter $\alpha = 0.1$ is assumed. The initial void ratio, $e_0 = 1.5$, and the vertical and horizontal effective stress profiles of the clay layer are part of the input data that must be supplied to the finite element program for this coupled

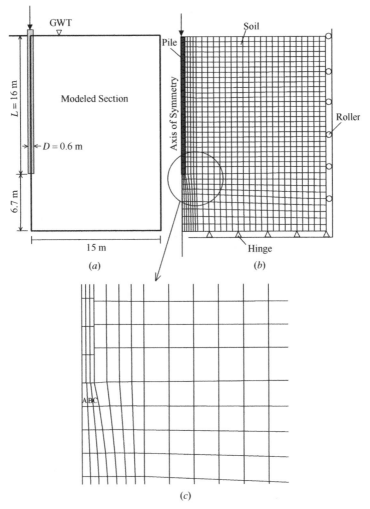

FIGURE 8.7   Concrete-filled pipe pile embedded in a thick homogeneous clay layer: (*a*) problem geometry; (*b*) finite element discretization; (*c*) enlarged mesh near the end of the pile.

(consolidation) analysis. In this analysis, the initial horizontal effective stress is assumed to be 50% of the vertical effective stress.

The problem is run in two steps. In step 1 the effective self-weight ($= \gamma' z$, where $\gamma' = \gamma_{\text{sat}} - \gamma_w$ and z is the depth below the ground surface) of the clay layer is applied using the "body-force" option. Note that the groundwater table is coincident with the ground surface (Figure 8.7*a*). As mentioned earlier, the clay layer is assumed to be elastoplastic, obeying the cap model. In general, using such a model is essential for limit equilibrium analysis. We are concerned with the ability of the clay layer to withstand the end-bearing stresses and skin shear stresses caused by the pile, and models such as the cap model, the Cam clay model, and the Mohr–Coulomb model can detect failure within the clay layer. During step 1, the "geostatic" command is invoked to make sure that equilibrium is satisfied within the clay layer. The geostatic option makes sure that the initial stress condition in any element within the clay layer falls within the initial yield surface of the cap model.

In step 2 a coupled (consolidation) analysis is invoked and the pile load is applied using the vertical displacement boundary condition to force the top surface of the pile to move downward a distance of 30 cm at a constant rate (3 cm/s). This high rate of loading is used to make sure that the soil will behave in an undrained manner. The pile load versus settlement curve obtained from the finite element analysis is shown in Figure 8.8. It is noted from the figure that the settlement increases as the load is increased in an approximately linear manner up to about a 570-kN pile load, at which a vertical lateral displacement of about 1.5 cm is encountered. Shortly after that, the pile plunges in a fast downward descent, indicating that the load capacity of the pile has been reached. For comparison, the pile load capacity of 699 kN, predicted by the α-method (Example 8.1), is also shown in Figure 8.8. It is noted that the finite element prediction of pile load capacity is approximately 15% smaller than the bearing capacity predicted by the α-method (i.e., the finite element analysis is more conservative in this particular case).

Figure 8.9 shows the evolution of excess pore water pressure in three soil elements located immediately below the pile tip (Figure 8.7c). The increase in excess pore water pressure during pile loading is expected since the loading is purposely made undrained.

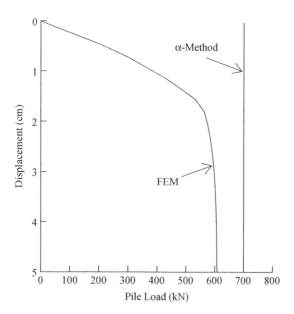

FIGURE 8.8 Pile load versus settlement curve: α-method versus FEM.

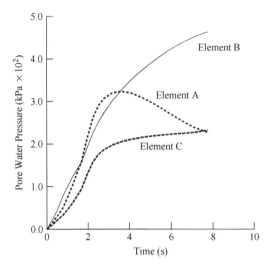

FIGURE 8.9 Evolution of excess pore water pressure (during undrained loading) in elements A, B, and C located immediately below the pile tip (Figure 8.7c).

8.3.2 β-method

This method can be used for both cohesive and cohesionless soils. The method is based on effective stress analysis and is suited for short- and long-term analyses of pile load capacity.

Friction Capacity: β-Method   Consider a pile embedded in a thick homogeneous soil that is fully saturated. Define σ'_h as the average lateral effective stress exerted on the pile by the surrounding soil. This stress can be taken as the lateral effective stress at the pile midpoint. The friction stress, f_s, between the pile and the surrounding soil can be calculated by multiplying the friction factor, μ, between the pile and soil with σ'_h. Thus, $f_s = \mu \sigma'_h$. But $\sigma'_h = K_0 \sigma'_v$, where σ'_v is the vertical effective stress at the pile midpoint and K_0 is the lateral earth pressure coefficient at rest. Therefore, $f_s = \mu K_0 \sigma'_v$.

Now we can calculate the skin friction force between the pile surface and soil:

$$Q_f = f_s(\text{contact area}) = \mu K_0 \sigma'_v \times \text{perimeter} \times \text{length}$$
$$= \beta \sigma'_v \times \text{perimeter} \times \text{length} \tag{8.8}$$

where

$$\beta = \mu K_0 \tag{8.9}$$

In general, the lateral earth pressure coefficient at rest is given by

$$K_0 = (1 - \sin \phi')(\text{OCR})^{0.5} \tag{8.10}$$

where OCR is the overconsolidation ratio. Recall that OCR $= 1$ for normally consolidated clays and OCR > 1 for overconsolidated clays.

For a pile with a variable diameter that is embedded in a layered system containing n layers, (8.8) can be modified as follows:

$$Q_f = \sum_{i=1}^{i=n} \left[\beta_i (\sigma'_v)_i \times \text{perimeter}_i \times \text{length}_i \right] \tag{8.11}$$

Note that $(\sigma'_v)_i$ is the vertical effective stress at the center of soil layer i.

In clays, the value of β can be estimated from (8.9) and (8.10) with $\mu = \tan \frac{2}{3}\phi'$ (Burland, 1973). For sands, however, McClellend (1974) suggested values of β ranging from 0.15 to 0.35. Meyerhof (1976) suggested values of $\beta = 0.44$, 0.75, and 1.2 for $\phi' = 28°, 35°$, and $37°$, respectively.

End-Bearing Capacity: β-Method   Using Terzaghi's bearing capacity equation, the bearing capacity at the base of the pile can be calculated:

$$f_b = (\sigma'_v)_b N_q + c'_b N_c \tag{8.12}$$

where $(\sigma'_v)_b$ is the vertical effective stress at the base of the pile, c'_b is the cohesion of the soil under the base of the pile, and N_q and N_c are bearing capacity coefficients. The corresponding load capacity, Q_b, is

$$Q_b = f_b A_b = [(\sigma'_v)_b N_q + c'_b N_c] A_b \tag{8.13}$$

where A_b is the cross-sectional area of the base of the pile.

Janbu (1976) presented equations to estimate N_q and N_c for various soils:

$$N_q = \left(\tan \phi' + \sqrt{1 + \tan^2 \phi'} \right)^2 \exp(2\eta\tan\phi') \tag{8.14}$$

$$N_c = \left(N_q - 1 \right) \cot\phi' \tag{8.15}$$

where η is an angle defining the shape of the shear surface around the tip of a pile as shown in Figure 8.10. The angle η ranges from $\pi/3$ for soft clays to 0.58π for dense sands.

Ultimate Load Capacity   The ultimate load capacity of a pile is the sum of its friction capacity and end-bearing capacity:

$$Q_{\text{ult}} = Q_f + Q_b \tag{8.16}$$

Example 8.3   Figure 8.11 shows a closed-end pipe pile with a diameter $D = 0.6$ m and embedded length $L = 16$ m in a thick homogeneous layer of saturated clay with a friction angle $\phi' = 30°$ and cohesion intercept $c' = 0$. The groundwater table is coincident with the top surface of the soil. Assume that $\eta = \pi/3$ and a

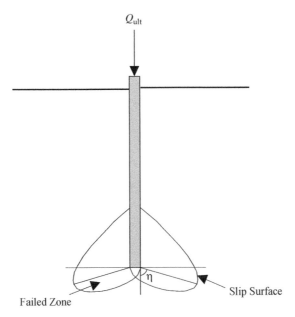

FIGURE 8.10 Shear surface around the tip of a pile: definition of the angle η. (Adapted from Janbu, 1976.)

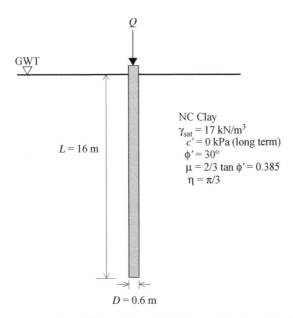

FIGURE 8.11 Pile configuration for Examples 8.3 and 8.4

soil–pile friction factor $\mu = 0.385$. Calculate the total load capacity of the pile using the β-method.

SOLUTION: *Friction capacity*   First, calculate the vertical effective stress at the pile midpoint:

$$\sigma'_v = (17 - 9.81)\left(\frac{16}{2}\right) = 57.52 \text{ kPa}$$

Eq. (8.10): $K_0 = (1 - \sin 30°)(1)^{0.5} = 0.5$

Eq. (8.9): $\beta = (0.385)(0.5) = 0.1925$

Eq. (8.8): $Q_f = (0.1925)(57.52)[\pi(0.6)](16) = 334 \text{ kN}$

End-bearing capacity   First, calculate the vertical effective stress at the tip of the pile:

$$(\sigma'_v)_b = (17 - 9.81)(16) = 115 \text{ kPa}$$

Eq. (8.14): using $\eta = \frac{\pi}{3} \rightarrow N_q = 10.05$

Eq. (8.15): $N_c = 15.7$

Eq. (8.13): $Q_b = [(115)(10.05) + (0)(15.7)][\pi(0.3^2)] = 327 \text{ kN}$

Total load capacity

$$\text{Eq. (8.16): } Q_{total} = Q_f + Q_b = 334 + 327 = 661 \text{ kN}$$

Example 8.4   Using the finite element method, calculate the long-term load capacity of a pipe pile with diameter $D = 0.6$ m and embedded length $L = 16$ m in a thick homogeneous clay layer as shown in Figure 8.11 (same as Example 8.3). The groundwater table is coincident with the top surface of the soil. The saturated clay has a friction angle $\phi' = 30°$ and a cohesion intercept $c' = 0$. Assume a soil–pile friction factor $\mu = 0.385$. Compare the results of the finite element analysis with the analytical solution obtained in Example 8.3 (using the β-method).

SOLUTION: *Finite element solution* (filename: Chapter8_Example4.cae) In this example a limit equilibrium solution is sought for a thick layer of clay loaded in drained conditions by a 16-m-long concrete-filled pipe pile with $D = 0.6$ m. The problem geometry, boundary conditions, and materials are identical to those of Example 8.3, providing a direct means to compare the finite element analysis results with the analytical solution obtained in Example 8.3 using the β-method.

The axisymmetric finite element mesh for this problem is identical to the one used in Example 8.2 (Figure 8.7). The material parameters used for this example are also identical to those of Example 8.2. The pile is initially in perfect contact with the soil. The interaction between the pile and the soil is simulated using a penalty-type interface between the pile and the soil with a friction factor of 0.385. This type of interface is capable of describing the frictional interaction between the pile surface and the soil in contact.

The problem is run in two steps. In step 1 the effective self-weight of the clay layer is applied using the "body-force" option. During this step, the "geostatic" command is invoked to make sure that equilibrium is satisfied within the clay layer. The geostatic option makes sure that the initial stress condition in any element within the clay layer falls within the initial yield surface of the cap model. This is identical to step 1 used in Example 8.2.

The main difference between this example and Example 8.2 is the loading rate applied in step 2: An undrained condition indicates a high loading rate (Example 8.2), and a drained condition indicates a low loading rate (this example). Also, the top surface of the clay layer in this example is made permeable to allow the excess pore water pressure to dissipate. Again, the present example involves long-term (drained) loading conditions, indicating that the pile should be loaded very slowly to prevent the generation of excess pore water pressure anywhere within the finite element mesh. Thus, in step 2 a coupled (consolidation) analysis is invoked and the pile load is applied using the vertical displacement boundary condition to force the top surface of the pile to move downward a distance of 30 cm at a very small constant rate $= 3 \times 10^{-13}$ cm/s. This small loading rate is used to make sure that the soil will behave in a drained manner. To confirm that, the excess pore water pressures in elements A, B, and C (shown in Figure 8.7c) are plotted in Figure 8.12. It is noted from the figure that the excess pore water pressure is essentially nil in these elements during loading, indicating that it is indeed a drained loading condition.

The pile load versus settlement curve obtained from the finite element analysis is shown in Figure 8.13. It is noted from the figure that the settlement increases linearly as the load is increased up to about a 400-kN pile load, at which a pile settlement of about 1.1 cm is encountered. Shortly after that, the pile plunges in a fast downward descent, indicating that the load capacity of the pile (530 kN) has been reached. For comparison, the pile load capacity of 661 kN, predicted by the β-method (Example 8.3), is also shown in Figure 8.13. It is noted that the finite element prediction of pile load capacity is approximately 15% smaller than the bearing capacity predicted by the β-method (i.e., the finite element analysis is more conservative in this particular case).

8.4 PILE GROUPS

Piles are generally used in groups. A square pile group arrangement is shown in Figure 8.14. Other arrangements, such as rectangular and circular, are possible. The spacing, s, between two piles center to center should be greater than $2D$, where D is the pile diameter. A concrete cap is generally used to connect the heads of the

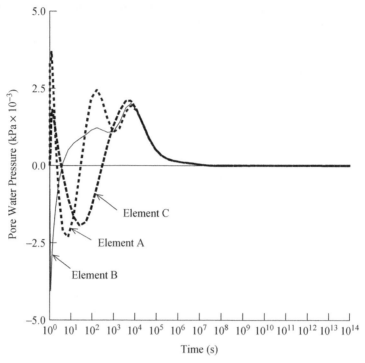

FIGURE 8.12   Evolution of excess pore water pressure (during drained loading) in elements A, B, and C located immediately below the pile tip (Figure 8.7c).

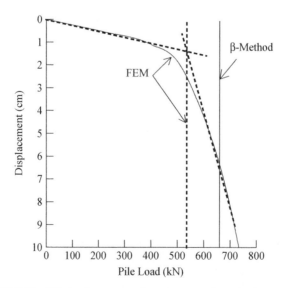

FIGURE 8.13   Pile load versus settlement curve: β-method versus FEM.

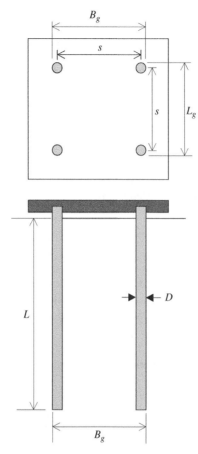

FIGURE 8.14 Pile group.

piles in a pile group. Loads are applied to the cap that transfer them to the piles. Two mechanisms of pile group failure are possible:

1. *Single-pile failure mechanism.* In this mechanism each single pile in the group fails individually, and the failure of all piles occurs simultaneously. In this case the pile group capacity, $(Q_{ult})_{npiles}$, is equal to nQ_{ult}, where n is the number of piles in the group and Q_{ult} is the load capacity of a single pile. Q_{ult} for a single pile can be calculated using the α-method and/or the β-method described above.

2. *Block failure mechanism.* In this mechanism the pile group, along with the soil between the piles, fail as a monolith (big block) that has the dimensions $B_g \times L_g \times L$ defined in Figure 8.14. The group load capacity for this failure mechanism can be calculated using the α-method and/or the β-method applied to a "mammoth" pile having the dimensions of the failing block.

8.4.1 α-Method

Recall that the α-method is based on the undrained shear strength of cohesive soils, and it is suited for short-term pile load capacity calculations. The following equation is based on the α-method and can be used to calculate pile group capacity assuming block failure mechanism:

$$(Q_{ult})_{block} = \left\{ \sum_{i=1}^{i=n} \left[\alpha_i (c_u)_i \times \text{perimeter}_{ig} \times \text{length}_i \right] \right\} + (c_u)_b N_c (A_b)_g \quad (8.17)$$

where $(Q_{ult})_{block}$ is the pile group capacity with block failure mechanism, perimeter_{ig} is the perimeter of the pile group $= 2(B_g + L_g)$, and $(A_b)_g$ is the cross-sectional area of the pile group $= B_g L_g$.

8.4.2 β-Method

Recall that the β-method can be used for both cohesive and cohesionless soils. The method is based on effective stress analysis and is suited for short- and long-term analyses of pile load capacity. The following equation is based on the β-method and can be used to calculate pile group capacity assuming block failure mechanism:

$$(Q_{ult})_{block} = \left\{ \sum_{i=1}^{i=n} \left[\beta_i (\sigma'_v)_i \times \text{perimeter}_{ig} \times \text{length}_i \right] \right\} + [(\sigma'_v)_b N_q + c'_b N_c](A_b)_g$$

$$(8.18)$$

where $(Q_{ult})_{block}$ is the pile group capacity with block failure, $\text{perimeter}_{ig} = 2(B_g + L_g)$, and $(A_b)_g = B_g \times L_g$.

Example 8.5   Consider a pile group consisting of four concrete piles with a square cross section 0.6×0.6 m^2 and positioned as shown in Figure 8.15. The embedded length of the piles is $L = 9.15$ m in a thick homogeneous layer of saturated clay with an average undrained shear strength $c_u = 14$ kPa and a friction angle $\phi' = 30°$. The clay layer underlying the base of the piles has an undrained shear strength $c_u = 21.3$ kPa and a friction angle $\phi' = 30°$. Calculate the total load capacity of the pile group. Use the α-method and the β-method. All necessary parameters are included in the figure.

SOLUTION:

$$\text{Single-pile perimeter} = 4(0.6) = 2.4 \text{ m}$$

$$A_b = (0.6)^2 \text{m}^2 = 0.36 \text{ m}^2$$

$$\text{Perimeter}_g = (4)(1.8 + 0.3 + 0.3) = 9.6 \text{ m}$$

$$(A_b)_g = (1.8 + 0.3 + 0.3)^2 = 5.76 \text{ m}^2$$

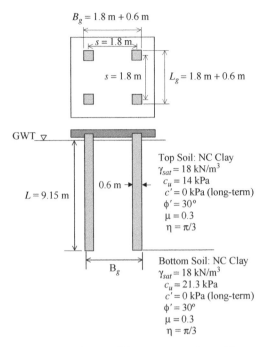

FIGURE 8.15 Pile group consisting of four concrete piles with square cross sections.

Using the α-method, we have the following results:
Block failure mechanism Equation (8.2) yields $\alpha = 1$ since $c_u < 25$ kPa.

$$Q_f = \alpha c_u (\text{perimeter}_g) L = (1)(14)(9.6)(9.15) = 1230 \text{ kN}$$

$$Q_b = 9 c_u (A_b)_g = (9)(21.3)(5.76) = 1106 \text{ kN}$$

$$(Q_{\text{ult}})_{\text{block}} = 1230 + 1106 = 2336 \ kN$$

Single-pile failure mechanism

$$Q_f = \alpha c_u (\text{perimeter})(L) = (1)(14)(2.4)(9.15) = 307 \text{ kN}$$

$$Q_b = 9 c_u A_b = (9)(21.3)(0.36) = 69 \text{ kN}$$

$$Q_{\text{ult}} = 307 + 69 = 376 \text{ kN}$$

$$(Q_{\text{ult}})_{\text{npiles}} = (4)(376) = 1504 \ kN$$

Using the β-method, we have:
Block failure mechanism

$$\beta = \mu(1 - \sin \phi')(\text{OCR})^{0.5} = 0.3(1 - \sin 30°)(1)^{0.5} = 0.15$$

$$\sigma'_v = (18 - 9.81)\left(\frac{9.15}{2}\right) = 37.5 \text{ kPa (at the midpoint of the pile)}$$

$$(\sigma'_v)_b = (18 - 9.81)(9.15) = 74.9 \text{ kPa (at the base of the pile)}$$

$$N_q = \left(\tan 30° + \sqrt{1 + \tan^2 30°}\right)^2 \exp\left(2\frac{\pi}{3}\tan 30°\right) = 10.05$$

$$N_c = (10.05 - 1)\cot 30° = 15.7$$

$$Q_f = \beta\sigma'_v \times \text{perimeter}_g \times \text{length} = (0.15)(37.5)(9.6)(9.15) = 494 \text{ kPa}$$

$$Q_b = \left[(\sigma'_v)_b N_q + c'_b N_c\right](A_b)_g = [(74.9)(10.05) + (0)(15.7)](5.76)$$

$$= 4336 \text{ kPa}$$

$$(Q_{ult})_{block} = 494 + 4336 = 4830 \ kN$$

Single-pile failure mechanism

$$Q_f = (0.15)(37.5)(2.4)(9.15) = 123 \text{ kN}$$

$$Q_b = [(74.9)(10.05) + (0)(15.7)](0.36) = 271 \text{ kN}$$

$$Q_{ult} = 123 + 271 = 395 \text{ kN}$$

$$(Q_{ult})_{npiles} = (4)(395) = 1578 \ kN$$

Compare the load capacities calculated. You will note that the single-pile failure mechanism using the α-method gives the lowest load capacity, 1504 kN. Thus, the ultimate load capacity of this pile group is 1504 kN. Assuming a safety factor of 3, the allowable load on this pile group is (1504 kN)/3 = 501 kN.

Example 8.6   Using the finite element method, calculate the total load capacity of the pile group described in Example 8.5. Assume both (a) undrained and (b) drained conditions. Compare your answers with those obtained using the α- and β-methods, respectively (Example 8.5). All necessary parameters are included in Figure 8.15.

SOLUTION: *Finite element solution* (filename: Chapter8_Example6_alpha.cae, Chapter8_Example6_beta.cae)   (a) *Undrained loading condition*   In this example a limit equilibrium solution is sought for a thick layer of clay loaded in undrained conditions by a pile group consisting of four 9.15-m-long concrete piles. Each pile has a square cross section of $0.6 \times 0.6 \text{ m}^2$. The problem geometry, boundary conditions, and materials are identical to those of Example 8.5, providing a direct means to compare the finite element analysis results with the analytical solution obtained in Example 8.5 using the α- and β-methods.

Pile group problems are three-dimensional by nature and are treated as such in the following finite element analysis. The pile group in this example has a square arrangement as shown in Figure 8.15; therefore, the finite element mesh of the pile group and the surrounding soil can take advantage of this symmetric condition. Only one-fourth of the geometry is considered, as shown in Figure 8.16. The finite element mesh of this fourth is shown in Figure 8.17. This simplification, however,

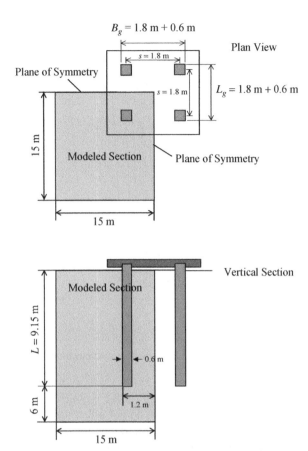

FIGURE 8.16   Simplifying the problem of Example 8.6 using symmetry planes.

cannot be used for pile groups that are loaded laterally. Such problems are not symmetrical and the entire group should be analyzed together.

As in any soil–pile system, interface elements that are capable of simulating the frictional interaction between the pile surface and the soil must be used. Note that in a finite element modeling of a pile group it is not possible to enforce a specific mode of failure: single–pile failure mechanism or block failure mechanism. Any of these two mechanisms could occur, depending on problem configuration, properties of the piles and the surrounding soil, and loading conditions (drained or undrained loading). In this analysis, the piles are assumed to be embedded in the soil before applying pile loads. Excess pore water pressures caused by pile driving are assumed to have been dissipated completely before the application of pile loads.

The three-dimensional finite element mesh (Figure 8.17) comprises two parts: the concrete pile and the soil. Interface elements between the two parts are used. Note that the mesh takes advantage of symmetry about two orthogonal planes as shown in Figures 8.16 and 8.17. The mesh is 15 m long (in the x-direction), 15 m

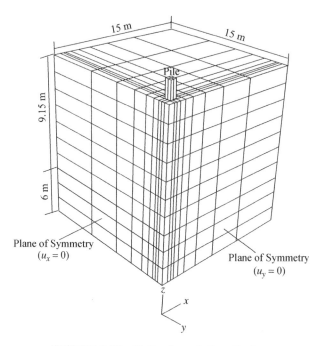

FIGURE 8.17   Finite element discretization.

wide (in the y-direction), and 15.15 m high (in the z-direction). Mesh dimensions should be chosen such that the boundaries do not affect the solution. This means that the mesh must be extended in all three dimensions. It is accepted that the length and the width of the mesh should be greater than $30D$, where D is the diameter of the pile; and the height of the mesh should be greater than $2L$, where L is the embedded length of the pile. The present mesh is used only for illustration and does not conform to these values.

In the analysis, the pile is assumed to be in perfect contact with the soil at the start. The interaction between the pile and the soil is simulated using penalty-type interface elements between the pile and the soil with a friction factor of 0.3. This type of interface is capable of describing the frictional interaction (Coulomb type) between the pile surface and the soil in contact.

Eight-node linear brick elements with reduced integration and without pore water pressure are used for the pile. The elements used for the soil are eight-node pore fluid/stress elements with trilinear displacement, trilinear pore pressure, and reduced integration. The base of the clay layer is fixed in the x, y, and z directions. There are two planes of symmetry as shown in Figure 8.17. The first plane allows sliding in the y and z directions, but prevents displacement in the x-direction. The second symmetry plane allows sliding in the x and z directions but prevents displacement in the y-direction. It is noted that the mesh is finer in the vicinity of the pile since that zone is the zone of stress concentration. No mesh convergence studies have been performed.

The elastic response of the clay is assumed to be linear and isotropic, with a Young's modulus of 68.9×10^3 kPa and a Poisson ratio of 0.3. The cap model, with the parameters $d = 0$ and $\beta = 50.2°$ that were matched to the Mohr–Coulomb failure criterion parameters $c' = 0$ and $\phi' = 30°$, is used to simulate the drained and undrained behavior of the clay. This procedure was explained in detail in Section 8.2. The undrained shear strength of the clay layer, as estimated by this procedure, yields $c_u \approx 14$ kPa at the pile midheight and $c_u \approx 21$ kPa at the pile tip. These values are consistent with the undrained shear strength of the top and bottom clay layers shown in Figure 8.15.

The cap eccentricity parameter is chosen as $R = 0.4$. The initial cap position (which measures the initial consolidation of the specimen) is taken as $\varepsilon^{pl}_{vol(0)} = 0.0$, and the cap hardening curve is assumed to be a straight line passing through two points ($p' = 0.57$ kPa, $\varepsilon^{pl}_{vol} = 0.0$) and ($p' = 103$ kPa, $\varepsilon^{pl}_{vol} = 0.0032$). The transition surface parameter $\alpha = 0.1$ is assumed. The initial void ratio $e_0 = 1.5$, and the vertical and horizontal effective stress profiles of the clay layer are part of the input data that must be supplied to the finite element program for this coupled (consolidation) analysis. In this analysis, the initial horizontal effective stress is assumed to be 50% of the vertical effective stress.

The problem is run in two steps. In step 1 the effective self-weight ($= \gamma'z$, where $\gamma' = \gamma_{sat} - \gamma_w$, and z is the depth below the ground surface) of the clay layer is applied using the "body-force" option. Note that the groundwater table is coincident with the ground surface (Figure 8.15). As mentioned earlier, the clay layer is assumed to be elastoplastic obeying the cap model. In general, using such a model is essential for limit equilibrium analysis. We are concerned with the ability of the clay layer to withstand the end-bearing stresses and skin shear stresses caused by the pile, and models like the cap model, the Cam clay model, and the Mohr–Coulomb model can detect failure within the clay layer. During step 1, the "geostatic" command is invoked to make sure that equilibrium is satisfied within the clay layer. The geostatic option makes sure that the initial stress condition in any element within the clay layer falls within the initial yield surface of the cap model.

In step 2 a coupled (consolidation) analysis is invoked and the pile load is applied using the vertical displacement boundary condition to force the top surface of the pile to move downward a distance of 30 cm at a constant rate (0.003 cm/s). This loading rate is found sufficient to invoke undrained soil behavior. Figure 8.18 shows the evolution of excess pore water pressure in a soil element located immediately below the pile tip. The substantial increase in the excess pore water pressure during pile loading is indicative that the loading is undrained.

The pile group capacity ($=$ pile load $\times$ 4 because of symmetry) versus settlement curve obtained from the finite element analysis is shown in Figure 8.19. It is noted from the figure that the settlement increases as the load is increased up to about a 3000-kN pile load, at which a pile group settlement of about 4 cm is encountered. Shortly after that, the pile moves downward at a greater rate, indicating that the load capacity of the pile has been reached. For comparison, the pile group capacity of 1504 kN, predicted by the α-method with the single-pile failure mechanism

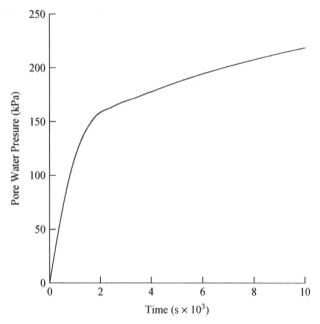

FIGURE 8.18   Evolution of excess pore water pressure (during undrained loading) in a soil element located immediately below the pile tip.

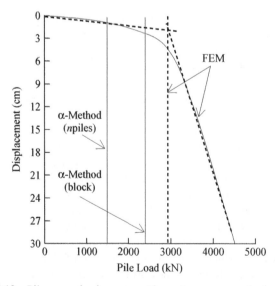

FIGURE 8.19   Pile group load versus settlement curve: α-method versus FEM.

(Example 8.5), is shown in Figure 8.19. The pile group capacity of 2336 kN, predicted by the α-method with the block failure mechanism (Example 8.5), is also shown in the figure. It is noted that the finite element prediction of pile group load capacity is greater than the bearing capacity predicted by the α-method with both failure mechanisms.

(b) *Drained loading condition* The problem geometry, boundary conditions, and materials are identical to those of the undrained loading condition described in part (a). The finite element solution for the drained loading condition will be compared with the analytical solution obtained in Example 8.5 using the β-method. The finite element mesh for this problem is identical to the one used in part (a). The material parameters used for this example are also identical to those of part (a). The problem is run in two steps. In step 1 the effective self-weight of the clay layer is applied in a manner identical to part (a).

The main difference is the loading rate applied in step 2: A drained condition indicates a low loading rate. Also, the top surface of the clay layer in this part is made permeable to allow the excess pore water pressure to dissipate. The present example involves long-term (drained) loading conditions, indicating that the pile should be loaded very slowly to prevent the generation of excess pore water pressure anywhere within the finite element mesh. Thus, in step 2 a coupled (consolidation) analysis is invoked and the pile load is applied using the vertical displacement boundary condition to force the top surface of the pile to move downward a distance of 30 cm at a very small constant rate $= 3 \times 10^{-13}$ cm/s. This small loading rate is used to make sure that the soil will behave in a drained manner. To confirm that, the excess pore water pressure in an element located immediately below the pile tip is plotted in Figure 8.20. It is noted from the

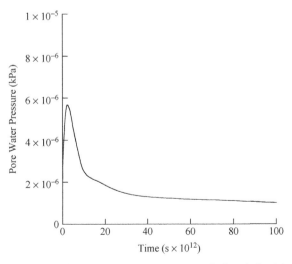

FIGURE 8.20 Evolution of excess pore water pressure (during drained loading) in a soil element located immediately below the pile tip.

figure that the excess pore water pressure is essentially zero in this element during loading, indicating that it is indeed a drained loading condition.

The pile group load versus settlement curve obtained from the finite element analysis is shown in Figure 8.21. It is noted from the figure that the settlement increases as the load is increased up to about 2000 kN, at which a pile group settlement of about 2.5 cm is encountered. Following that, the pile group suddenly moves downward at a much higher rate, indicating that the load capacity of the pile group has been reached. For comparison, the pile group capacity of 1578 kN, predicted by the β-method with the single-pile failure mechanism (Example 8.5), is shown in Figure 8.21. The pile group capacity of 4830 kN, predicted by the β-method with the block failure mechanism (Example 8.5), is also shown in the figure. It is noted that the finite element prediction of pile group load capacity is slightly greater than the bearing capacity predicted by the α-method with the single-pile failure mechanism, but much smaller than the load capacity predicted by the α-method with the block failure mechanism.

8.5 SETTLEMENTS OF SINGLE PILES AND PILE GROUPS

Working loads cause piles and pile groups to settle. Calculating such settlements is usually difficult because of the many factors that must be considered. These factors include the working load magnitude, the pile shape and dimensions, the group shape and spacing, the soil–pile interface characteristics, the soil strata stiffness and strength, and the stiffness of the pile(s). Fortunately, the settlements of the pile head and the cap of a pile group rarely exceed 10 mm when they are

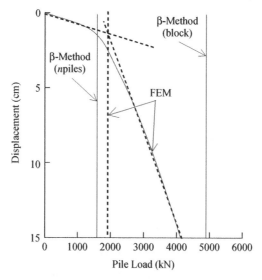

FIGURE 8.21   Pile group load versus settlement curve: β-method versus FEM.

subjected to working loads. Note that the working load is several times smaller than the ultimate capacity of a single pile or a pile group.

There are several methods available to estimate pile and pile group settlements: empirical, semiempirical, analytical, and numerical methods. Except for the numerical method, these methods are beyond the scope of this book. But for a rough estimate of the settlement, S, of a single pile subjected to a working axial load, the following empirical equations can be used: For bored piles, $0.003D \leq S \leq 0.01D$, where D is the diameter of the pile; for driven piles, $0.008D \leq S \leq 0.012D$.

In the realm of finite element analysis, calculating the settlements of single piles and pile groups at any loading level is readily available as part of the solution. Let's take, for instance, the 16-m-long pile analyzed in Example 8.4. Let's calculate the settlement of the pile head corresponding to the allowable load, Q_{all}, assuming a safety factor FS = 3.

The ultimate load $Q_{ult} = 530$ kPa was estimated from the load versus settlement curve shown in Figure 8.13, which was obtained from the finite element analysis described in Example 8.4. The allowable load Q_{all} can be calculated by dividing the ultimate load by the safety factor: $Q_{all} = Q_{ult}/\text{FS} = 530/3 = 177$ kPa. The pile head settlement $S \approx 4$ mm that corresponds to a load $= 177$ kPa can be estimated with the help of Figure 8.13.

8.6 LATERALLY LOADED PILES AND PILE GROUPS

Fully and partially embedded piles and drilled shafts can be subjected to lateral loads as well as axial loads in various applications, including sign posts, power poles, marine pilings, and post-and-panel retaining walls. Piles and drilled shafts resist lateral loads via shear, bending, and earth passive resistance. Thus, their resistance to lateral loads depends on (1) pile stiffness and strength (pile configuration, in particular the pile length-to-diameter ratio, plays an important role in determining pile stiffness, hence its ability to resist shear and bending moments); (2) soil type, stiffness, and strength; and (3) end conditions: fixed end versus free end.

Several analytical approaches are available for the design of laterally loaded piles and drilled shafts. These approaches can be divided into three categories: elastic approach, ultimate load approach, and numerical approach. The elastic approach is used to estimate the response of piles subjected to working loads assuming that the soil and the pile behave as elastic materials. Ultimate loads cannot be calculated using this approach. (Why?) Matlock and Reese (1960) proposed a method for calculating moments and displacements along a pile embedded in a cohesionless soil and subjected to lateral loads and moments at the ground surface. They used a simple Winkler's model that substitutes the elastic soil that surrounds the pile with a series of independent elastic springs. Using the theory of beams on an elastic foundation, they were able to obtain useful equations that allow the calculation of lateral deflections, slopes, bending moments, and shear forces at any point along the axis of a laterally loaded pile. A similar elastic solution by Davisson and Gill

(1963) is also available for laterally loaded piles embedded in cohesive soils. Note that this approach requires that the coefficient of subgrade reaction at various depths be known. Best results can be obtained if this coefficient is measured in the field, but that is rarely done. Several methods that use the ultimate load approach are available for the design of laterally loaded piles and drilled shafts (e.g., Broms, 1965; and Meyerhof, 1995). These methods provide solutions in the form of graphs and tables that are easy for students and engineers to use.

8.6.1 Broms' Method

The ultimate load approach embodied in Broms' method is suitable for short and long piles, for restrained- and free-headed piles, and for cohesive and cohesionless soils. A short pile will rotate as one unit when it is subjected to lateral loads as shown in Figure 8.22a. The soil in contact with the short pile is assumed to fail in shear when the ultimate lateral load is reached. On the other hand, a long pile is assumed to fail due to the bending moments caused by the ultimate lateral load; that is, the shaft of the pile will fail at the point of maximum bending moment, forming a plastic hinge as shown in Figure 8.22b. Also, the term *restrained-headed pile* indicates that the head of the pile is connected to a rigid cap that prevents the head of the pile from rotation. Broms' method assumes that the pile is equivalent to a beam on an elastic foundation.

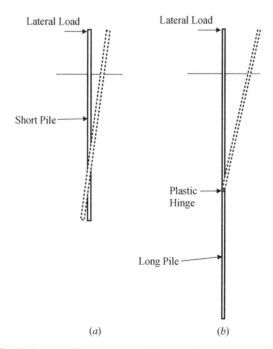

FIGURE 8.22 Definition of (*a*) short and (*b*) long piles in terms of lateral loading.

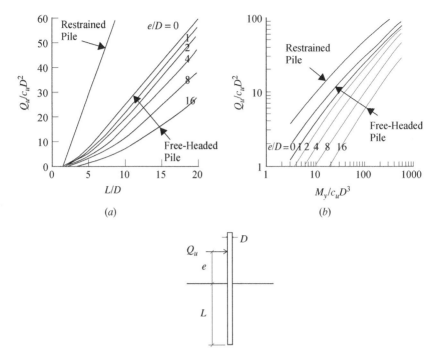

FIGURE 8.23 Broms' method solution for piles embedded in cohesive soils: (*a*) short piles; (*b*) long piles.

Broms' method presents the solution for short piles embedded in cohesive soils with a set of curves shown in Figure 8.23*a*. The curves relate the pile's embedment length-to-diameter ratio, L/D, to the normalized ultimate lateral force, $Q_u/c_u D^2$, for various e/D ratios. In general, piles having a length-to-diameter ratio L/D greater than 20 are long piles. Figure 8.23*b* can be used for long piles embedded in cohesive soils. The curves in this figure relate the normalized ultimate lateral force, $Q_u/c_u D^2$, to the normalized yield moment of the pile, $M_{\text{yield}}/c_u D^3$, for various e/D ratios. These curves are used only when $L/D > 20$ and when the moment generated by the ultimate lateral load is greater than the yield moment of the pile.

For short piles embedded in cohesionless soils, Broms' method provides the curves given in Figure 8.24*a*, which relate the pile's embedment length-to-diameter ratio L/D to the normalized ultimate lateral force $Q_u/K_p D^3 \gamma$ for various e/D ratios. Note that K_p is the passive lateral earth pressure coefficient and γ is the unit weight of the soil around the pile.

Figure 8.24*b* can be used for long piles embedded in cohesionless soils. The curves in this figure relate the normalized ultimate lateral force $Q_u/K_p D^3 \gamma$ to the normalized yield moment of the pile $M_{\text{yield}}/K_p D^4 \gamma$ for various e/D ratios. These curves are used only when $L/D > 20$ and when the moment generated by the ultimate lateral load is greater than the yield moment of the pile.

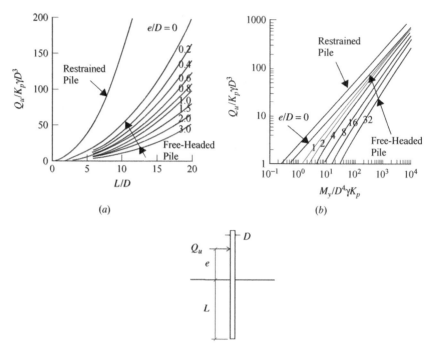

(a)   (b)

FIGURE 8.24 Broms' method solution for piles embedded in cohesionless soils: (a) short piles; (b) long piles.

Example 8.7 Using Broms' method, calculate the ultimate lateral load capacity of the pile shown in Figure 8.25. The embedded potion of the pile is 3.66 m long and is made of a 0.75-m-diameter concrete cylinder (bored pile). The aboveground portion of the pile is 3.96 m long and is made of an HP steel section. The lateral

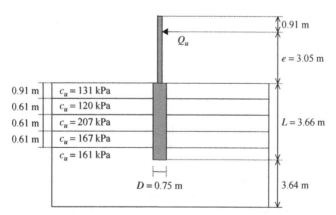

FIGURE 8.25 Problem configuration for Example 8.7: calculation of the ultimate lateral load capacity of a pile.

load is applied 3.05 m above the ground level. The soil consists of five layers with varying undrained shear strengths as indicated in the figure.

SOLUTION: Let's calculate the average undrained shear strength of the five layers:

$$c_u = \frac{131 + 120 + 207 + 167 + 161}{5} = 157 \text{ kPa}$$

$$\frac{L}{D} = \frac{3.66}{0.75} = 4.88 < 20 \rightarrow \text{short pile}$$

$$\frac{e}{D} = \frac{3.05}{0.75} = 4$$

From Figure 8.23a (Broms' method for short piles in cohesive soils) we obtain the normalized ultimate lateral force $Q_u/c_u D^2 \approx 3$. Thus, the ultimate lateral load capacity is

$$Q_u \approx 3c_u D^2 \approx 3(157)(0.75)^2 \approx 265 \text{ kN}$$

8.6.2 Finite Element Analysis of Laterally Loaded Piles

Finite element analysis of laterally loaded piles is similar to that of axially loaded piles except for loads that are applied in the horizontal direction at or above the ground level. These loads can be concentrated loads or moments or a combination of the two. In the case of an axially loaded pile, the finite element mesh of the pile and the surrounding soil can take advantage of axisymmetry. Unfortunately, this simplification cannot be taken advantage of for piles that are loaded with lateral loads. Such piles should be treated using three-dimensional analysis as shown in the following example.

Example 8.8   Using the finite element method, calculate the ultimate lateral load capacity of the pile described in Example 8.7. Assume undrained loading conditions. Compare your answer with that obtained using Broms' method (Example 8.7). Also, compare the lateral load–displacement curve from your finite element analysis with the field test results obtained by the Ohio Department of Transportation (ODOT) on the same pile. Actually, the pile configuration, soil strata, and soil properties assumed in Examples 8.7 and 8.8 are taken from an experiment conducted by ODOT.

SOLUTION: *Finite element solution* (filename: Chapter8_Example8.cae)   In this example a limit equilibrium solution is sought for clay strata loaded in undrained conditions by a single pile with a lateral load applied above ground level. The problem geometry, boundary conditions, and materials are identical to those of Example 8.7, providing a direct comparison of the finite element analysis results with the analytical solution obtained in Example 8.7 using Broms' method.

Piles with lateral loads are three-dimensional by nature and will be treated as such in the following finite element analysis. Note that this problem is symmetrical about a plane that contains the axis of the pile and the line of action of the lateral load (Figure 8.26). Thus, the finite element mesh of half of the pile and the surrounding soil is considered as shown in Figure 8.27. Interface elements that

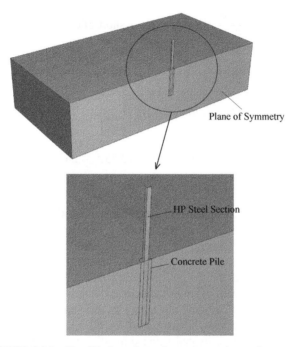

FIGURE 8.26 Simplified configuration using a plane of symmetry.

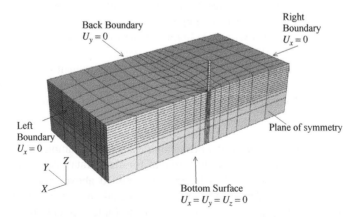

FIGURE 8.27 Finite element discretization.

are capable of simulating the frictional interaction between the pile surface and the soil are used. Since this is a nondisplacement (bored) pile, the excess pore water pressure after pile installation is assumed to be zero in this finite element analysis.

The three-dimensional finite element mesh (Figure 8.27) comprises two parts: the concrete pile with the aboveground HP section and the soil. Note that the mesh takes advantage of symmetry about a vertical plane as indicated in Figure 8.27. The mesh is 30 m long (in the x-direction), 15 m wide (in the y-direction), and 7.3 m high (in the z-direction). Mesh dimensions should be chosen such that the boundaries do not affect the solution. This means that the mesh must be extended in all three dimensions, and that is considered in this mesh.

In the analysis, the pile is assumed to be in perfect contact with the soil at the start. Interaction between the pile and the soil is simulated using penalty-type interface elements between the pile and the soil with a friction factor of 0.3. This type of interface is capable of describing the frictional interaction (Coulomb type) between the pile surface and the soil in contact.

Eight-node linear brick elements with reduced integration and without pore water pressure are used for the pile. The elements used for the soil are eight-node pore fluid/stress elements with trilinear displacement, trilinear pore pressure, and reduced integration. The base of the clay strata is fixed in the x, y, and z directions. There is one plane of symmetry that allows sliding in the x and z directions, as shown in Figure 8.27. It is noted that the mesh is finer in the vicinity of the pile since that zone is the zone of stress concentration. No mesh convergence studies have been performed.

The elastic response of the clay layers is assumed to be linear and isotropic, with a Young's modulus that is a function of the undrained shear strength of each layer as indicated in Table 8.1. The cap model parameters d and β are given in the same table for each soil layer. These parameters were matched to the Mohr–Coulomb failure criterion parameters c_u (Table 8.1) and $\phi_u = 0$ to simulate the undrained behavior of the clay layers.

The cap eccentricity parameter is chosen as $R = 0.1$ for all clay layers. The initial cap position (which measures the initial consolidation of the specimen) is taken as $\varepsilon^{pl}_{vol(0)} = 0.0$, and the cap hardening curve is assumed to be a straight line passing through two points [($p' = 35$ kPa, $\varepsilon^{pl}_{vol} = 0.0$) and ($p' = 1034$ kPa, $\varepsilon^{pl}_{vol} = 0.0464$)]. The transition surface parameter $\alpha = 0$ is assumed. The initial void

TABLE 8.1 Undrained Soil Properties

Soil Layer	Depth (m)	c_u (kPa)	E (MPa)	d (kPa)	β
1 (top)	0–0.91	131	32.7	226	0
2	0.91–1.52	120	30	208	0
3	1.52–2.13	207	51.7	358	0
4	2.13–2.74	167	41.9	290	0
5 (bottom)	2.74–7.31	161	40.3	280	0

ratio $e_0 = 1.5$ and the vertical and horizontal effective stress profiles of the clay strata are part of the input data that must be supplied to the finite element program for this coupled (consolidation) analysis. In this analysis, the initial horizontal effective stress is assumed to be 50% of the vertical effective stress.

The problem is run in two steps. In step 1 the effective self-weight ($= \gamma'z$, where $\gamma' = \gamma_{sat} - \gamma_w$ and z is the depth below the ground surface) of the clay layer is applied using the "body-force" option. During step 1, the "geostatic" command is invoked to make sure that equilibrium is satisfied within the clay layer. The geostatic option makes sure that the initial stress condition in any element within the clay strata falls within the initial yield surface of the cap model.

In step 2 a coupled (consolidation) analysis is invoked and the pile lateral load is applied using the concentrated load option. A loading rate of 0.891 kN/s is used and found sufficient to invoke undrained soil behavior (because of the low permeability of the soil). The pile lateral load capacity ($=$ pile load $\times 2$ because of symmetry) versus displacement curve obtained from the finite element analysis is shown in Figure 8.28. It is noted from the figure that the horizontal displacement increases as the lateral load is increased up to about a 475-kN pile load, at which a pile lateral displacement of about 5 cm is encountered. Shortly after that, the pile moves laterally at a greater rate, indicating that the lateral load capacity of the pile has been reached. For comparison, the pile lateral load capacity of 265 kN predicted by Broms' method (Example 8.7) is shown in Figure 8.28. It is noted

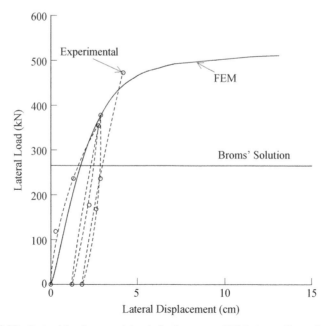

FIGURE 8.28 Lateral load versus lateral displacement: FEM versus Broms' solution and experimental data.

that the finite element prediction of pile lateral load capacity is about two times greater than the capacity predicted by Broms' method.

Figure 8.28 includes the load–displacement curve obtained from the field test conducted on the same pile by ODOT. The finite element results are in relatively good agreement with the measured results. Note that the ODOT test included two cycles of loading and unloading, and that the pile test was terminated at a lateral load approaching 480 kN. It can be concluded from the figure that the ultimate lateral load of the pile predicted by Broms' method grossly underestimated the measured lateral load capacity of the pile.

Figure 8.29 shows the distribution of the horizontal stresses within the soil strata that corresponds to the ultimate lateral load of the pile (≈ 475 kN). This stress distribution is consistent with the stress distribution expected for a short pile with lateral loading. A short pile will rotate, acting as one unit, when subjected to a lateral load as indicated in Figure 8.22a. Intuitively, this type of rotation would cause a passive earth pressure distribution similar to the one shown in Figure 8.29.

Example 8.9 A post-and-panel wall is a wall type that has gained some use because it offers advantages under certain conditions. A typical post design consists of a steel H section set in a column of concrete as shown in Figure 8.30. The column of concrete extends from the final ground surface to the base elevation computed

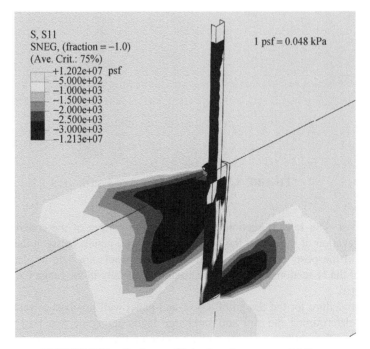

FIGURE 8.29 Contours of lateral earth pressure at failure.

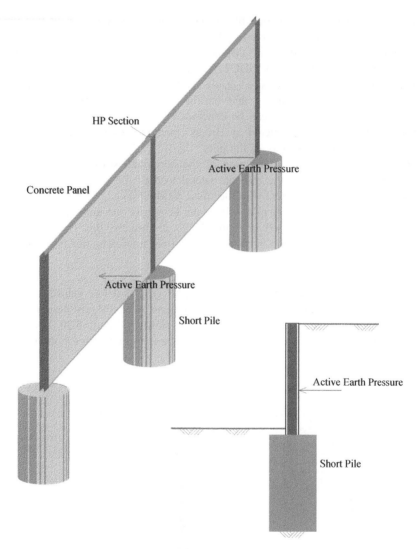

FIGURE 8.30   Post-and-panel wall system.

for the post. The steel H section extends from the bottom of the concrete to the design elevation of the top of the wall. Installing wall panels between the exposed sections of the posts completes construction. The panels are held in place by the flanges of the H sections. The concrete column is usually in the range 0.6 to 1.2 m in diameter.

The procedure for the design of a post-and-panel wall involves selecting a post spacing, determining the soil and surcharge loads acting on that post and then determining the optimum length and diameter of the post necessary to develop passive soil pressures sufficient to resist loads acting on the post. A post-and-panel wall may be designed either as a cantilever system or as a tieback system. This type of wall may be used either for conventional "bottom-up" construction or to retain existing facilities by "top-down" construction.

It is proposed that the concrete column be eliminated and replaced by a steel plate of the same width and length as the concrete column, as shown in Figure 8.31. This plate would be welded to the H section and then the composite unit would be driven into the ground to the plan base elevation required. The remainder of the construction would proceed without change. This alternative post system (termed

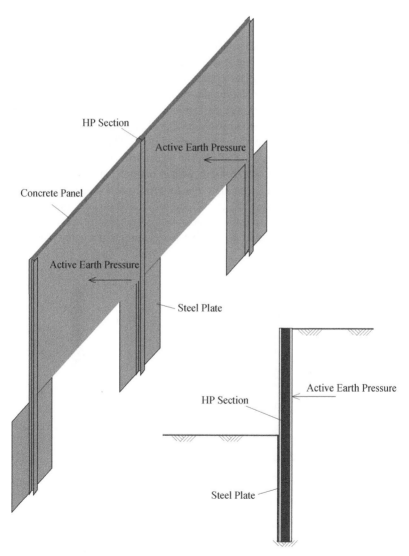

FIGURE 8.31 Plate system alternative to the post-and-panel wall system.

"the plate system" henceforth) offers benefits such as ease of construction, reduced construction time, and lower wall costs. Although this system seems feasible, there are concerns regarding its performance, in particular the amount of bending in the post and the defection of the wall due to active earth pressures exerted by the retained soil.

The objective of this example is to assess the feasibility of the plate system. If the system is deemed feasible, one can use finite element analyses to develop design criteria (not part of this exercise) for the plate system based on exposed wall heights, applied soil loads, post dimensions, and parameters of the retained soil and the foundation soil.

SOLUTION: *Finite element solution* (filename: Chapter8_Example9_concrete.cae, Chapter8_Example9_steel.cae)   In this example we carry out a preliminary feasibility analysis of the proposed plate system using the finite element code that was verified in Example 8.8 by simulating the behavior of a full-scale field test of a post with a concrete column that was performed by the Ohio DOT. The feasibility analysis consists of an objective comparison of the behavior of a conventional post with a concrete column and a post with a welded plate. The two post systems analyzed herein are shown in Figure 8.32. In the figure the conventional post system consists of a 5.3-m-long concrete column with a diameter of 0.91 m. The aboveground H section is 5.3 m long. The plate system proposed consists of a 5.3-m-long, 0.91-m-wide, 25-mm-thick steel plate welded to a 10.6-m-long post with an H section (5.3 m aboveground). In both systems, the lateral load is applied at the one-third point measured from ground level. This is to simulate the active force caused

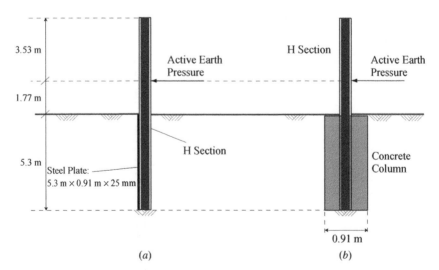

FIGURE 8.32 Problem configuration for Example 8.9: (*a*) post with plate system; (*b*) post with concrete column system.

by the triangular distribution of the lateral earth pressure exerted by the retained soil on the concrete panel (see Figure 8.30). The foundation soil is assumed to be a medium-dense sand with $c' = 0$ and $\phi' = 37°$. The three-dimensional finite element meshes of the two systems are not shown here but resemble the mesh used in Example 8.8.

Figure 8.33 shows the predicted horizontal displacement versus applied lateral load for both post systems. In the figure the horizontal displacement is the displacement of the post at ground level. It is clear from the figure that the conventional post with concrete column system is much stiffer than the plate system proposed. With reference to Figure 8.31, for a 5.3-m-high concrete panel spanning 3 m between two posts center to center, the active lateral force exerted on the concrete panel is approximately 200 kN. From Figure 8.33 this lateral load will cause a horizontal displacement of 4 mm in the conventional post with a concrete column system. By contrast, the same load will cause 25 mm of horizontal displacement in the plate system proposed. This large difference in displacement is attributed to the large flexural stiffness of the concrete column in a conventional post system compared to the flexural stiffness of the 25-mm-thick steel plate in the plate system proposed. Using typical values of Young's moduli for steel and concrete ($E_{\text{steel}} = 206$ GPa and $E_{\text{concrete}} = 28.7$ GPa), we can calculate the stiffness of the concrete column as

$$\frac{E_{\text{concrete}} I_{\text{column}}}{L_{\text{column}}} = \frac{E_{\text{concrete}} \left(\frac{1}{4}\pi r^4\right)}{L_{\text{column}}} = \frac{28.7 \left[\frac{1}{4}\pi(0.457)^4\right]}{5.3} = 186 \times 10^3 \text{ kN} \cdot \text{m}$$

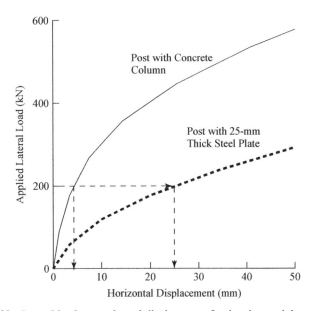

FIGURE 8.33   Lateral load versus lateral displacement for the plate and the post-and-panel wall systems.

and the stiffness of the steel plate is

$$\frac{E_{steel}I_{plate}}{L_{plate}} = \frac{E_{steel}\left(bt^3/12\right)}{L_{plate}} = \frac{(206)(0.91)[(0.0254)^3/12]}{5.3} = 48.3 \text{ kN} \cdot \text{m}$$

This means that the stiffness of the concrete column is approximately 4000 times greater than the stiffness of the steel plate for this specific example!

A question arises now: *Is 25 mm of horizontal displacement tolerable for the plate system proposed under a working load of 200 kN?* If the answer is no, the plate system proposed needs to be improved. In such a case one can increase the stiffness of the plate and the stiffness of the post (H section). Note that this discussion is based on only a single analysis, and many other cases need to be considered before offering remedies for the plate system proposed. Nevertheless, if more stiffness is needed, one can increase the thickness, and possibly the width, of the plate. Also, plate stiffeners can be used for added stiffness (stiffeners are welded steel sheets that are orthogonal to the plate). Anchors can also be used to reduce lateral displacements. Another option is to increase the embedded length of the post and the welded plate. All these options can be investigated easily using the finite element mesh provided with this example (with minor modifications).

PROBLEMS

8.1 A concrete-filled pipe pile (Figure 8.34) with an external diameter $D = 0.5$ m and length $L = 18.5$ m is driven into a thick homogeneous NC clay layer (the embedded length of the pile is 18 m). The undrained shear strength of the NC clay layer varies with depth: $c_u = (10 + 5z)$ kPa. The groundwater table is 3 m below the ground surface, the bulk unit weight of the soil is 18 kN/m³, and the saturated unit weight is 19 kN/m³. Calculate the ultimate load capacity of the pile using the α-method.

8.2 Redo Problem 8.1 using the finite element method. Since the pile has a circular cross section and is axially loaded, it can be assumed axisymmetric. If the cap model is used, the cap parameters β and d can be calculated using the short-term strength parameters c_u and $\phi_u = 0$ (divide the clay layer into several layers to account for the variation of c_u with depth). The cap eccentricity parameter for this soil is $R = 0.5$. The initial cap position (which measures the initial consolidation of the specimen) is $\varepsilon^{pl}_{vol(0)} = 0.0$, and the cap hardening curve is a straight line passing through two points $[(p' = 1 \text{ kPa}, \varepsilon^{pl}_{vol} = 0.0)$ and $(p' = 500 \text{ kPa}, \varepsilon^{pl}_{vol} = 0.012)]$. The transition surface parameter is $\alpha = 0.05$. Compare the results of the finite element analysis with the analytical solution obtained in Problem 8.1.

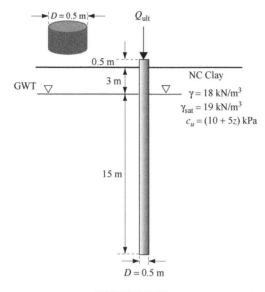

FIGURE 8.34

8.3 An open-ended steel tube pile (Figure 8.35) with an external diameter $D = 0.5$ m and a wall thickness of 6 mm is driven into a thick homogeneous layer of saturated clay with a friction angle $\phi' = 33°$ and cohesion intercept $c' = 0$. The groundwater table is coincident with the top surface of the soil. Assume that $\eta = \pi/3$ and a soil–pile friction factor $\mu = 0.35$. Using the β-method, calculate the required length of the pile to carry an ultimate load of 900 kN.

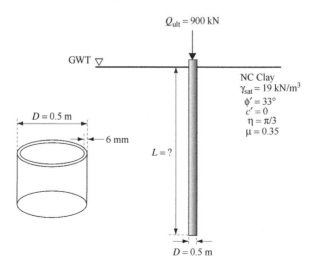

FIGURE 8.35

8.4 Using the finite element method with the cap model, calculate the ultimate load capacity of a 20-m-long open-ended steel tube pile (Figure 8.35) with an external diameter $D = 0.5$ m and a wall thickness of 6 mm. The pile is driven into a thick homogeneous layer of saturated clay with a friction angle $\phi' = 33°$ and cohesion intercept $c' = 0$. The groundwater table is coincident with the top surface of the soil. Assume a soil–pile friction factor $\mu = 0.35$.

The cap parameters β and d can be calculated using the long-term strength parameters $\phi' = 33°$ and $c' = 0$. The cap eccentricity parameter for this soil is $R = 0.5$. The initial cap position (which measures the initial consolidation of the specimen) is $\varepsilon^{pl}_{vol(0)} = 0.0$, and the cap hardening curve is a straight line passing through two points $[(p' = 1 \text{ kPa}, \varepsilon^{pl}_{vol} = 0.0)$ and $(p' = 500 \text{ kPa}, \varepsilon^{pl}_{vol} = 0.012)]$. The transition surface parameter is $\alpha = 0.05$.

8.5 Redo Problem 8.4 using the Cam clay model. The Cam clay parameter M can be calculated from the long-term strength parameter $\phi' = 33°$. The soil has $\lambda = 0.12$, $\kappa = 0.02$, $e_0 = 1.42$, and OCR = 1.2.

8.6 Calculate the total settlement (elastic settlement + consolidation settlement) of the pile in Problem 8.4. Use a working load that is four times smaller than the ultimate load capacity calculated in Problem 8.4.

8.7 A concrete-filled pipe pile (Figure 8.36) with an external diameter $D = 0.8$ m and length $L = 25$ m is driven into a 10-m-thick soft clay layer underlain by a stiff clay layer. The undrained shear strength of the normally consolidated soft clay layer is $c_u = 50$ kPa, its effective strength parameters are $\phi' = 18°$ and $c' = 0$, and its saturated unit weight is 18.5 kN/m³. On the

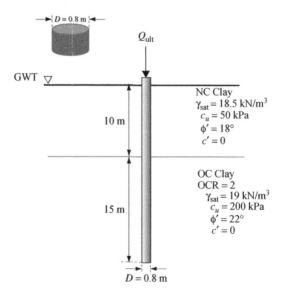

FIGURE 8.36

other hand, the overconsolidated clay layer has an overconsolidation ratio of 2, an undrained shear strength of 200 kPa, effective strength parameters of $\phi' = 22°$ and $c' = 0$, and a saturated unit weight of 19 kN/m^3. The groundwater table is coincident with the ground surface. Calculate the short- and long-term ultimate load capacity of the pile using the α- and β-methods.

8.8 Consider a pile group consisting of four concrete piles with a square cross section 0.9×0.9 m^2 and positioned as shown in Figure 8.37. The embedded length of the piles is $L = 20$ m. The piles are driven into a thick homogeneous layer of saturated clay with an average undrained shear strength $c_u = 74$ kPa and a friction angle $\phi' = 23°$. The clay layer under the base of the piles has an undrained shear strength $c_u = 121$ kPa and a friction angle $\phi' = 27°$. Calculate the ultimate load capacity of the pile group. Use the α- and β-methods. All parameters needed are included in the figure.

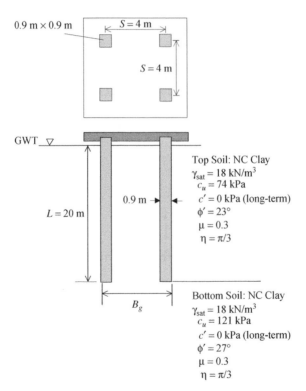

FIGURE 8.37

8.9 Redo Problem 8.8 using the finite element method. Assume both undrained and drained conditions. For drained conditions, the cap parameters β and d can be calculated using the long-term strength parameters ϕ' and c' for each soil layer. For undrained conditions they can be calculated using the undrained shear strength c_u for each soil layer. The top soil layer has the following parameters: the cap eccentricity parameter $R = 0.4$, the initial cap position (which measures the initial consolidation of the specimen) is $\varepsilon_{vol(0)}^{pl} = 0.0$, the cap hardening curve is a straight line passing through two points $[(p' = 1 \text{ kPa}, \varepsilon_{vol}^{pl} = 0.0)$ and $(p' = 500 \text{ kPa}, \varepsilon_{vol}^{pl} = 0.025)]$, and the transition surface parameter is $\alpha = 0.1$. The bottom soil layer has $R = 0.4$, $\varepsilon_{vol(0)}^{pl} = 0.002$, the cap hardening curve is a straight line passing through two points $[(p' = 1 \text{ kPa}, \varepsilon_{vol}^{pl} = 0.0)$ and $(p' = 500 \text{ kPa}, \varepsilon_{vol}^{pl} = 0.01)]$, and $\alpha = 0.1$.

8.10 Calculate the total settlement (elastic settlement + consolidation settlement) of the pile group in Problem 8.9. Use a working load that is three times smaller than the ultimate load capacity calculated in Problem 8.9.

8.11 Using Broms' method, calculate the ultimate lateral load capacity of the 0.75-m-diameter concrete (bored) pile shown in Figure 8.38. The embedded portion of the pile is 25 m long. The aboveground portion of the pile is 5 m long. The lateral load is applied 3 m above ground level. The soil consists of two layers: a soft clay layer, underlain by a stiff clay layer, as shown in the figure. The average undrained shear strength of the top soil layer is $c_u = 74$ kPa and its friction angle is $\phi' = 23°$. The bottom clay has an undrained shear strength $c_u = 121$ kPa and a friction angle $\phi' = 27°$.

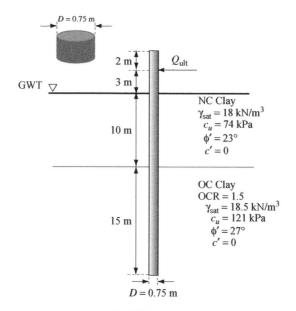

FIGURE 8.38

8.12 Using the finite element method, calculate the ultimate lateral load capacity of the pile described in Problem 8.11. Assume undrained loading conditions. The cap parameters β and d can be calculated using the undrained shear strength c_u for each soil layer. The top soil layer has the following parameters: the cap eccentricity parameter $R = 0.4$, the initial cap position (which measures the initial consolidation of the specimen) is $\varepsilon_{vol(0)}^{pl} = 0.0$, the cap hardening curve is a straight line passing through two points $[(p' = 1 \text{ kPa}, \varepsilon_{vol}^{pl} = 0.0)$ and $(p' = 500 \text{ kPa}, \varepsilon_{vol}^{pl} = 0.025)]$, and the transition surface parameter is $\alpha = 0.1$. The bottom soil layer has: $R = 0.4$, $\varepsilon_{vol(0)}^{pl} = 0.002$, the cap hardening curve is a straight line passing through two points $[(p' = 1 \text{ kPa}, \varepsilon_{vol}^{pl} = 0.0)$ and $(p' = 500 \text{ kPa}, \varepsilon_{vol}^{pl} = 0.01)]$, and $\alpha = 0.1$. Compare your answer with that obtained using Broms' method (Problem 8.11). What is the lateral displacement of the pile at the point of load application when the applied load is one-third of the ultimate load?

CHAPTER 9

PERMEABILITY AND SEEPAGE

9.1 INTRODUCTION

Soils have interconnected voids that form many tortuous tiny tubes that allow water to flow. The average size of these tubes depends on soil porosity, which in turn determines how easy (or difficult) it is for water to seep through the soil. Being able to calculate the quantity of water flowing through a soil and the forces associated with this flow is crucial to the design of various civil engineering structures, such as earth dams, concrete dams, and retaining walls.

The *coefficient of permeability* (or *permeability*) in soil mechanics is a measure of how easily a fluid (water) can flow through a porous medium (soil). Soils with coarser grains have larger voids; therefore, their permeabilities are larger. It follows that gravels are more permeable than sands, and sands are more permeable than silts. Because of their extremely small permeabilities, clays are used to construct the cores of earth dams that act as water barriers. In the environmental engineering literature, *hydraulic conductivity* is often used instead of *coefficient of permeability*, but they have the same meaning.

The flow of water through soils, called *seepage*, occurs when there is a difference in the water level (energy) on the two sides of a structure such as a dam or a sheet pile, as shown in Figure 9.1. The side with a higher water level is called *upstream*, and the side with a lower water level is called *downstream*. Water seeps through the soil under a dam from the zone of high energy (upstream) to the zone of low energy (downstream) in accordance with Bernoulli's equation. The flow velocity is governed by Darcy's equation, which requires knowledge of the coefficient of permeability. It is often necessary to calculate the quantity of seepage (e.g., through

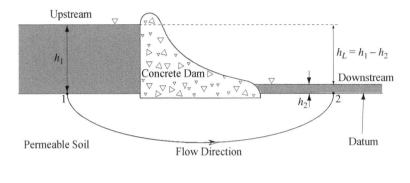

FIGURE 9.1 Seepage through a porous medium.

an earth dam, underneath a concrete dam, or around a sheet pile), and a good estimate of the soil's coefficient of permeability is needed.

The coefficient of permeability k is a soil parameter that depends on the average size of the pores in the soil and is related to the distribution of particle sizes, particle shape, and soil structure. Permeabilities vary widely depending on soil type. The ratio of the permeability of a typical sand to that of a typical clay can be on the order of 10^6. A small fraction of fine material in a coarse-grained soil can lead to a significant reduction in its permeability.

9.2 BERNOULLI'S EQUATION

For a steady-state flow of a nonviscous incompressible fluid, *Bernoulli's equation* (9.1) calculates the total head at a point as the summation of three components: pressure head, elevation head, and velocity head [i.e., total head (h) = pressure head (h_p) + elevation head (h_e) + velocity head (h_v)]:

$$h = h_p + h_e + h_v = \frac{u}{\gamma_w} + z + \frac{v^2}{2g} \qquad (9.1)$$

where u is the fluid pressure, v the velocity at a point within the fluid, and g the gravitational acceleration. The total head and its three components have length units. The elevation head is measured with respect to an arbitrarily selected datum as shown in Figure 9.2. Note that the selected datum must be a horizontal line. If a point such as point A in Figure 9.2 is above the datum, its elevation head is positive. If a point is below the datum then its elevation head is negative. At point A, the fluid pressure u is the pressure felt by an observer (diver) located at that point. The pressure head at point A is given as $h_p = u/\gamma_w$. The velocity head is given as $h_v = v^2/2g$. If we position the tip of a standpipe (piezometer) at point A, the

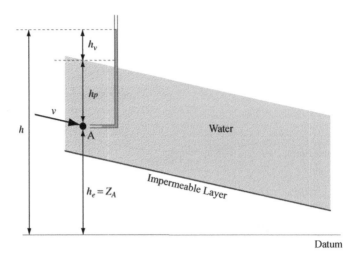

FIGURE 9.2   Steady-state flow of a non viscous incompressible fluid.

water will rise inside it to a height equal to the sum of the pressure head and the velocity head, as shown in Figure 9.2.

Water flowing through soil usually has a very small flow velocity. This means that the velocity head in (9.1) can be neglected. Thus, Bernoulli's equation for flow through soils reduces to

$$h = h_p + h_e = \frac{u}{\gamma_w} + z \tag{9.2}$$

In this equation the total head h at a given point represents the energy possessed by the fluid at that point. While flowing through soil from a point of higher total head (h_1) to a point of a lower total head (h_2), water loses some energy, due to soil resistance (friction). This loss of energy, expressed as the total head loss h_L, is the difference in water levels between upstream and downstream $(h_L = h_1 - h_2)$, as shown in Figure 9.1. In soil mechanics the fluid pressure u is called *pore water pressure or pore pressure*. The pore pressure at any point in the flow region is given as

$$u = h_p \gamma_w \tag{9.3}$$

The hydraulic gradient (i) is defined as the total head loss per unit length. Consider a particle of water flowing from point A to point B as shown in Figure 9.3. The total head at point A is $h_A = u_A/\gamma_w + Z_A$, and the total head at point B is $h_B = u_B/\gamma_w + Z_B$. Note that the total head at A is greater than that at B, which is why the water particle travels from A to B. The head loss between A and B is $h_A - h_B$. The average hydraulic gradient between A and B is the head loss between

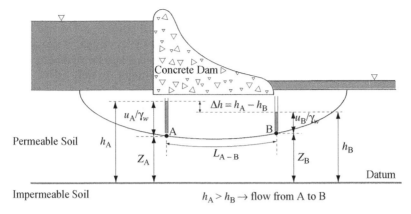

FIGURE 9.3 Definition of head loss and hydraulic gradient.

A and B divided by the length AB along the flow path:

$$i_{A\text{-}B} = \frac{h_A - h_B}{L_{A\text{-}B}} = \frac{(u_A/\gamma_w + Z_A) - (u_B/\gamma_w + Z_B)}{L_{A\text{-}B}} \tag{9.4}$$

Example 9.1 A 0.5-m-long soil specimen is subjected to a steady-state flow with a constant head loss $h_L = 1.5$ m as shown in Figure 9.4. Calculate the total head at point C.

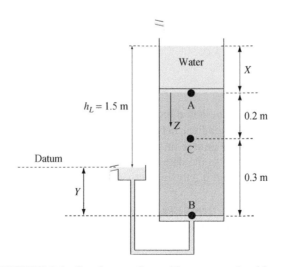

FIGURE 9.4 Steady-state flow with a constant head loss.

SOLUTION: Let us assume that the datum is coincident with the downstream water as shown in Figure 9.4. To calculate the total head at point C, we would need the value of the pressure head at that point [equation (9.2)], and that is not available. Instead, let us calculate the total heads at points A and B and the hydraulic gradient i_{A-B}. Knowing that $i_{A-B} = i_{A-C}$, we can calculate h_C with the help of (9.4).

To simplify the solution we will assume the height of the water in the upper reservoir to be X (above point A), as shown in Figure 9.4. Also, we assume the height of the water in the lower reservoir to be Y (above point B) as shown in the same figure. At point A,

$$h_p = X$$

$$h_e = 1.5 \text{ m} - X$$

$$h_A = h_p + h_e = X + 1.5 \text{ m} - X = 1.5 \text{ m}$$

At point B,

$$h_p = Y$$

$$h_e = -Y$$

$$h_B = h_p + h_e = Y - Y = 0 \text{ m}$$

From (9.4) we have

$$i_{A-B} = \frac{h_A - h_B}{L_{A-B}} = \frac{1.5 \text{ m} - 0 \text{ m}}{0.5 \text{ m}} = 3$$

but $i_{A-B} = i_{A-C}$; therefore,

$$i_{A-C} = \frac{h_A - h_C}{L_{A-C}} = \frac{1.5 \text{ m} - h_C}{0.2 \text{ m}} = 3 \rightarrow h_C = 0.9 \text{ m}$$

Example 9.2 Consider the one-dimensional flow condition shown in Figure 9.5. Determine the hydraulic gradient between points A and B and the flow direction knowing that the pore pressures at points A and B are 2.943 and 11.772 kPa, respectively. With respect to the datum shown in the figure, the elevation heads of points A and B are 0.5 and 0.2 m, respectively.

SOLUTION: Given $u_A = 2.943$ kPa, $u_B = 11.772$ kPa, $Z_A = 0.5$ m, and $Z_B = 0.2$ m. The water will flow from A to B only if h_A is greater than h_B. Applying (9.2) to point A yields

$$h_A = h_p + h_e = \frac{u_A}{\gamma_w} + z_A = \frac{2.943 \text{ kPa}}{9.81 \text{ kN/m}^3} + 0.5 \text{ m} = 0.8 \text{ m}$$

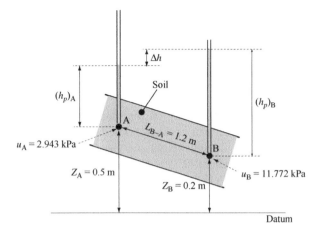

FIGURE 9.5   One-dimensional seepage.

At point B,

$$h_B = h_p + h_e = \frac{u_B}{\gamma_w} + z_B = \frac{11.772 \text{ kPa}}{9.81 \text{ kN/m}^3} + 0.2 \text{ m} = 1.4 \text{ m}$$

Since h_B is greater than h_A, the water will flow from B to A (up the slope). The hydraulic gradient from B to A is

$$i_{B\text{-}A} = \frac{h_B - h_A}{L_{A\text{-}B}} = \frac{1.4 \text{ m} - 0.8 \text{ m}}{1.2 \text{ m}} = 0.5$$

9.3   DARCY'S LAW

When the flow through soil is laminar, *Darcy's law* (Darcy, 1856) applies:

$$v = ki \tag{9.5}$$

In this equation the flow velocity (v) is proportional to the hydraulic gradient (i). The parameter k is the coefficient of permeability. Since i is dimensionless, k must have velocity units. In geotechnical engineering k is commonly given in cm/s or m/s. The coefficient of permeability is strongly affected by the density of packing of the soil particles, which can be expressed simply using void ratio e or porosity n. In coarse-grained soils, the effective grain size D_{10} is well correlated with permeability. For uniform sands with $C_u < 5$ and a D_{10} value of 0.1 to 3 mm, k is calculated using *Hazen's equation* (Hazen, 1892):

$$k(\text{cm/s}) = D_{10}^2(\text{mm}) \tag{9.6}$$

For laminar flow in saturated coarse-grained soils, the *Kozeny–Carman equation* (Kozemy, 1927 and Carmen, 1956) can be used to estimate k:

$$k = \frac{1}{k_0 k_T S_s^2} \frac{e^3}{1+e} \frac{\gamma_w}{\eta} = C \frac{e^3}{1+e} \tag{9.7}$$

where k_0 and k_T are factors depending on the shape and tortuosity of the pores, respectively; S_s is the surface area of the solid particles per unit volume of solid material; and γ_w and η are the unit weight and viscosity of the pore water.

The Kozeny–Carman equation (9.7) is not well suited for silts and clays. The following equation (Taylor, 1948) can be used for clays:

$$\log k = \log k_0 - \frac{e_0 - e}{C_k} \tag{9.8}$$

where C_k is the permeability change index (for natural clays use $C_k \approx e_0/2$, where e_0 is the in situ void ratio) and k_0 is the in situ permeability (at $e = e_0$). Typical permeability values for common soil types are: 100 to 1 cm/s for clean gravel, 1 to 0.01 cm/s for coarse sand, 0.01 to 0.001 cm/s for fine sand, 0.001 to 0.00001 cm/s for silty clay, and less than 0.000001 cm/s for clay.

9.4 LABORATORY DETERMINATION OF PERMEABILITY

The coefficient of permeability of a coarse-grained soil can be determined in the laboratory using a constant-head permeability test. The test includes a cylindrical soil specimen that is subjected to a constant head as shown in Figure 9.6. The length of the soil specimen is L and its cross-sectional area is A. The total head loss (h_L) along the soil specimen is equal to the constant head, which is the difference in elevation between the water levels in the upper and lower reservoirs as shown in the figure. A constant head implies that we have reached a steady-state condition in which the flow rate is constant (i.e., does not vary with time). Using a graduated flask, we can collect a volume of water (Q) in a period of time (t). From this we can calculate the flow rate $q(= Q/t)$.

The flow velocity is given by $v = q/A$, and the hydraulic gradient along the soil specimen is $i = h_L/L$. Applying Darcy's law, we can write

$$v = ki \rightarrow \frac{q}{A} = \frac{Q}{tA} = k\frac{h_L}{L}$$

Therefore,

$$k = \frac{QL}{h_L A t} \tag{9.9}$$

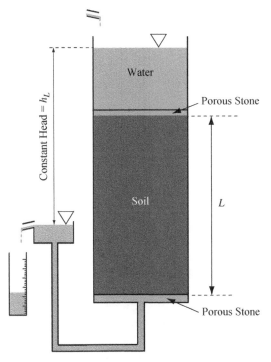

FIGURE 9.6 Constant-head permeability test.

The constant-head permeability test cannot be used for fine-grained soils because of their low permeabilities; it will take a long time to collect a measurable quantity of water to calculate the flow rate. The falling-head laboratory test is used instead. A schematic diagram of the falling-head permeability test is shown in Figure 9.7. The length of the cylindrical soil specimen is L and its cross-sectional area is A. The inside cross-sectional area of the standpipe is a. The elapsed time t needed for the height of the water column inside the standpipe to drop from h_1 to h_2 is recorded.

Using Darcy's law and equating the flow rate in the standpipe and the soil specimen, it can be shown that

$$k = \frac{aL}{At} \ln \frac{h_1}{h_2} \qquad (9.10)$$

For more representative values, the coefficient of permeability can be measured in the field using pump-in or pump-out tests on a well. In these tests, the flow rate required to maintain the water table at a constant height is measured and the coefficient of permeability is calculated using simple analytical expressions (see, e.g., Das, 2004).

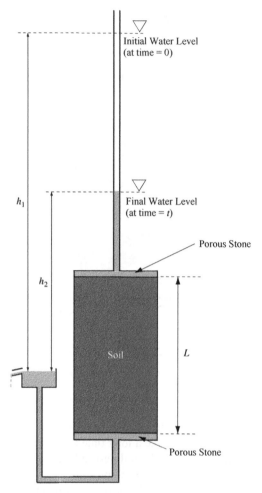

h_1

h_2

FIGURE 9.7 Falling-head permeability test.

9.5 PERMEABILITY OF STRATIFIED SOILS

Consider a stratified soil having horizontal layers of thickness $H_1, H_2, H_3, \ldots, H_n$ with coefficients of permeability $k_1, k_2, k_3, \ldots, k_n$, as shown in Figure 9.8a. For flow perpendicular to soil stratification, as shown in the figure, the flow rate q through area A of each layer is the same. Therefore, the head loss across the n layers is given as

$$h_L = \frac{H_1 q}{k_1 A} + \frac{H_2 q}{k_2 A} + \frac{H_3 q}{k_3 A} + \cdots + \frac{H_n q}{k_n A} = \left(\frac{H_1}{k_1} + \frac{H_2}{k_2} + \frac{H_3}{k_3} + \cdots + \frac{H_n}{k_n} \right) \frac{q}{A}$$

$$(9.11)$$

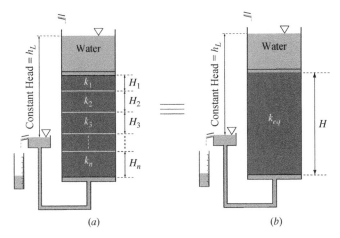

FIGURE 9.8   Flow perpendicular to soil stratification.

Figure 9.8*b* shows an "equivalent" soil layer that can replace the stratified system shown in Figure 9.8*a*. The thickness of the equivalent soil layer is $H(= H_1 + H_2 + H_3 + \cdots + H_n)$ and its permeability is k_{eq}. For this equivalent system, using Darcy's law we can write

$$h_L = \frac{qH}{Ak_{eq}} \tag{9.12}$$

Equating (9.11) with (9.12), we obtain the equivalent, or average, coefficient of permeability k_{eq}:

$$k_{eq} = \frac{H}{H_1/k_1 + H_2/k_2 + H_3/k_3 + \cdots + H_n/k_n} \tag{9.13}$$

For a flow that is parallel to soil stratification, such as the one shown in Figure 9.9, the head loss h_L over the same flow path length L will be the same for each layer. Thus, $i_1 = i_2 = i_3 = \cdots = i_n$. The flow rate through a layered system (with width = 1 unit) is

$$q = k_1 i_1 H_1 \times 1 + k_2 i_2 H_2 \times 1 + k_3 i_3 H_3 \times 1 + \cdots + k_n i_n H_n \times 1$$
$$= (k_1 H_1 + k_2 H_2 + k_3 H_3 + \cdots + k_n H_n)i \tag{9.14}$$

For the equivalent system shown in Figure 9.9*b*, we have

$$q = k_{eq} H i \tag{9.15}$$

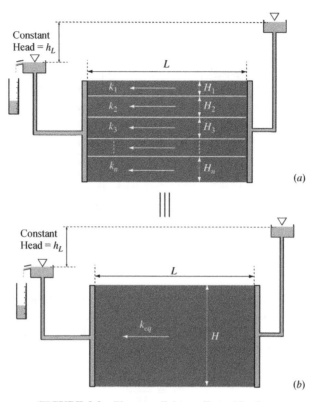

FIGURE 9.9   Flow parallel to soil stratification.

Equating (9.14) with (9.15), we obtain the equivalent, or average, coefficient of permeability k_{eq}:

$$k_{eq} = \frac{k_1 H_1 + k_2 H_2 + k_3 H_3 + \cdots + k_n H_n}{H} \tag{9.16}$$

9.6   SEEPAGE VELOCITY

Consider a flow through a soil specimen with a cross-sectional area A normal to the direction of flow (Figure 9.10). Flow velocity can be calculated using Darcy's law (9.5), which relates flow velocity to the hydraulic gradient. The flow rate q through the cross-sectional area A is calculated as the product of flow velocity v and total cross-sectional area of the soil specimen:

$$q = vA \tag{9.17}$$

In reality, water flows through a cross-sectional area A_v (area of voids) that is smaller than the total area A. Recall that water follows a tortuous path through the

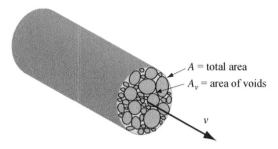

FIGURE 9.10   Seepage velocity.

pores. The seepage velocity v_s is the average velocity of the water flowing through the pores. It can be estimated by dividing the flow rate q by the average area of voids A_v on a cross section normal to the direction of flow:

$$v_s = \frac{q}{A_v} \tag{9.18}$$

Soil porosity n is related to the volume of voids:

$$n = \frac{V_v}{V} = \frac{A_v}{A} \tag{9.19}$$

where V_v is the volume of voids and V is the total volume of a soil specimen. Substituting (9.19) into (9.18), we have

$$v_s = \frac{v}{n} \tag{9.20}$$

9.7   STRESSES IN SOILS DUE TO FLOW

The effective stress depends on the "flow" condition that may occur in a soil specimen. Three possible conditions for the same soil specimen are shown in Figure 9.11. Let's consider the case of a homogeneous soil layer in a container as shown in Figure 9.11a. The thickness of the soil layer is H_2. Above the soil there is a layer of water H_1 thick. There is another reservoir that can be used to create an upward flow (upward seepage) through the soil sample. We assume that the valve leading to the upper reservoir is closed, so no water is flowing through the soil sample. This is the no-flow condition.

Upward seepage can be induced by opening the valve leading to the upper reservoir as shown in Figure 9.11b. The upper reservoir causes the water to flow upward through the soil sample. This steady-state upward flow occurs in the field as a result of artesian pressure when a less permeable layer is underlain by a permeable layer connected through the ground to a water source providing pressures higher

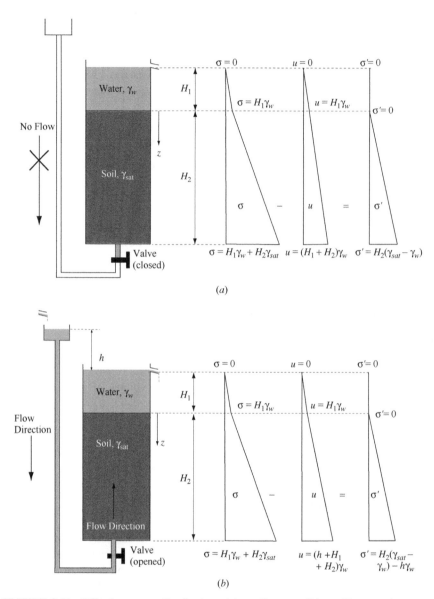

FIGURE 9.11 Effective stress distribution: (a) no-flow condition; (b) upward-seepage condition; (c) downward-seepage condition.

than local hydrostatic pressures. Downward seepage can be induced by lowering the upper reservoir as shown in Figure 9.11c. This steady-state downward flow occurs in the field when water is pumped at a constant rate from an underground aquifer. Pore pressures are then lower than hydrostatic pressures.

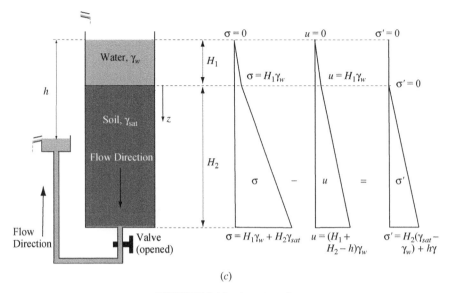

(c)

FIGURE 9.11 (*continued*)

In reference to Figure 9.11, the soil specimen has the same total vertical stress in the three conditions. The total vertical stresses, pore water pressures, and effective stresses at depth z are:

No-flow condition:

$$\sigma_v = \gamma_w H_1 + \gamma_{sat} z, \qquad u = \gamma_w(H_1 + z), \qquad \sigma'_v = \gamma' z$$

Upward-flow condition:

$$\sigma_v = \gamma_w H_1 + \gamma_{sat} z, \qquad u = \gamma_w(H_1 + z) + iz\gamma_w, \qquad \sigma'_v = \gamma' z - iz\gamma_w$$

Downward-flow condition:

$$\sigma_v = \gamma_w H_1 + \gamma_{sat} z, \qquad u = \gamma_w(H_1 + z) - iz\gamma_w, \qquad \sigma'_v = \gamma' z + iz\gamma_w$$

In the upward-flow condition the pore water pressure increases and the effective stress decreases with depth. On the other hand, when the flow is downward, the pore water pressure decreases and the effective stress increases. In reference to Figure 9.11b (upward-flow condition), if the hydraulic gradient induced by the head difference is large enough ($i = i_{cr}$), the upward seepage force will cause the effective stress within the soil to become zero, thus causing a sudden loss of soil strength in accordance with the effective stress principle. This condition resembles that of the exit soil element on the downstream side of the sheet pile, as discussed in Example 9.4. If the hydraulic gradient in the exit element is large ($i_{exit} = i_{cr}$), the exit element becomes unstable—the upward seepage force is large enough to cause the exit element to "float."

The soil specimen in Figure 9.11b will be totally destabilized when the effective stress within the soil specimen becomes zero. We can obtain this condition if we set the effective stress at the bottom of the soil layer equal to zero:

$$H_2\gamma' - h\gamma_w = 0 \tag{9.21}$$

or

$$\frac{h}{H_2} = \frac{\gamma'}{\gamma_w} \tag{9.22}$$

The hydraulic gradient through the soil specimen is given by

$$i = i_{cr} = \frac{h}{H_2} \tag{9.23}$$

where the hydraulic gradient i is equal to the "critical" hydraulic gradient i_{cr} because it causes the soil specimen to be destabilized. Substituting (9.22) into (9.23) yields

$$i_{cr} = \frac{\gamma'}{\gamma_w} \tag{9.24}$$

It can be shown (from the phase diagram) that

$$i_{cr} = \frac{\gamma'}{\gamma_w} = \frac{G_s - 1}{1 + e} \tag{9.25}$$

9.8 SEEPAGE

Two-dimensional steady-state flow of the incompressible pore fluid is governed by Laplace's equation, which can be derived based on Darcy's law and the concept of flow continuity. Let us consider the two-dimensional flow problem shown in Figure 9.12 involving a row of sheet piles embedded into a permeable soil. The sheet pile is impermeable and its role is to separate the high water level (upstream) from the lower level (downstream). Water will migrate through the porous soil from the high-energy zone to the low-energy zone around the embedded sheet pile. We assume that flow occurs only in the x–z plane, an acceptable assumption if the row of sheet piles is very long in the y-direction. No flow occurs in the y-direction.

For the rectangular soil element with dimensions dx, dz, and unit thickness (Figure 9.12), the rate of flow into the element in the x-direction is given as

$$(dq_x)_{in} = v_x \, dz \times 1$$

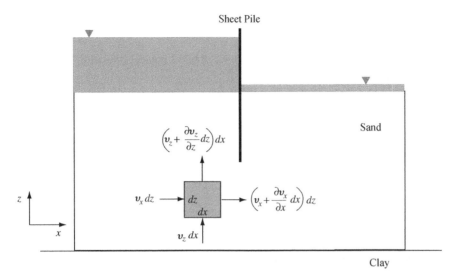

FIGURE 9.12 Two-dimensional steady-state flow of an incompressible pore fluid.

The rate of flow out of the element in the x-direction is

$$(dq_x)_{\text{out}} = \left(v_x + \frac{\partial v_x}{\partial x} dx \right) dz \times 1$$

Similar expressions can be written for the z-direction:

$$(dq_z)_{\text{in}} = v_z \, dx \times 1$$

$$(dq_z)_{\text{out}} = \left(v_z + \frac{\partial v_z}{\partial z} dz \right) dx \times 1$$

A balance of flow requires that

$$(dq_x)_{\text{out}} + (dq_z)_{\text{out}} = (dq_x)_{\text{in}} + (dq_z)_{\text{in}}$$

Therefore,

$$\left(v_x + \frac{\partial v_x}{\partial x} dx \right) dz + \left(v_z + \frac{\partial v_z}{\partial z} dz \right) dx = v_x \, dz + v_z \, dx$$

or

$$\frac{\partial v_x}{\partial x} + \frac{\partial v_z}{\partial z} = 0 \tag{9.26}$$

Using Darcy's law and assuming an isotropic condition in which $k = k_x = k_y$, we can write

$$v_x = k_x i_x = k \frac{\partial h}{\partial x} \tag{9.27}$$

and

$$v_z = k_z i_z = k \frac{\partial h}{\partial z} \tag{9.28}$$

Substituting (9.27) and (9.28) into (9.26), we obtain Laplace's equation for isotropic flow conditions:

$$\frac{\partial^2 h}{\partial x^2} + \frac{\partial^2 h}{\partial z^2} = 0 \tag{9.29}$$

In three dimensions, Laplace's equation becomes

$$\frac{\partial^2 h}{\partial x^2} + \frac{\partial^2 h}{\partial y^2} + \frac{\partial^2 h}{\partial z^2} = 0 \tag{9.30}$$

Two-dimensional steady flow of the incompressible pore fluid is governed by Laplace's equation (9.29), which indicates that any inequity in flow into and out of a soil element in the x-direction must be compensated by a corresponding opposite inequity in the z-direction. Laplace's equation can be solved analytically, graphically, or numerically. We will present the analytical solution of Laplace's equation for a simple one-dimensional boundary value flow problem. The example shows how analytical solution of the one-dimensional Laplace's equation satisfies the boundary conditions. Unfortunately, the analytical solution for two-dimensional problems is extremely tedious because of the complexity of the boundary conditions involved. In this case, use of the graphical or numerical solution is more practical.

Example 9.3   A 0.5-m-long soil specimen is subjected to steady-state flow under a constant head $= 1.5$ m as shown in Figure 9.4 (same as Example 9.1). Using Laplace's equation, calculate the total head at point C.

SOLUTION: Let us adopt the one-dimensional coordinate system shown in Figure 9.4. Laplace's equation for this one-dimensional flow condition becomes

$$\frac{\partial^2 h}{\partial z^2} = 0$$

The solution of this equation is of the form $h = A_1 z + A_2$, where A_1 and A_2 are constants to be determined from the boundary conditions of the flow domain:

- At the top of the soil specimen (point A), we have $z = 0$ and $h = 1.5$ m (see Example 9.1) considering the datum being located at downstream; therefore, 1.5 m $= A_1(0) + A_2$, or $A_2 = 1.5$ m.
- At the bottom of the soil specimen (point B), $z = 0.5$ m and $h = 0$; therefore, $0 = A_1 (0.5 \text{ m}) + 1.5$ m, or $A_1 = -3$ (dimensionless).

Substituting A_1 and A_2 into the solution of Laplace's equation, we get $h = -3z + 1.5$ m. Finally, at point C we have $z = 0.2$ m $\rightarrow h = -3(0.2 \text{ m}) + 1.5$ m $= 0.9$ m.

9.9 GRAPHICAL SOLUTION: FLOW NETS

The graphical solution to Laplace's equation for two-dimensional seepage problems is presented by a *flow net* consisting of two families of lines: equipotential lines and flow lines. These two sets of lines are orthogonal and must form quadrilateral flow elements (curvilinear squares) as shown in Figure 9.13a. An *equipotential line* connects points of equal total head h (potential), and a *flow line* represents the path traveled by a drop of water, indicating the direction of seepage down a hydraulic gradient.

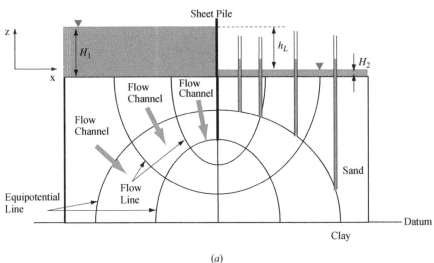

(a)

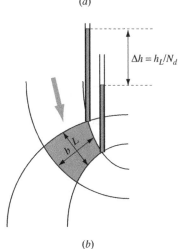

(b)

FIGURE 9.13   Graphical solution of Laplace's equation: (*a*) flow net; (*b*) flow element.

If we position the tips of several standpipes (piezometers) along a single equipotential line, the water would rise to the same level in each standpipe. This means that the total head is the same along an equipotential line since the total head is the sum of the pressure head (height of water in a standpipe) and the elevation head measured from an assumed datum, as shown in Figure 9.13a. There is a potential change (decrease) between two consecutive equipotential lines (Figure 9.13b). This is called the *potential drop* Δh and is calculated as

$$\Delta h = \frac{h_L}{N_d} \tag{9.31}$$

where h_L is the total head loss ($h_L = H_1 - H_2$) and N_d is the number of potential drops. Knowing that there is a potential drop between any two equipotential lines, it follows that N_d is equal to the total number of equipotential lines minus 1. Also, there is a flow channel between two consecutive flow lines. Thus, the number of flow channels N_f is equal to the total number of flow lines minus 1.

9.9.1   Calculation of Flow

Consider the flow element shown in Figure 9.13b. This flow element is bounded by two flow lines and two equipotential lines. Let us assume that the width of the flow element is b (the distance between the two flow lines) and the length of the element is L. As the pore-water traverses along the flow channel a distance L it encounters a potential drop Δh. Therefore, the average hydraulic gradient along the element is $i = \Delta h/L$.

Using Darcy's law we can calculate the flow rate within a flow channel assuming that the width of the channel is 1 unit of length in the y-direction:

$$\Delta q = kb\frac{\Delta h}{L} \tag{9.32}$$

Substituting (9.31) into (9.32) yields

$$\Delta q = kb\frac{h_L}{LN_d} \tag{9.33}$$

It is preferred when sketching flow nets to use curvilinear square elements with $L \approx b$. If all flow elements in a flow channel are curvilinear squares, then (9.33) becomes

$$\Delta q = k\frac{h_L}{N_d} \tag{9.34}$$

The total flow rate for a flow net with N_f flow channels is given as

$$q = N_f\Delta q = kh_L\frac{N_f}{N_d} \tag{9.35}$$

9.9.2 Flow Net Construction

A two-dimensional cross section of the seepage problem should be sketched to scale in both the x and z directions. Figure 9.14a shows a sheet pile embedded in a 19-m-thick sand layer underlain by a virtually impermeable clay layer. The embedment length of the sheet pile is 8 m. The problem is drawn to scale, keeping in mind that the graphical solution is only relevant to the flow domain, which includes the 19-m-thick sand layer and the embedded part of the sheet pile. The water on both sides of the sheet pile is used to establish the boundary conditions at the top surface of the sand layer, as explained next. The bottom surface of the sand layer has an impermeable boundary condition.

We would start drawing a flow net for a particular two-dimensional seepage problem by establishing the first and last flow lines, which usually correspond

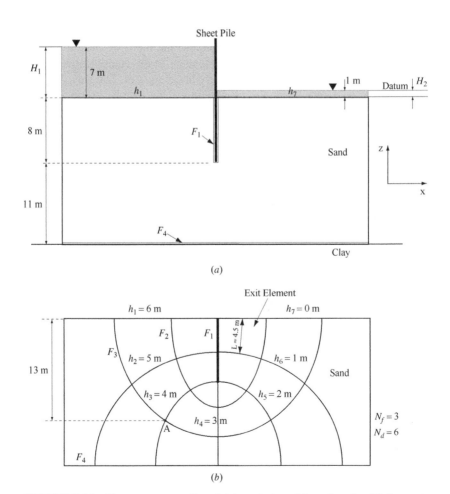

FIGURE 9.14 Flow net construction: (a) boundaries of flow domain; (b) flow net.

to impermeable boundary conditions. Then we draw a few "parallel" flow lines between the first and last lines. Note that a surface across which there is no flow (e.g., an impermeable soil layer or an impermeable wall) is a flow line. In Figure 9.14a the perimeter of the sheet pile [line F_1 (F for "flow")] is a flow line, and the bottom impermeable boundary of the sand layer (line F_4) is also a flow line. Let's consider line F_1 as the first flow line and line F_4 as the last flow line. In between these two lines, we can draw two flow lines marked F_2 and F_3, as shown in Figure 9.14b.

Next, we would establish the first and last equipotential lines, which usually correspond to permeable boundary conditions. Then we draw a few "parallel" lines in between. Note that a surface on which the total head is constant (e.g., from the level of a river) is an equipotential line. In Figure 9.14a, line h_1 (h for "total head"), is the first equipotential line having $h = 6$ m, and line h_7 is the last equipotential line with $h = 0$ m, assuming the datum to be coincident with the downstream water level as shown in the figure. Between lines h_1 and h_7 let's draw a few equipotential lines: h_2, h_3, h_4, h_5, and h_6, as shown in Figure 9.14b.

While drawing the flow lines and the equipotential lines, we need to remember two rules: (1) the flow lines are always orthogonal to the equipotential lines, and (2) all resulting flow elements need to be curvilinear squares. The way you judge a flow element to be a curvilinear square is by inscribing a circle inside the element that touches the four curved sides at the same time. Note that drawing a flow net is a trial-and-error process in which we do not know beforehand how many flow lines and equipotential lines are required to establish a correct flow net. So you will need to use a pencil and make frequent use of the eraser! If you are able successfully to establish two families of lines (flow net) that obey the aforementioned two rules, your flow net will represent the graphical solution of Laplace's equation, (9.29).

Example 9.4 Figure 9.14 shows a sheet pile embedded in a 19-m-thick sand layer underlain by a virtually impermeable clay layer. The embedment length of the sheet pile is 8 m. The figure also shows the flow net associated with this seepage problem. (a) Calculate the flow rate through the sand layer per unit length (in the y-direction) knowing that the coefficient of permeability of the sand layer is $k = 10^{-3}$ m/s. (b) Calculate the pore water pressure at point A. (c) Calculate the hydraulic gradient of the exit element and its safety factor against heaving. The saturated unit weight of the sand is $\gamma_{sat} = 19$ kN/m^3.

SOLUTION:

(a) The total head loss is $h_L = H_1 - H_2 = 7$ m $- 1$ m $= 6$ m. From the flow net shown in Figure 9.14b we have $N_f = 3$ and $N_d = 6$. The flow rate per unit length can be calculated using (9.35):

$$q = kh_L \frac{N_f}{N_d} = 10^{-3} \text{ m/s(6m)} \left(\frac{3}{6}\right) = 3 \times 10^{-3} \text{ m}^3/\text{s per meter}$$

(b) Let us calculate the total head associated with the equipotential lines in Figure 9.14b. The first equipotential line has a total head of $h_1 = h_L = 6$ m, with the datum located as shown in the figure. The number of potential drops is $N_d = 6$; therefore, the potential drop between two adjacent equipotential lines is $\Delta h = h_L/N_d = (6$ m$)/6 = 1$ m. Now we can calculate the total head in subsequent equipotential lines as follows:

$$h_1 = h_L = 6\text{m}$$

$$h_2 = h_L - 1 \times \Delta h = 6\text{m} - 1 \times 1\text{m} = 5\text{m}$$

$$h_3 = h_L - 2 \times \Delta h = 6\text{m} - 2 \times 1\text{m} = 4\text{m}$$

$$h_4 = h_L - 3 \times \Delta h = 6\text{m} - 3 \times 1\text{m} = 3\text{m}$$

$$h_5 = h_L - 4 \times \Delta h = 6\text{m} - 4 \times 1\text{m} = 2\text{m}$$

$$h_6 = h_L - 5 \times \Delta h = 6\text{m} - 5 \times 1\text{m} = 1\text{m}$$

$$h_7 = h_L - 6 \times \Delta h = 6\text{m} - 6 \times 1\text{m} = 0\text{m}$$

Point A is located on the third equipotential line, which has a total head of 4 m. According to Bernoulli's equation, (9.2), the total head at any point is the sum of the pressure head and the elevation head: $h = h_p + h_e = u/\gamma_w + z$. The elevation head at point A is -14 m because point A is below the assumed datum. So $h_p = u/\gamma_w = h - h_e = 4\text{m} - (-14\text{m}) = 18\text{m}$. Therefore, the pore pressure at point A is $u = (18$ m$)\gamma_w = (18\text{m})(9.81\text{kN/m}^3) = 177$ kPa.

(c) The exit element on the downstream side of the sheet pile is shown in Figure 9.14b. The flow net indicates that the exit element is subject to a total head loss of 1 m ($= h_6 - h_7$). As the pore water flows from the bottom of the exit element toward the top, a distance L of approximately 4.5 m, it encounters a head loss of 1 m. Therefore, the *exit hydraulic gradient* can be calculated as

$$i_{\text{exit}} = \frac{h_6 - h_7}{L} = \frac{\Delta h}{L}$$

Let us define the safety factor against heaving as

$$\text{FS} = \frac{i_{\text{cr}}}{i_{\text{exit}}}$$

in which $i_{\text{cr}} = \gamma'/\gamma_w$ (9.24). When this safety factor is 1, the exit hydraulic gradient is equal to the critical hydraulic gradient, and the exit element is in the state of incipient failure. To prevent that, this safety factor should be equal to or greater than 1.5.

Is the exit element in Figure 9.14*b* safe? To answer that, we need to calculate its safety factor as follows:

$$FS = \frac{i_{cr}}{i_{exit}} = \frac{\gamma'/\gamma_w}{\Delta h/L} = \frac{\gamma'}{\gamma_w}\frac{L}{\Delta h}$$

Using $\gamma_{sat} = 19$ kN/m^3, $L = 4.5$ m, and $\Delta h = 1$ m in the equation above, we get

$$FS = \left(\frac{19 - 9.81}{9.81}\right)\left(\frac{4.5}{1}\right) = 4.2 > 1.5$$

and the exit element is safe.

9.10 FLOW NETS FOR ANISOTROPIC SOILS

Many natural sedimentary soils are anisotropic with respect to permeability; their horizontal permeabilities are significantly greater than their vertical permeabilities. Let us consider an anisotropic soil with permeability k_x and k_z in the x and z directions, respectively. The equation governing seepage in such a soil is given as

$$k_x\frac{\partial^2 h}{\partial x^2} + k_z\frac{\partial^2 h}{\partial z^2} = 0 \qquad (9.36)$$

This elliptic equation (not Laplace's equation) can be simplified by applying the following transformation on the x-dimension:

$$x_T = x\sqrt{\frac{k_z}{k_x}}$$

Substituting the equation above into (9.36), we get

$$\frac{\partial^2 h}{\partial x_T^2} + \frac{\partial^2 h}{\partial z^2} = 0$$

The equation above has the same form as Laplace's equation for isotropic flow conditions, (9.29). This means that for an anisotropic flow condition we can sketch a flow net using the same procedure for isotropic flow as described earlier provided that the structure and the flow domain are stretched in the x-direction by multiplying their x-dimension by $\sqrt{k_z/k_x}$. Calculations of flow are done using an equivalent permeability $k_T = \sqrt{k_x k_z}$; thus,

$$q = \sqrt{k_x k_z}\,h_L\frac{N_f}{N_d} \qquad (9.37)$$

9.11   FLOW THROUGH EMBANKMENTS

Earth dams are constructed of well-compacted soils. A homogeneous earth dam consists of one type of soil but may contain a drainage blanket to collect seeping water. A zoned earth dam has several zones made of different materials, typically a locally available shell with a watertight clay core. Water seeps through the body of an earth dam and through its foundation. The velocity and quantity of seepage, especially through the earth dam, need to be controlled. If left uncontrolled, it can slowly erode soil from the body of the dam or its foundation, resulting in catastrophic failure of the dam. Erosion of the soil starts at the downstream side of the dam and advances progressively toward the reservoir, creating a direct "tunnel" to the reservoir. This phenomenon is known as *piping*.

In Section 9.8 we discussed how to estimate seepage under hydraulic structures using flow nets. In this section we learn how to estimate seepage through an earth dam (or embankment dam). Seepage through an earth dam presents an additional difficulty in that it is an unconfined flow problem where the flow domain is bounded at the top by a *phreatic surface* which represents the top flow line (Figure 9.15*a*). The pressure head along the phreatic surface is zero because the soil above it is assumed to be dry. Because the pressure head along the phreatic surface is zero, it follows that the total head changes and elevation head changes are equal. Thus, for equal total head intervals Δh between equipotential lines, there will be equal vertical distances between the points of intersection of equipotential lines with the phreatic surface as shown in Figure 9.15*b*.

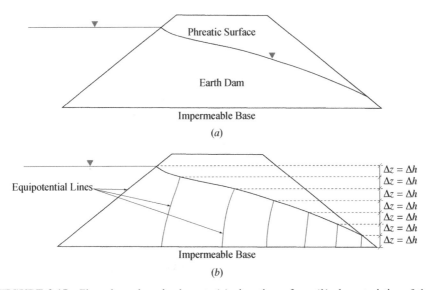

FIGURE 9.15   Flow through embankment: (*a*) phreatic surface; (*b*) characteristics of the phreatic surface.

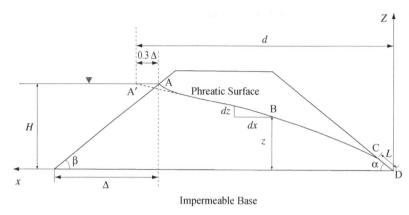

FIGURE 9.16 Construction of a phreatic surface in a homogeneous earth dam.

To calculate the flow rate through the dam and to sketch the flow net, we need to determine the shape of the phreatic surface for a particular embankment dam. As in any seepage problem, the flow net is very useful, especially for calculating the exit hydraulic gradient and the exit safety factor against piping. Figure 9.16 presents a homogeneous earth dam with an impervious foundation soil. The permeability of the dam is k (isotropic). The phreatic surface of this simple flow condition is assumed to be a parabolic surface (A'BC) with its focus at point D. The phreatic surface is slightly adjusted at the entrance (upstream) to start at point A instead of point A'. At the exit, the length L is calculated as

$$L = \frac{d}{\cos \alpha} - \sqrt{\frac{d^2}{\cos^2 \alpha} - \frac{H^2}{\sin^2 \alpha}} \tag{9.38}$$

where H is the height of the water in the reservoir, α the slope of the downstream face of the embankment dam, and the distance d, defined in Figure 9.16, can be calculated based on the geometry of the embankment dam (Schaffernak, 1917). The flow rate through the dam is given by

$$q = k \frac{\sin^2 \alpha}{\cos \alpha} L = k \frac{\sin^2 \alpha}{\cos \alpha} \left(\frac{d}{\cos \alpha} - \sqrt{\frac{d^2}{\cos^2 \alpha} - \frac{H^2}{\sin^2 \alpha}} \right) \tag{9.39}$$

9.12 FINITE ELEMENT SOLUTION

Let us consider the general case of anisotropic soil with permeability k_x and k_z in the x and z directions, respectively. Equation 9.36 governs seepage in such a soil:

$$k_x \frac{\partial^2 h}{\partial x^2} + k_z \frac{\partial^2 h}{\partial z^2} = 0$$

Developing a closed-form solution of 9.36 can be a very difficult task because the boundary conditions are difficult to satisfy even in the simplest two-dimensional flow problems. But with the finite element method, an approximate solution can be obtained in a simple manner. Without getting into much detail, it can be shown that discretization of (9.36) gives the following global equation that is applicable to both confined and unconfined flow problems:

$$[K]\{H\} = \{Q\} \tag{9.40}$$

where $[K]$ is the global stiffness matrix, $\{H\}$ a vector containing the nodal total heads (unknowns), and $\{Q\}$ a vector containing nodal flow. Note that each node has only one degree of freedom, which is the total head.

The global stiffness matrix $[K]$ is an assembly of the element stiffness matrices of the entire flow domain. The element stiffness matrix is given as

$$[K]^e = \int_{v_e} [B]^T [P][B] \, dv \tag{9.41}$$

where $[B]$ is derived from the shape function $[N]$, the matrix $[B]^T$ is the transpose of the matrix $[B]$, and $[P]$ is the element permeability matrix, defined as

$$[P] = \begin{bmatrix} k_x & 0 \\ 0 & k_z \end{bmatrix} \tag{9.42}$$

The numerical solution of (9.36) depends largely on the boundary conditions of a particular flow problem.

Boundary Conditions for Confined Flow For confined flow problems such as the one shown in Figure 9.17, the boundary conditions are as follows:

- Line AB is an equipotential line; therefore, all the nodal points along this line have a constant total head $H = H_1 + Z_1$ assuming the datum is located at the bottom of the flow domain as shown in the figure. We are using the symbol H for the total head, to be consistent with (9.40).

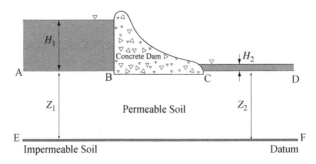

FIGURE 9.17 Boundary conditions for confined flow.

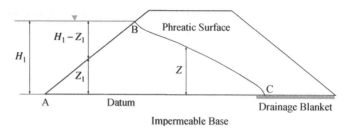

FIGURE 9.18 Boundary conditions for unconfined flow.

- Line CD is also an equipotential line having a total head $H = H_2 + Z_2$.
- Lines BC and EF represent impermeable boundaries along which $\partial H / \partial z = 0$.

Boundary Conditions for Unconfined Flow For unconfined flow problems, such as the problem of flow through an embankment dam shown in Figure 9.18, the boundary conditions are as follows:

- Line AB is an equipotential line with a constant total head $H = H_1$, considering that the datum is located at the interface between the embankment dam and the impermeable foundation soil.
- Line BC is the phreatic surface along which the pressure head is zero, which means that the total head is equal to the elevation head along this surface. Thus, the boundary condition along line BC is $H = Z$.
- Line AC represents impermeable boundary along which $\partial H / \partial z = 0$.

***Example 9.5*:** *Sheet Pile Embedded in Isotropic Soil* Figure 9.19*a* shows a row of sheet piles embedded in a 9.2-m-thick isotropic silty sand layer ($k_x = k_z = 5 \times 10^{-5}$ m/s) underlain by an impermeable clay layer. The sheet pile is assumed to be of infinite length in the *y*-direction. The embedment length of the sheet pile is 4.6 m. (a) Construct the flow net associated with this seepage problem using the finite element method. (b) Calculate the flow rate through the silty sand layer per unit length (in the *y*-direction).

SOLUTION: (files: Chapter 9_Example5_equipotential.cae, Chapter 9_Example5_ flowlines.cae) The finite element discretization of the flow domain is shown in Figure 9.19*b*. The 3-m-deep water on top of the soil on the upstream side is replaced by a pore water pressure boundary condition with $u = 30$ kPa, as shown in the figure. The pore water pressure along the soil surface on the downstream side is set equal to zero. The two vertical sides and the bottom side of the flow domain are impermeable. There is no need to specify these boundaries as impermeable boundary conditions since the pore fluid flow formulation does not allow a flow of fluid across the surface of the domain. Instead of modeling the sheet pile with

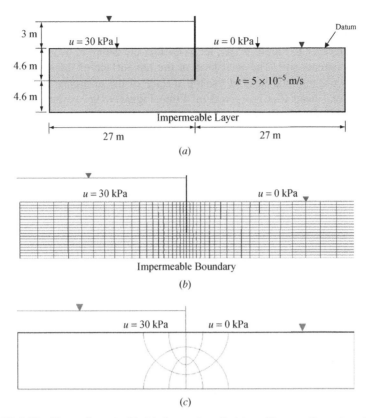

FIGURE 9.19   Sheet pile embedded in isotropic soil: (*a*) problem configuration; (*b*) finite element discretization; (*c*) flow net.

structural elements, a narrow V-shaped gap is created. The surfaces of the gap are regarded as external surfaces of the flow domain therefore they are impermeable.

In this example (and all the following examples) we consider fluid flow only; deformation of the soil is ignored. A plane-strain coupled pore fluid flow–deformation element is used in the current analysis. This type of element is generally used to calculate stresses and deformations associated with seepage forces. Nonetheless, in the present analysis we constrain all displacement degrees of freedom since we are only interested in establishing the flow net.

(a) Isotropic permeability of the soil is used with $k = 5 \times 10^{-5}$ m/s. The weight of the water is applied by gravity loading. A steady-state analysis is performed to obtain the equipotential lines shown in Figure 9.19c. To obtain the flow lines, the boundary conditions of the flow domain are altered as follows: The top surface of the soil on both sides of the sheet pile is made impermeable. To invoke flow, the embedded surface of the sheet pile is assigned a constant pore pressure $u = 0$, while the bottom boundary of the flow domain is assigned a higher pore pressure, $u = 30$ kPa. The resulting pore water pressure contours are the flow lines. The flow lines are placed over the equipotential lines to form the flow net shown in Figure 9.19c.

(b) In reference to Figure 9.19a, let's consider the datum to be at the top surface of the downstream soil. The total head at the top surface of the downstream soil is $h = 0$, while the total head at the top surface of the upstream soil is $h = 3$ m. Thus, the total head loss is $h_L = 3$ m. From Figure 9.19c we have $N_f = 3$ and $N_d = 6$. The flow rate is calculated as

$$q = kh_L \frac{N_f}{N_d} = 5 \times 10^{-5} \ m/s(3m)\left(\frac{3}{6}\right) = 7.5 \times 10^{-5} \ m^3/s \ \text{per meter}$$

$$= 6.48 \ m^3/\text{day per meter}$$

Example 9.6: *Sheet Pile Embedded in Anisotropic Soil*   Figure 9.20a shows a sheet pile embedded in a 9.2-m-thick anisotropic silty sand layer ($k_x = 15 \times 10^{-5}$ m/s and $k_z = 5 \times 10^{-5}$ m/s) underlain by an impermeable clay layer. The embedded length of the sheet pile is 4.6 m. Construct the flow net associated with this seepage problem using the finite element method.

SOLUTION: (files: Chapter9_Example6_equipotential.cae, Chapter9_Example6_flowlines.cae)   Hand-drawing of anisotropic flow nets is possible following the procedure discussed in Section 9.10. The procedure can be tedious, especially for relatively complicated geometries. In this example we use the finite element method

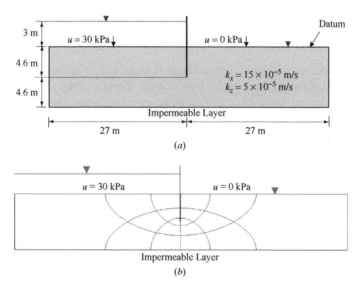

FIGURE 9.20   Sheet pile embedded in anisotropic soil: (*a*) problem configuration; (*b*) flow net.

to construct a flow net for anisotropic flow domain. The finite element procedure to solve this problem is identical to that of Example 9.5. The only exception is that we use the orthotropic soil permeability option herein with $k_x = 15 \times 10^{-5}$ m/s and $k_z = 5 \times 10^{-5}$ m/s. The flow net is constructed in the same manner as in Example 9.5. Figure 9.20b shows the resulting anisotropic flow net.

Let's consider the datum to be at the top surface of the downstream soil (Figure 9.20a). The total head at the top surface of the downstream soil is $h = 0$, and the total head at the top surface of the upstream soil is $h = 3$ m; therefore, the total head loss is $h_L = 3$ m. From Figure 9.20b we have $N_f = 3$ and $N_d = 6$. The flow rate for this anisotropic flow condition is calculated as

$$q = \sqrt{k_x k_z} h_L \frac{N_f}{N_d} = \sqrt{(15 \times 10^{-5})(5 \times 10^{-5})} \text{m/s}(3\text{m}) \left(\frac{3}{6}\right)$$

$$= 1.3 \times 10^{-4} \text{ m}^3/s \text{ per meter}$$

$$= 11.22 \text{ m}^3/\text{day per meter}$$

Example 9.7: *Asymmetric Sheet Pile Problem*   Two parallel rows of sheet piles are embedded in an isotropic silty sand layer underlain by an impermeable layer as shown in Figure 9.21a. The silty sand layer has a nonuniform thickness, and the embedded lengths of the two sheet piles are different, as indicated in the figure. Using the finite element method, construct a flow net for the flow domain and calculate the flow rate, per unit length, of the water seeping into the trench between the two rows. The sheet piles are assumed to be of infinite length in the y-direction.

SOLUTION: (files: Chapter9_Example7_equipotential.cae, Chapter9_Example7_flowlines.cae)   Figure 9.21b presents the finite element mesh used for the analysis. The 3-m-deep water on top of the soil on both sides of the trench is replaced by a pore water pressure boundary condition with $u = 30$ kPa as shown in the figure. The pore water pressure on the soil surface located at the bottom of the trench is set equal to zero. The two vertical sides and the bottom side of the flow domain are impermeable. The sheet piles are simulated using narrow V-shaped gaps whose surfaces are impermeable.

A plane-strain coupled pore fluid flow-deformation element is used in the current analysis. Isotropic permeability of the soil is used with $k = 5 \times 10^{-5}$ m/s. The weight of the water is applied by gravity loading. A steady-state analysis is performed to obtain the equipotential lines shown in Figure 9.21c. To obtain the flow lines, the boundary conditions of the flow domain are altered as follows: The top surfaces of the soil on both sides of the trench are made impermeable, as is the soil surface located at the bottom of the trench. To invoke flow, the embedded surfaces of the two sheet piles are assigned a constant pore pressure $u = 0$, and the bottom boundary of the flow domain is assigned a higher pore pressure $u = 30$ kPa. The resulting pore water pressure contours are the flow lines. The flow lines are placed over the equipotential lines to form the flow net shown in Figure 9.21c.

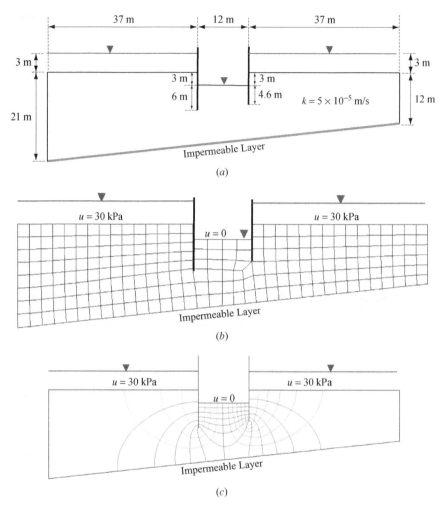

FIGURE 9.21   Asymmetric sheet pile: (*a*) problem configuration; (*b*) finite element discretization; (*c*) flow net.

In reference to Figure 9.21*c*, we note that the flow net is asymmetric. We have six flow channels on the right-hand side and only four flow channels on the left-hand side. Assuming the datum to be at the bottom of the trench, the total head at the top surface of the flow domain would be 6 m. Thus, the total head loss between the top surface of the flow domain and the bottom surface of the trench is 6 m. The flow rate is calculated as

$$q = kh_L \frac{N_f}{N_d} = 5 \times 10^{-7} \text{m/s(6m)} \left(\frac{4+6}{12} \right) = 2.5 \times 10^{-6} \text{m}^3\text{/s per meter}$$

$$= 0.216 \text{ m}^3\text{/day per meter}$$

Example 9.8: *Three-Dimensional Sheet Pile Problem*   Four sheet piles are driven into a silty sand in a square cofferdam formation to a depth of 6.1 m. The cofferdam is 9.2 m × 9.2 m in plan. Figure 9.22*a* shows one-fourth of the problem configuration. The soil is excavated to a depth of 4.6 m as shown in the figure. The water level outside the cofferdam is 3 m above the ground surface (not shown in the figure). Inside the cofferdam the water level is kept at the bottom of the excavation (using a pump). Using the finite element method, construct a three-dimensional flow net for the flow domain. Note that there is an impermeable soil layer under the silty sand at a depth of 15.1 m.

SOLUTION: (files: Chapter9_Example8_equipotential.cae, Chapter9_Example8_flowlines.cae)   Figure 9.22*b* presents the three-dimensional finite element mesh used for the analysis. Note that only one-fourth of the geometry is considered because of symmetry. The 3-m-deep water on top of the soil outside the cofferdam is replaced by a pore water pressure boundary condition with $u = 30$ kPa as shown in the figure. The pore water pressure on the soil surface located inside the cofferdam is set equal to zero. All vertical sides and the bottom side of the flow domain are impermeable. The sheet piles are simulated using narrow V-shaped gaps whose surfaces are also impermeable.

A three-dimensional coupled pore fluid flow-deformation element is used in the current analysis. Isotropic permeability of the soil is used with $k = 5 \times 10^{-5}$ m/s. The weight of the water is applied by gravity loading. A steady-state analysis is performed to obtain the equipotential lines (surfaces) shown in Figure 9.22*c*. To obtain the flow lines, the boundary conditions of the flow domain are altered as follows: The top surfaces of the soil outside the cofferdam are made impermeable, as is the soil surface located at the bottom of the cofferdam. To invoke flow, the embedded surfaces of the sheet piles are assigned a constant pore pressure $u = 0$, and the bottom boundary of the flow domain is assigned a higher pore pressure $u = 30$ kPa. The resulting pore water pressure contours are the flow lines (surfaces). The flow lines are shown in Figure 9.22*d*.

Example 9.9: *Symmetric Concrete Dam Problem*   Calculate the flow rate of seepage per unit length under the concrete dam shown in Figure 9.23*a*. The dam is 20 m high and has a 60-m-wide base. The dam is underlain by a 90-m-thick soil with $k = 0.03$ cm/s (isotropic). The water in the reservoir is 12 m high. The concrete dam is assumed to be very long in the y-direction.

SOLUTION: (files: Chapter9_Example9_equipotential.cae, Chapter9_Example9_flowlines.cae)   The finite element mesh of the flow domain is shown in Figure 9.23*b*. Note that the mesh is made finer in the vicinity of the dam. The top surface of the soil on the upstream side of the dam is assigned a pore water pressure boundary condition with $u = 117.7$ kPa to account for the pressure caused by the 12-m-high upstream water. The pore pressure boundary condition on the top surface of the soil on the downstream side is assigned $u = 0$. The bottom surface of the concrete dam is an impermeable boundary condition. The vertical sides

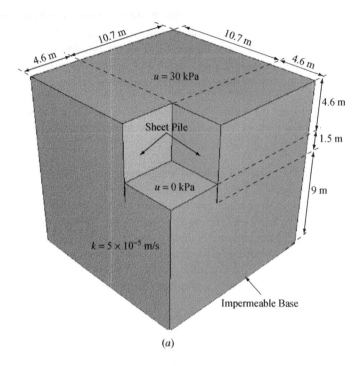

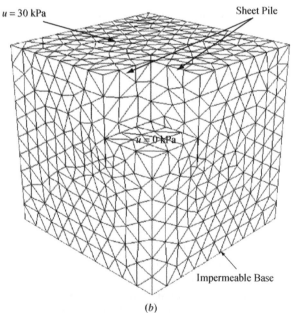

FIGURE 9.22 Three-dimensional sheet pile problem: (*a*) configuration; (*b*) finite element discretization; (*c*) equipotentials; (*d*) flow lines.

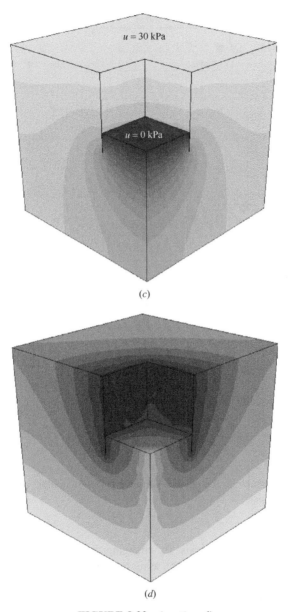

(c)

(d)

FIGURE 9.22   (*continued*)

and the bottom side of the flow domain are also assigned impermeable boundary conditions.

The element used in this analysis is a plane-strain coupled pore fluid flow-deformation element. Soil permeability is assumed to be isotropic with $k = 0.03$ cm/s. The equipotential lines shown in Figure 9.23*c* are obtained by

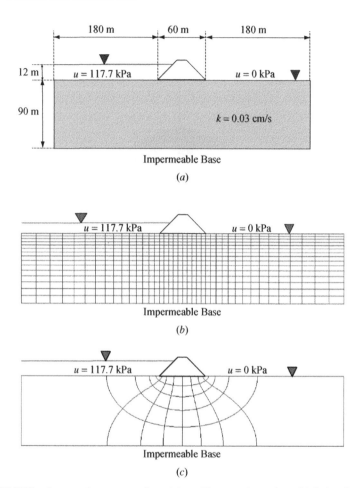

FIGURE 9.23 Symmetric concrete dam: (*a*) problem configuration; (*b*) finite element discretization; (*c*) flow net.

performing a steady-state analysis. By changing the boundary conditions, we can obtain the flow lines. This is done by assigning impermeable boundary conditions to the top surfaces of the soil on both sides of the dam. Also, the bottom surface of the dam is assigned a constant pore pressure $u = 0$, while the bottom boundary of the flow domain is assigned a higher pore pressure $u = 117.7$ kPa. The flow lines, shown in Figure 9.23*c*, are obtained by plotting contour lines of pore pressure within the flow domain. The flow lines and the equipotential lines are superimposed to form the flow net shown in Figure 9.23*c*.

Considering datum to be on the top surface of the soil on the downstream side, the total head at the top surface of the soil on the downstream side is equal to zero. The total head at the top surface of the soil on the upstream side is 12 m. Thus, the total head loss is 12 m. Also, from the flow net shown in Figure 9.23*c* we have

$N_f = 5$ and $N_d = 8$. The flow rate is calculated as

$$q = kh_L \frac{N_f}{N_d} = 0.0003\text{m/s}(12\text{m}) \left(\frac{5}{8}\right) = 2.25 \times 10^{-3} \text{ m}^3/\text{s per meter}$$

$$= 194.4 \text{ m}^3/\text{day per meter}$$

The calculated flow rate seems too high! This is partially due to the high permeability of the soil. Let's include a sheet pile at the heel of the concrete dam to see if we can reduce the flow rate (next example).

Example 9.10: *Asymmetric Concrete Dam Problem* Construct a flow net for the flow domain under the concrete dam shown in Figure 9.24a. The dam is 20 m

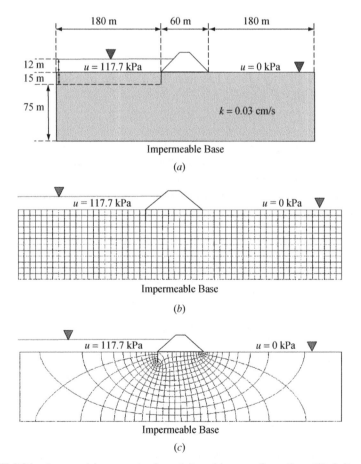

FIGURE 9.24 Asymmetric concrete dam: (*a*) problem configuration; (*b*) finite element discretization; (*c*) flow net.

high and has a 60-m-wide base. The concrete dam includes a row of sheet piles with an embedded length of 15 m as shown in the figure. The dam is underlain by a 90-m-thick soil with $k = 0.03$ cm/s (isotropic). The water in the reservoir is 12 m high.

SOLUTION: (files: Chapter9_Example10_equipotential.cae, Chapter9_Example10 _flowlines.cae) The finite element procedure to solve this problem is similar to that of Example 9.9. The only difference is the presence of the 15-m-long sheet pile at the heel of the concrete dam. The sheet pile is simulated by a narrow V-shaped gap, the surfaces of which are regarded as external surfaces of the flow domain; therefore, they are impermeable. Figure 9.24b presents the finite element mesh used in the analysis. The resulting flow net is shown in Figure 9.24c. From the figure we have $N_f = 13$ and $N_d = 24$. The flow rate is calculated as

$$q = kh_L \frac{N_f}{N_d} = 0.0003\text{m/s}(12\text{m})\left(\frac{13}{24}\right) = 1.95 \times 10^{-3} \text{ m}^3/\text{s per meter}$$

$$= 168.5 \text{ m}^3/\text{day per meter}$$

This flow rate is slightly (13%) less than the flow rate of a dam without a sheet pile (Example 9.9).

Example 9.11   Figure 9.25a shows a long horizontal drain located at 4.5 m below the ground surface. The drain is 0.6 m in diameter. The 9-m-thick sandy clay layer has $k = 5 \times 10^{-7}$ m/s and is underlain by an impermeable stratum. (a) Draw the flow net for the groundwater flow. (b) Calculate the discharge through the drain in m^3/day per meter length of drain when the water level is 3 m above the ground surface as shown in the figure.

SOLUTION: (files: Chapter9_Example11_equipotential.cae, Chapter9_Example11 _flowlines.cae) (a) The finite element discretization of the flow domain is shown in Figure 9.25b. The 3-m-deep water on top of the soil is replaced by a pore water pressure boundary condition with $u = 30$ kPa at the top surface of the flow domain. The pore water pressure along the perimeter of the drain is set equal to zero. The two vertical sides and the bottom side of the flow domain are impermeable. The natural boundary condition in the pore fluid flow formulation provides no flow of fluid across the surface of the domain, therefore, no further specification is needed on these surfaces.

The permeability of the soil is 5×10^{-7} m/s. The weight of the water is applied by gravity loading. A steady-state analysis is performed to obtain the equipotential lines shown in Figure 9.25c. To obtain the flow lines, the boundary conditions of the flow domain are altered (reversed) as follows: The top surface ABC and the perimeter of the drain are made impermeable (Figure 9.25b). To invoke flow, line BD is assigned a constant pore pressure $u = 0$, while lines AF, CH, and EG are

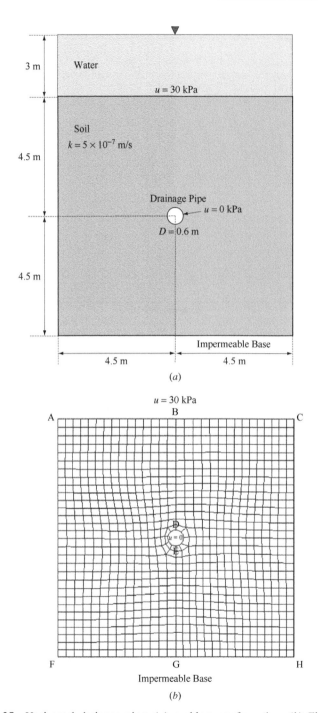

FIGURE 9.25 Horizontal drainage pipe: (*a*) problem configuration; (*b*) Finite element mesh; (*c*) flow net.

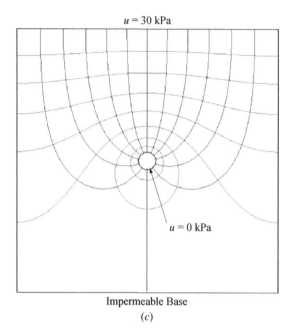

$u = 30$ kPa

$u = 0$ kPa

Impermeable Base

(c)

FIGURE 9.25 (*continued*)

assigned a higher pore pressure $u = 30$ kPa. The resulting pore pressure contours are actually the flow lines. These are superimposed to the equipotential lines to form the flow net shown in Figure 9.25c.

(b) Consider the datum at the drain level. This means that the total head along the perimeter of the drain is approximately zero. The total head at the top of the flow domain (the ground surface) is therefore 7.5 m. (Why?) The head loss (h_L) from the surface of the flow domain to the center of the drain is 7.5 m. From the flow net shown in Figure 9.25c we have $N_f = 12$ and $N_d = 8$:

$$q = kh_L \frac{N_f}{N_d} = 5 \times 10^{-7} \text{m/s}(7.5\text{m}) \left(\frac{12}{8}\right) = 5.625 \times 10^{-6} \text{m}^3/\text{s per meter}$$

$$= 0.486 \text{ m}^3/\text{day per meter}$$

Example 9.12   Using the finite element method, establish the phreatic surface of the homogeneous earth dam shown in Figure 9.26a. The dam is 24 m high and it is filled to one-half of its height. The foundation soil is impermeable. A 9-m-wide drainage blanket is used to control seepage through the dam.

SOLUTION: (file: Chapter9_Example12_equipotential.cae)   This example illustrates the use of the finite element method to solve for the flow through a homogeneous embankment dam in which fluid flow is occurring in a gravity field and

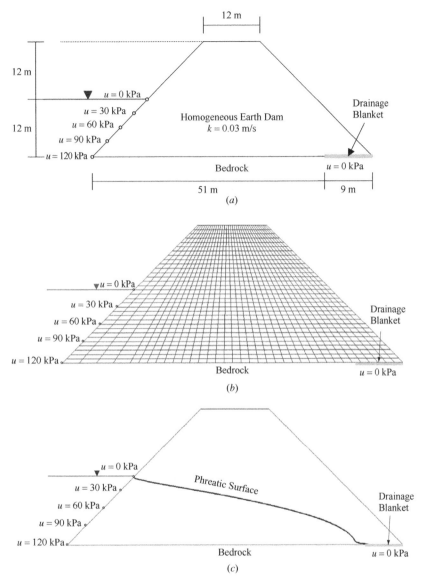

FIGURE 9.26 Phreatic surface of a homogeneous earth dam: (*a*) problem configuration; (*b*) finite element mesh; (*c*) calculated phreatic surface.

the location of the phreatic surface is a part of the solution. The phreatic surface in the dam is the locus of points at which the pore fluid pressure, u, is zero. In this problem we consider fluid flow only; deformation of the dam is ignored. A fully coupled pore fluid flow-deformation element will be used. This type of element is generally used to calculate stresses and deformations associated with seepage forces. Nonetheless, in the present analysis we constrain all displacement degrees of freedom since we are only interested in establishing the phreatic surface.

The geometry of this homogeneous earth dam is shown in Figure 9.26a. The dam is filled to one-half of its height. The earth dam includes a drainage blanket at its base. Since the dam is assumed to be long, we use coupled pore pressure/displacement plane strain elements. The finite element mesh is shown in Figure 9.26b. Half of the upstream face of the dam is subject to water pressure as shown in Figure 9.26a. The pore pressure on this face varies with depth: $u = (H_1 - Z_1)\gamma_w$, where H_1 ($= 12$ m) is the elevation of the water surface and Z_1 is elevation as indicated in Figure 9.18. Part of the bottom of the dam is assumed to rest on an impermeable foundation. Since the natural boundary condition in the pore fluid flow formulation provides no flow of fluid across a surface of the model, no further specification is needed on this surface. The drainage blanket boundary is assigned zero pore pressure ($u = 0$).

Again, the phreatic surface in the dam is determined as the locus of points at which the pore water pressure is zero. Above this surface the pore water pressure is assumed to be zero in this particular analysis (i.e., the soil above the phreatic surface is assumed to be dry). In reality, the pore pressure above the phreatic surface is negative. The capillary tension causes the fluid to rise against the gravitational force, thus creating a capillary zone. The effect of capillary tension on the location of the phreatic surface is minimal and can be ignored, as is done in the present analysis.

The permeability of the fully saturated soil of the dam is 3×10^{-2} m/s. The initial void ratio of the soil is 1.0. Initially, the embankment dam is assumed to be saturated with water up to the level of the water in the reservoir. This means that the initial pore pressure varies between zero at the upstream water level and a maximum of 120 kPa at the base of the dam. The weight of the water is applied by gravity loading. A steady-state analysis is performed in five increments to allow the numerical algorithm to resolve the high degree of nonlinearity in the problem. The phreatic surface is shown in Figure 9.26c. This phreatic surface is established by plotting the contour line along which the pore water pressure is zero.

PROBLEMS

9.1 Points A and B are located on the same flow line as shown in the Figure 9.3. The distance between the two points is 35 m and the average hydraulic gradient between the two points is 0.1. Knowing that the pressure head at point A is 70 kPa, calculate the pore water pressure at point B. What is the height of water in a standpipe piezometer positioned at point B?

9.2 Two 0.5-m-thick soil layers are subjected to a steady-state flow condition with a constant head $h_L = 1.5$ m as shown in Figure 9.27. The top layer

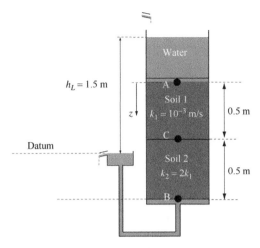

FIGURE 9.27

has $k_1 = 0.001$ m/s and the bottom layer has $k_2 = 2k_1$. Calculate the pore water pressure at point C located at the interface between the two layers.

9.3 Three 0.2-m-thick soil layers are subjected to a steady-state flow condition with a constant head $h_L = 1.5$ m. The top layer has $k_1 = 0.001$ m/s, the middle layer has $k_2 = 2k_1$, and the bottom layer has $k_3 = 3k_1$. Calculate the equivalent permeability for the case of water flowing perpendicular to soil stratification (Figure 9.8) and for the case of water flowing parallel to soil stratification (Figure 9.9).

9.4 Refer to Figure 9.11b. Calculate the head h that will cause the 2-m-thick soil layer to heave (total loss of strength). The saturated unit weight of the soil is 17.9 kN/m³.

9.5 The seepage force per unit volume is given as $i\gamma_w$. Calculate the average seepage force in the exit element shown in Figure 9.14b.

9.6 Calculate the pore water pressure distribution on both sides of the sheet pile shown in Figure 9.14b.

9.7 As shown in Figure 9.28, a row of sheet piles is embedded in a 4-m-thick soil layer with $k = 10^{-3}$ cm/s. The embedment length of the sheet piles is d. The row of sheet piles is very long in the direction perpendicular to the figure. Establish a flow net and calculate the flow rate per unit length and the exit hydraulic gradient (**a**) with $d = 1$ m, (**b**) with $d = 2$ m, and (**c**) with $d = 3$ m. Which case has the maximum flow rate per unit length? Which case has the maximum exit hydraulic gradient?

9.8 A row of sheet piles is embedded in a two-layer soil system as shown in Figure 9.29. The embedment length of the sheet piles is 2 m. The row of

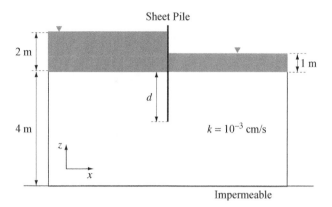

FIGURE 9.28

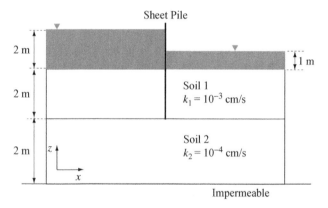

FIGURE 9.29

sheet piles is very long in the direction perpendicular to the figure. The top soil layer has $k_1 = 0.001$ cm/s, and the bottom layer has $k_2 = 0.1k_1$. Using the finite element method, establish a flow net for this layered soil system. Calculate the flow rate per unit length and the exit hydraulic gradient.

9.9   Calculate the flow rate and the exit hydraulic gradient for the circular cofferdam shown in Figure 9.30. The radius of the cofferdam is 2 m and its embedment length is 2 m. The soil layer is 4 m thick and its permeability is 10^{-3} m/s. Use the finite element method to establish the flow net. [*Hint*: The problem is axisymmetric.]

9.10 The concrete dam shown in Figure 9.31 is very long in the direction normal to the figure. The permeable soil under the dam is 25 m thick with $k = 10^{-3}$ cm/s and underlain by an impermeable layer. To control seepage through the soil under the dam, a row of sheet piles is embedded at the toe of the dam with an embedment length of 10 m. Calculate **(a)** the flow

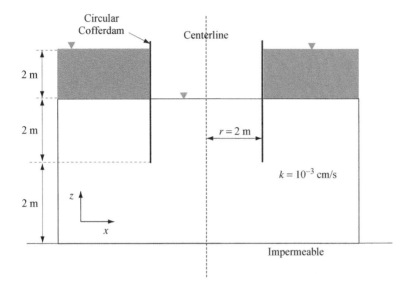

FIGURE 9.30

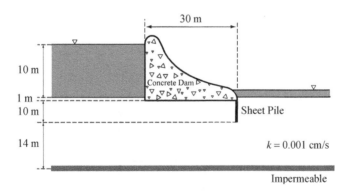

FIGURE 9.31

rate per unit length through the soil under the dam, **(b)** the exit hydraulic gradient, and **(c)** the pore water pressure distribution exerted on the bottom surface of the dam.

9.11 Using the finite element method, establish the phreatic surface for each of the two homogeneous earth dams shown in Figure 9.32. The first dam includes a drainage blanket to control seepage. Both dams are constructed using the same soil having $k = 5 \times 10^{-3}$ cm/s. The dams are underlain by an impermeable (nonfissured) rock.

9.12 Two infinitely long parallel rows of sheet piles are embedded in a 6-m-thick permeable soil layer underlain by an impermeable layer (Figure 9.33). The

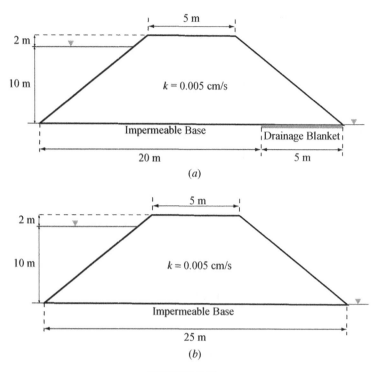

FIGURE 9.32

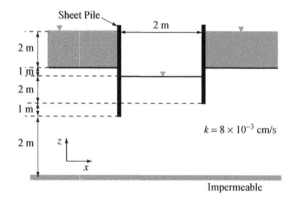

FIGURE 9.33

embedment length of the sheet pile row on the left-hand side is 4 m, while the embedment length of the sheet pile row on the right-hand side is 3 m. Calculate the flow rate into the trench per unit length. Also calculate the exit hydraulic gradient at each side of the trench.

REFERENCES

ABAQUS User's Manual (2002). ABAQUS/Standard Users's Manual, Volume II, Version 6.3. Hibbitt, Karlsson & Sorenson, Inc., Pawtucket, Rhode Island.

Ahlvin, R. G., and Ulery, H. H. (1962). *Tabulated Values for Determining the Complete Pattern of Stresses, Strains, and Deflections Beneath a Uniform Circular Load on a Homogeneous Half Space*, Highway Research Bulletin 342, Transportation Research Board, National Research Council, Washington, DC, pp. 1–13.

American Petroleum Institute (1984). *API Recommended Practice for Planning, Designing and Constructing Fixed Off-shore Platforms*, API, Washington, DC.

American Society for Testing and Materials (2004). *ASTM Standards*, Vol. 04.08, ASTM, West Conshohocken, PA.

Boussinesq, J. (1883). *Application des potentials a l'étude de l'équilibre et du mouvement des solides élastiques*, Gauthier-Villars, Paris.

Broms, B. B. (1965). Design of laterally loaded piles, *Journal of the Soil Mechanics and Foundation Division, ASCE*, Vol. 91, No. SM3, pp. 79–99.

Budhu, M. (1999). *Soil Mechanics and Foundations*, Wiley, Hoboken, NJ.

Budhu, M. (2007). *Soil Mechanics and Foundations*. 2nd ed., Wiley, Hoboken, NJ.

Burland, J. B. (1973). Shaft friction piles in clay—a simple fundamental approach, *Ground Engineering*, Vol. 6, No. 3, pp. 30–42.

Carman, P. C. (1956). *Flow of Gases Through Porous Media*, Butterworth Scientific Publications, London.

Casagrande, A. (1932). Research of Atterberg limits of soils, *Public Roads*, Vol. 1, No. 8, pp. 121–136.

Casagrande. A. 1936. Determination of the presconsolidation load and its practical signif-icance. p. 60-64. *In* Proc. Inter. Conf. on Soil Mechanics and Foundation Engineering. Vol. III, Harvard University, Cambridge, MA.

Casagrande, A., and Fadum, R. E. (1940). *Notes on Soil Testing for Engineering Purposes* School of Engineering Publication 8, Harvard University, Cambridge, MA.

Coulomb, C. A. (1776). Essai sur une application des règles de maximis et minimis a quelques problèmes de statique, relatifs a l'architecture, *Memories de la societe, Royale des Sciences, Paris*, Vol. 3.38.

Darcy, H. (1856). *Les fontaines publiques de la ville de Dijon*, Dalmont, Paris.

Das, B. M. (2004). *Principles of Foundation Engineering*, 5th ed., Brooks/Cole, Pacific grove, CA.

Davisson, M. T., and Gill, H. L. (1963). Laterally loaded piles in a layered soil system, *Journal of the Soil Mechanics and Foundation Division, ASCE*, Vol. 89, No. SM3, pp. 63–94.

Desai, C. S., and Siriwardane, H. J. (1984). *Constitutive Laws for Engineering Materials with Emphasis on Geologic Materials*, Prentice-Hall, Englewood cliffs, NJ.

Fleming, W. G. K., Weltman, A. J., Randolph, M. F., and Elson, W. K. (1985). *Piling Engineering*, Halsted Press, New York.

Gibson, R. E., Schiffman, R. L., and Pu, S. L. (1970). Plane strain and axially symmet-ric consolidation of a clay layer on a smooth impervious base, *Quarterly Journal of Mechanics and Applied Mathematics*, Vol. 23, Pt. 4, pp. 505–520.

Hazen, A. (1892). "Phycical properties of sands and gravels with reference to use in filtra-tion," *Report to Mass. State Board of Health*, 539.

Jaky, J. (1944). The coefficient of earth pressure at rest, *Journal of the Society of Hungarian Architects and Engineers*, Vol. 7, pp. 355–358.

Janbu, N. (Ed.) (1976). Static bearing capacity of friction piles, *Proceedings of the 6th Euro-pean Conference on Soil Mechanics and Foundation Engineering*, Vol. 1.2, pp. 479–488.

Kim, M. K., and Lade, P. V. (1988). Single hardening constitutive model for frictional materials: I. Plastic potential function, *Computers and Geotechnics*, Vol. 5, pp. 307–324.

Kozeny, J. (1927). Ueber kapillare Leitung des Wassers in Boden, Akademic der Wiss, exsehalften (Wien), Vol. 136, No. 2a, p. 271.

Lade, P. V. (2005). Overview of constitutive models for soils, Geotechnical special Pub-lication, N 128, Soil Constitutive Models–Evaluation, Selection and Calibration, pp. 1–34.

Lade, P. V., Jakobsen, K.P. 2002. Incrementalization of a single hardening constitutive model for frictional materials. Int. Journ. Num. and Anal. Methods in Geomech. 26, 647–659. DHI ref. 3/02.

Lade, P. V., and Kim, M. K. (1988). Single hardening constitutive model for frictional materials: II. Yield criterion and plastic work contours, *Computers and Geotechnics*, Vol. 6, pp. 13–29.

Lade, P. V., and Nelson, R. B. (1987). Modeling the elastic behavior of granular materials, *International Journal for Numerical and Analytical Methods in Geomechanics*, Vol. 11, pp. 521–542.

Lade, P. V. (1977). "Elasto-plastic stress strain theory for cohesionless soil with curved yield surface". International Journal of Solids and Structures, **13**, 1019–1035.

Matlock, H., and Reese, L. C. (1960). Generalized solution for laterally loaded piles, *Journal of the Soil Mechanics and Foundation Division, ASCE*, Vol. 86, No. SM5, pp. 63–91.

McClelland, B. (1974). Design of deep penetration piles for ocean structures, *Journal of the Geotechnical Engineering Division, ASCE*, Vol. 100, No. GT7, pp. 705–747.

Meyerhof, G. G. (1963). Some recent research on the bearing capacity of foundations, *Canadian Geotechnical Journal*, Vol. 1, No. 1, pp. 16–26.

Meyerhof, G. G. (1976). Bearing capacity and settlement of pile foundations, *Journal of the Geotechnical Engineering Division, ASCE*, Vol. 102, No. GT3, pp. 195–228.

Meyerhof, G. G. (1995). Behavior of pile foundations under special loading conditions: 1994 R. M. Hardy keynote address, *Canadian Geotechnical Journal*, Vol. 32, No. 2, pp. 204–222.

Rankine, W. M. J. (1857). On stability of loose earth, *Philosophical Transactions of the Royal Society, London*, P. I, pp. 9–27.

Roscoe, K. H., and Burland, J. B. (1968). On the generalized stress–strain behavior of "Wet" Clay, *In Engineering Plasticity*, J. Heyman and F. Leckie (Ed.), Cambridge University Press, Cambridge, pp. 535–609.

Schaffernak, F. (1917). Über die Standicherheit durchlaessiger geschuetteter Dämme, Allgemeine. Bauzeitung.

Schofield, A., and Wroth, C. P. (1968). *Critical State Soil Mechanics*, McGraw-Hill, London.

Semple, R. M., and Rigden, W. J. (1984). Shaft capacity of driven piles in clay, *Analysis and Design of Pile Foundations*, American Society of Civil Engineers, New York, pp. 59–79.

Skempton, A. W. (1953). The colloidal activity of clays, *Proceedings of the 3rd International Conference on Soil Mechanics and Foundation Engineering*, London, Vol. 1, pp. 57–61.

Skempton, A. W. (1959). Cast-in-situ bored piles in London clay, *Geotechnique*, Vol. 9, No. 4, pp. 153–173.

Taylor, D. W. (1948). *Fundamentals of Soil Mechanics*, Wiley, New York.

Terzaghi, K. (1925). *Erdbaumechanik auf Bodenphysikalischer Grundlager*, Deuticke, Vienna, Austria.

Terzaghi, K. (1936). Relation between soil mechanics and foundation engineering: presidential address, *Proceedings of the First International Conference on Soil Mechanics and Foundation Engineering*, Boston, MA, Vol. 3, pp. 13–18.

Terzaghi, K. (1943). *Theoretical Soil Mechanics*, Wiley, New York.

INDEX